AF182535

11 Springer Series in Chemical Physics

Edited by Fritz Peter Schäfer

Springer Series in Chemical Physics

Editors: V. I. Goldanskii R. Gomer F. P. Schäfer J. P. Toennies

Preface

This conference on liquid crystals of one- and two-dimensional order and their applications is the third in a series of European conferences devoted mainly to smectic liquid crystals. Its purpose was to bring together people working on the frontiers of the field of liquid crystals. Ordinary nematic liquid crystals were left out in order to limit the size of the meeting. The number of registered participants still reached 148.

The conference shed new light on the classification of smectic mesophases, especially through the interaction of the Halle (GDR) and Hull (England) groups. It saw lively discussions on the famous blue phase of cholesterics. There were illuminating presentations on lyotropic nematic liquid crystals, on reentrant nematics, mesomorphic polymer phases, and related subjects. Much room was given to bilayers, monolayers, and interfaces, mostly to further the use of the concepts and methods of liquid crystal physics in exploring biomembranes. Other topics were device applications of smectic and cholesteric liquid crystals and nematic polymers, both of which hold promise of technological breakthroughs, apart from their scientific interest.

The conference benefitted greatly from the willingness of many experts to present reviews of one of their specialities. It was sponsored by the European Physical Society, the "Gesellschaft Deutscher Chemiker", and the "Deutsche Physikalische Gesellschaft". We would like to thank the "Deutsche Forschungsgesellschaft" and the "Bayerisches Staatsministerium für Unterricht und Kultus" for financial support and the "Kurverwaltung of Garmisch-Partenkirchen" for their cooperation. Finally, the organisation of the conference and the compilation of the proceedings would not have been possible without the dedicated help of Frau H. Assmann and Frau J. Klingebiel who deserve our special thanks.

Berlin, July, 1980 *W. Helfrich G. Heppke*

Contents

Part II. *A and C Smectic Phases and Structures*

Part III. *Defects, Elasticity, and Rheology of Smectics*

* Invited lecturer

Part I

Ordered Smectic Phases and Structures

The Smectic Phases of the N-(4-n-Alkyloxybenzylidene)-4'-n-alkylanilines (n0.m's) - Some Problems of Phase Identification and Structure

J.W. Goodby[1], G.W. Gray[1], A.J. Leadbetter[2], and M.A. Mazid[2]

[1] Department of Chemistry, The University, Hull, HU6 7RX, England
[2] Department of Chemistry, The University, Exeter, EX4 4QD, England

The *N*-(4-n-alkoxybenzylidene)-4'-n-alkylanilines are simple Schiff's bases with the general structure (I).

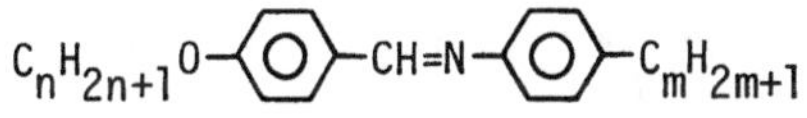

(I)

Even if we consider only combinations of the two alkyl groups such that n and m may have values between 1 and 10, a large number of compounds is involved, and most form liquid crystal phases. Moreover, many of these materials exhibit complex smectic polymorphic behaviour, and for this reason they have attracted considerable attention over the last few years. This group of compounds is also particularly well-known because one of its simpler members is MBBA (*N*-[4-methoxybenzylidene]-4'-n-butylaniline) one of the first materials found to give a nematic phase at room temperature [1]. The investigation of these Schiff's bases seems straightforward because of their ready accessibility from the appropriate aldehyde and amine, but obtaining the compounds in a highly pure state can pose problems. As Schiff's bases, they are liable to heterolytic cleavage by water, particularly under any conditions of acidic or basic catalysis. Care must therefore be exercised in handling the materials, and also in the process of their purification, particularly that of low melting, highly soluble members for which strong cooling, with consequent dangers from the condensation of moisture, must be used in any crystallisation process. Some discrepancies in earlier literature reports of the properties of these materials may well be attributable to matters relating to purity and degradation by moisture.

Comprehensive studies of the Schiff's bases (I) were made by Smith, Gardlund, and Curtis [2,3] and they suggested a simple and useful nomenclature scheme based on n0.m, where n represents the carbon content of the alkoxy group, and m the carbon content of the alkyl group. Thus, the compound

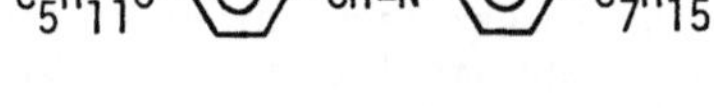

may be simply and unambiguously represented as 50.7.

The work of Smith, Gardlund, and Curtis [2,3], involving materials with values of n from 1 to 7 and of m from 4 to 8, was an important contribution and made useful reference to contemporary studies of the same system by Flannery and Haas [4], Fishel and Hsu [5], Knaak, Rosenberg, and Servé [6], Dietrich and Stieger [7], and Murase [8], drawing attention to discrepancies where they existed amongst the observations. Their values [2,3] for the transition temperatures have stood the test of time and their studies highlighted the varied and complex smectic polymorphism of these series. Remembering the imperfect understanding of smectic polymorphism at that time, and the fact that they did not employ miscibility methods, their conclusions were very enlightened. Not only were firm identifications of N, S_A, and S_B phases made, but also quite firm assignments of some smectic phases to the S_C category were proposed, and suggestions that other phases may be S_F in nature were put forward.

As this review will deal extensively with the smectic polymorphic behaviour of the compounds (I), it is appropriate that we should first make the following comments.

The nomenclature of smectic H and G phases

In 1971, Demus and Sackmann and their co-workers at Halle used S_G to describe a new phase observed in certain pyrimidines. For example, for 2-(4'-n-pentylphenyl)-5-(4"-n-pentyloxy)pyrimidine, they gave [9] the sequence of smectic phases observed as

G F C A Isotropic

In 1972, de Vries and Fishel [10] observed for 40.2 a smectic phase which had X-ray characteristics which were unique at that time, *viz*, a herringbone structure. 40.2 was defined as

K 41 H 51 N 66 Isotropic

This phase was then found in several other systems, including TBBA, and because it was mistakenly thought by many that these phases were *miscible* with orthogonal B phases, they became known as tilted B phases. The nomenclature system

S_B tilted ≡ S_{B_C} ≡ S_H

therefore developed in the literature.

However, from work carried out by Goodby and Gray over a number of years, and eventually published [11], it became clear that tilted S_B phases required a new code letter and we chose H after the work by de Vries and Fishel. Our conclusions regarding the immiscibility of S_{B_C} (S_H) and S_B phases were fully substantiated by Sackmann in his lecture at the International Conference at Bordeaux [12], and indeed it is now accepted from work by Doucet and Levelut [13] and de Jeu and de Poorter [14] that direct, reversible transitions between these two phases can occur.

However, unknown to us at that point in time, Richter, Demus, and Sackmann [15] had shown that the lowest temperature smectic phase of the pyrimidine [9] was *miscible* with the smectic phase of 40.2. As they had observed this phase first in the pyrimidine, and coded it as G, they naturally described 40.2 as

K 41 G 51 N 66 Isotropic

This duality of nomenclature for the same smectic phase was made more serious in its consequences because other workers had meantime used G to code the tilted analogue of the smectic E phase, which however the Halle group referred to as H. So a total inversion of nomenclature between H and G arose, causing anomalies in description such as those below:

TBBA:	(1)	H	G	C	A	N	Isotropic	(Halle group)
	(2)	G	H	C	A	N	Isotropic	(others)
Pyrimidine:	(1)	G	F	C	A	Isotropic		(Halle group)
	(2)	H	F	C	A	Isotropic		(others)

Following discussions between two of the authors (Gray and Goodby) and Sackmann and Demus during a recent visit to Halle, we decided that this confusing situation, which had arisen as an accident of history, compounded by some erroneous observations that need not be specified, should be put right. We therefore recommend strongly that a unified nomenclature system shall now be adopted and that this should be based on the historical priorities of the situation. Thus we recommend that:

> G be used for the phase originally described by many as a tilted B phase, and
>
> H be used for the phase that is the tilted analogue of an E phase

Consequently, the sequences denoted (1) above for both TBBA and the pyrimidine are now the accepted sequences, and the nomenclature used in this review is naturally based on this unified scheme.

Reasons for our interest in the nO.m series (I)

(1) From the work of Smith, Gardlund, and Curtis [2,3] it seemed that some members might exhibit F phases, which are always of interest, and that challenging and unsolved sequences of smectic phases were provided by several members. Thus, 50.6 appeared to give five smectic phases of which only the A and B, and possibly the C phases were identified with any certainty.

(2) Billard [16] reported the sequence for 70.7 as

G F C A N Isotropic

apparently confirming that the nO.m's were a source of F phases.

(3) Doucet and Levelut [13] reported examples of direct B to G transitions in the nO.m's and this was very relevant to our earlier discovery that these phases (then referred to as orthogonal and tilted B phases, respectively) were immiscible. In their work [13], 70.7 was assigned the following sequence of phases which compares/contrasts with the sequences quoted by the other workers listed overleaf. Note that in all cases, we have used the code letters employed by the authors at that time.

Doucet, Levelut [13]:		S_{B_C}	S_B	S_C	S_A	N	Isotropic
Smith *et al.* [2,3]:	(i)	S_4	S_B	S_C	S_A	N	Isotropic
	(ii)	S_4	S_F	S_C	S_A	N	Isotropic
Billard [16]:		S_G	S_F	S_C	S_A	N	Isotropic
Demus [17]:		S_G	S_B	S_C	S_A	N	Isotropic

Since S_G (Demus) ≡ S_{B_C} (Doucet and Levelut), it thus seemed that the sequence for 70.7 should be written as

G B C A N Isotropic

and that the suggested smectic F character of 70.7 must be incorrect. This then cast doubts on the potential of n0.m's to form F phases.

(4) Doucet and Levelut [13] had observed a double, rather than a single ring in the powder diffraction patterns corresponding to the Bragg reflections from the layers of the B phases of 70.7, 70.5, and 50.7, and it was of interest to see whether this arose with other homologues.

Plan of the Present Investigations

(a) Studies of the trends in smectic polymorphism in several homologous series of n0.m's using optical microscopy and miscibility methods.

(i) n0.1 series (n = 1 to 10)
(ii) n0.8 series (n = 1 to 7)
(iii) n0.4 series (n = 1 to 10)
(iv) 60.m series (m = 1 to 8)
(v) 50.m series (m = 1, 4, 6 to 8)

(b) Detailed structural studies of the smectic phases of selected members using X-ray diffraction methods and when possible incoherent neutron quasi elastic scattering (INQES). The homologues chosen were:

40.2	40.8	50.6	50.7	60.4
70.1	70.5	70.7	90.4	

The work involving the synthesis of the n0.m's and the microscopic and miscibility studies was carried out at Hull. The X-ray studies were made at Exeter, and the INQES studies by members of the Exeter group using facilities at ILL, Grenoble. Calorimetric measurements were made either at Hull or Exeter.

(a) Studies of homologous series

Data for the five series studied are tabulated (Tables 1 to 5) and also expressed in graphical form in Figs. 1 to 5.

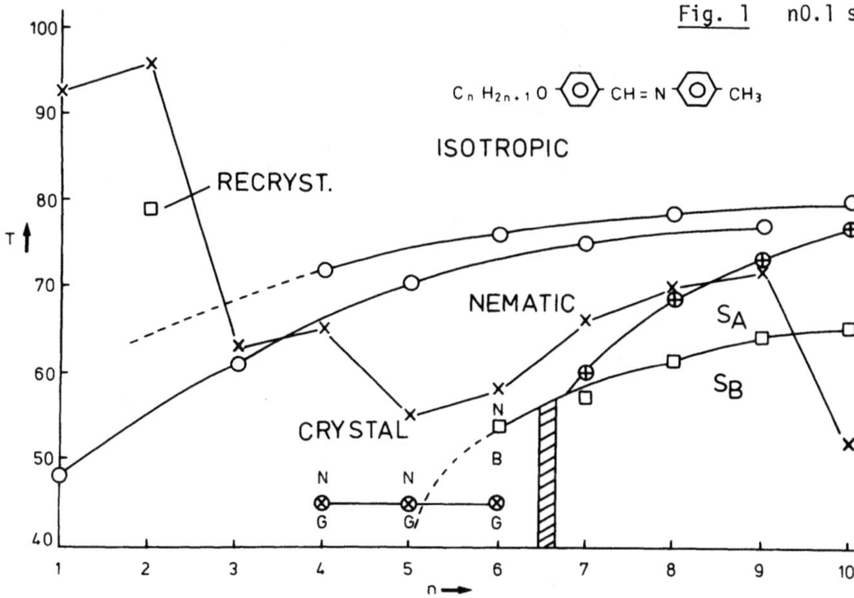

Fig. 1 n0.1 series

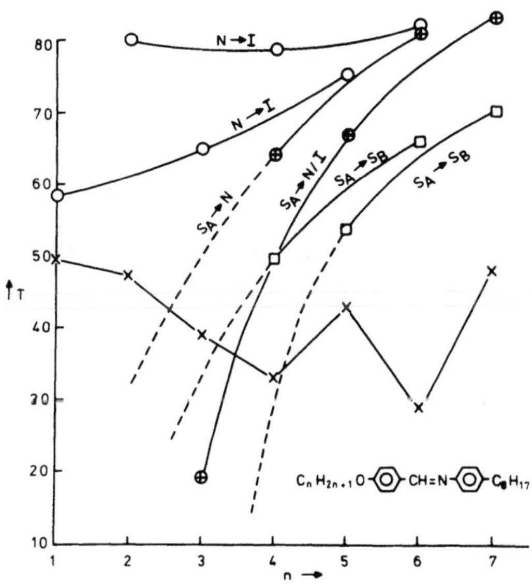

Fig. 2 n0.8 series

The main point of interest stemming from Figs. 1 to 5 is the complex way in which the smectic polymorphism varies with changes in the lengths of the alkyl chains at the ends of the molecule. Even within a given homologous series, quite dramatic changes in the smectic phase types that are exhibited may occur, sometimes between neighbouring homologues. These effects are at present not understood, and only the following general comments will be made on each of the five series.

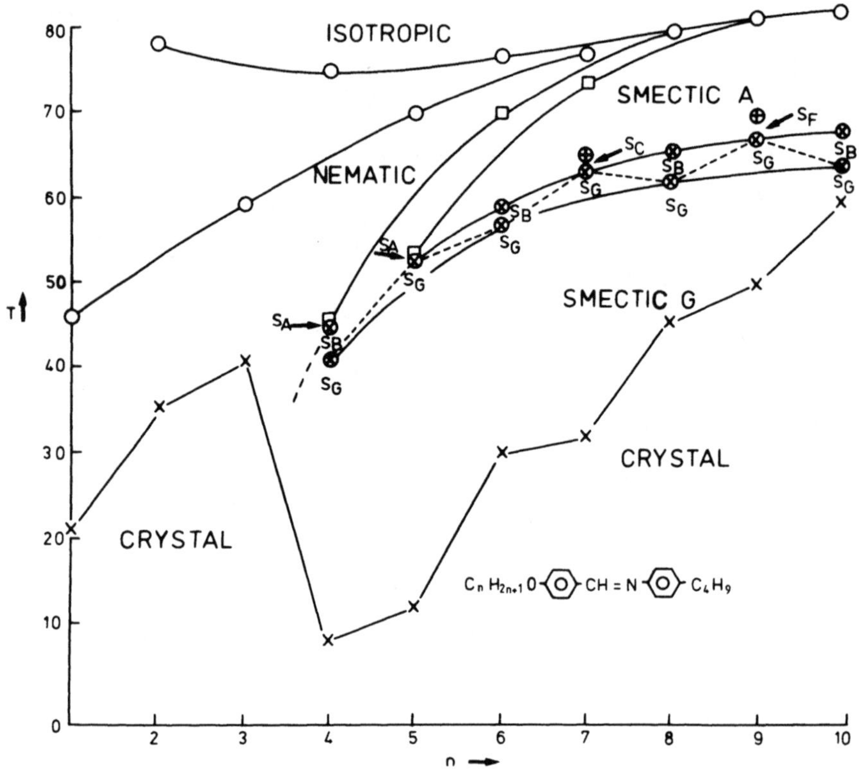

<u>Fig. 3</u> n0.4 series

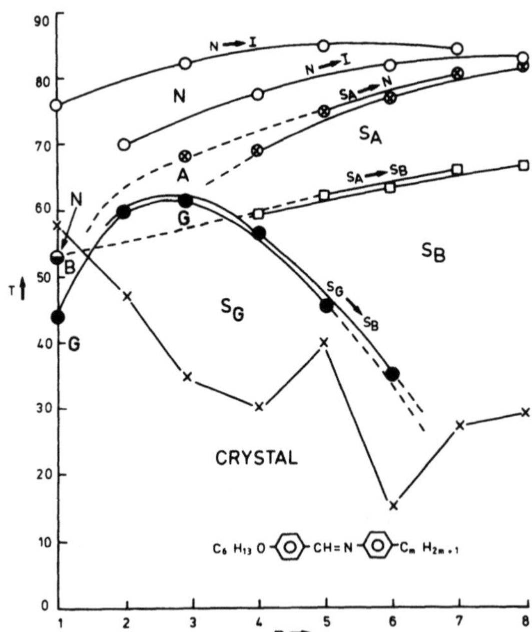

◄ <u>Fig. 4</u> 60.m series

Fig. 5 50.m series

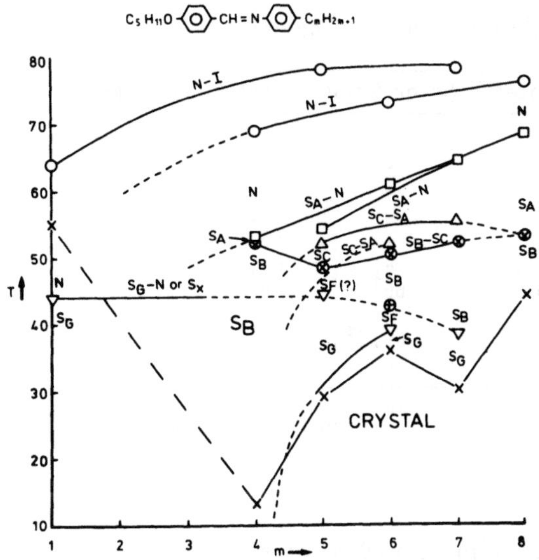

Table 1 Transition temperatures (^{O}C) for the n0.1 series

n	mp	S_G-S_B or N	S_B-S_A or N	S_A-N	N-I
1	92.5	-	-	-	-
2	96*	-	-	-	-
3	63	-	-	-	(61)
4	65	(45)	-	-	72
5	55	(44)	-	-	70.5
6	58	(44)	(53)	-	76
7	66	-	(57)	(60)	75
8	70	-	(61.5)	(69)	78.5
9	72	-	(64)	73	77.5
10	52	-	65	77	80

* temperature of recrystallisation, 79°C
( ) monotropic transition

Table 2 Transition temperatures (^{O}C) for the n0.8 series

n	mp	S_B-S_A	S_A-N	S_A or N-I
1	49.5	-	-	58.5
2	47.5	-	-	80.5
3	39	-	(19)	65
4	33	49.5	64.5	79
5	43	53.5	67	75.5
6	29	66	81.5	82.5
7	48	70	-	83

( ) monotropic transition

Table 3 Transition temperatures (^{o}C) for the n0.4 series

n	mp	S_G-S_B	S_B-S_A	Footnotes	S_C or S_F-S_A	S_A-N	S_A or N-I
1	21	-	-	-	-	-	46
2	35.5	-	-	-	-	-	78
3	41	-	-	-	-	-	59.5
4	8	41	45	-	-	45.5	75
5	12	-	-	53*	-	53.5	70
6	30	57.5	59	-	-	69.5	78
7	32	-	-	63.5**	65†	74	77
8	45	62	65.5	-	-	-	80
9	50	-	-	67††	69.5Φ	-	82
10	61	64	68	-	-	-	82.5

* G to A ** G to C † C to A †† G to F Φ F to A

Table 4 Transition temperatures (^{o}C) for the 60.m series

m	mp	S_G-S_B or S_A	S_B-S_A or N	S_G-N	S_A-N	N-I
1	58	(44)	(53)	-	-	76
2	47	-	-	58	-	70
3*	34	61	-	-	68	81
4	30	57.5	59	-	69.5	78
5	40	45	62.5	-	75	85
6	15	35	63	-	77	82
7	27	-	66	-	80.5	84
8	29	-	66.5	-	81.5	83

( ) monotropic transition

* Data from L Richter (private communication)

Table 5 Transition temperatures (^{o}C) for the 50.m series*

m	mp	S_G-S_B, S_F or N	S_F-S_B	S_B-S_C or S_A	S_C-S_A	S_A-N	N-I
1	55	(44)	-	-	-	-	64
4	13	-	-	52	-	53	69
6	36	38.5	42.5	50	51.8	60.5	73
7	30	38**	-	52	55	64	78
8	44	-	-	53	-	68	76

* 50.5 forms the phases G, B, C, A, N - L Richter (private communication)

** G to B

( ) monotropic transition

nO.1 Series (Fig. 1) - The main interest in this series is the occurrence of G phases for three members (40.1, 50.1, and 60.1), and the sudden replacement of these G phases by B and A phases as n is increased to 7 and above. Other features such as the rising N-I transition lines are normal.

nO.8 Series (Fig. 2) - In this series with an extended m chain, no G properties are observed, and we have purely nematic behaviour for the early members, a combination of nematic, A, and B properties for higher homologues, and the extinction of nematic properties after m = 6. The S_B-S_A, S_A-N, and N-I temperatures alternate with the same sense - which is unusual. The steep fall-off in the smectic curves as n decreases explains the suppression of smectic properties in the early homologues.

nO.4 Series (Fig. 3) - G properties are very pronounced for the higher homologues, and marked differences occur between the phase sequences exhibited by neighbouring homologues (emphasising the dangers in predicting phase behaviour on a basis of near-neighbour relations). Considering n = 4, 6, 8, and 10, each member has the sequence of phases

$$K \rightarrow S_G \rightarrow S_B \rightarrow S_A \rightarrow N/\text{Isotropic}$$

providing further examples of S_G-S_B transitions in pure compounds. However, the alternation senses for S_B-S_A transitions and S_G-S_B transitions are opposite. Since the thermal ranges of the B phases for even n values are narrow, B phases are extinguished for odd members (n = 5, 7, and 9), and the G phase changes, *not* to a B phase, but to *another* phase. For n = 5, an A phase is formed, but for n = 7, a C phase is formed, and for n = 9, an F phase. Thus, for 70.4 and 90.4, the G phase exerts its tendency to give another tilted phase on heating, and we get the injection of a C or an F phase before A. Smith *et al.* [2,3] were therefore correct in suggesting that S_2 for 70.4 is a C phase, and the liquid-like order within the layers of S_2 (S_C) was confirmed by the powder patterns obtained by Doucet and Levelut [13].

Compound 90.4 is the more interesting however, because of the F properties, and it provides the first example of a direct S_F-S_A transition. Miscibility results obtained with 90.4 have already been reported [18]. Subsequent X-ray diffraction studies [19] have shown that at the A to F transition, the S_A layers retain their orientation, while the molecules tilt (contrast 50.6 - see on). The diffraction pattern of the F phase shows it to have monoclinic symmetry with a hexagonal molecular packing in the plane normal to the long axes of the molecules which are tilted with respect to the layer planes. Lattice orientation and tilt direction have long range order, but the broadened diffraction peaks indicate that the molecular positions have relatively short range order (correlation length ca. 50Å). The F phase is therefore a weakly coupled 2-D system having long range bond orientational order, but short range positional order. Plates 1 to 3 show the microscopic textures obtained on cooling the isotropic liquid. Plate 2 shows the weak birefringence in the areas of S_F that were homeotropic in the A phase. This birefringence is more pronounced in the G phase. Plate 4 is a fine example of the arced or striped fans of the focal-conic F phase.

60.m Series (Fig. 4) - (m = 1, 2, 4 to 8) Here there are no unusual features, and the N-I, S_A-N, and S_A-S_B transition lines are typical. The G phases provide four further cases of direct S_G-S_B transitions in pure materials. Shorter chains appear to favour the thermal stability of the G phase, and for m = 7 and 8, no G phases occur.

Plate 1. S_A phase of 90.4

Plate 2. S_F phase of 90.4

Plate 3. S_G phase of 90.4

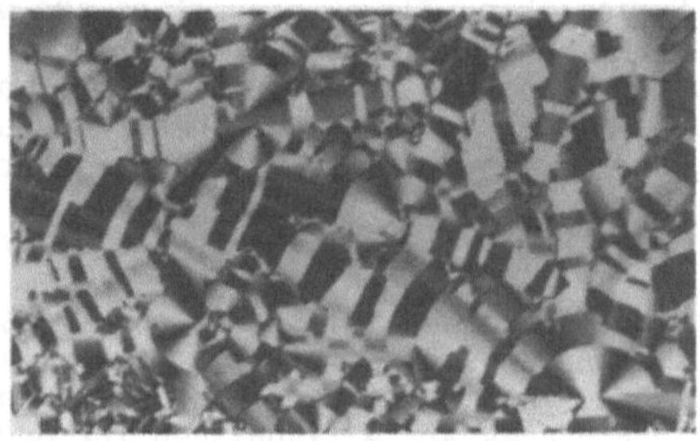

Plate 4. S_F phase of 90.4

50.m Series (Fig. 5) - (m = 1, 4, 6, 7, and 8) This series is much more complex, and full data are not available for m = 1 through to 8. Data from Smith *et al.* [2,3] for 50.5 are included in Fig. 5, but not in Table 5.

The N-I and S_A-N transition temperature curves are normal, except that there is a reversed sense of alternation for the two types of transition. That for the N-I transitions is normal. 50.1 exhibits a monotropic G phase, and 50.4 and 50.8 give B phases below the A phase. The situation is complicated for 50.5, 50.6, and 50.7 which give C phases. The 50.6 and 50.7 members give S_B-S_C transitions; 50.8 shows no C phase. The complexity is even greater however, because 50.7 exhibits a S_G-S_B transition. With an odd value of m (7), this transition may be related to the S_G-N transition for 50.1 - see partly dashed curve in Fig. 5. A S_G-S_B transition would therefore be expected for 50.3. It is noted that a transition for 50.5 also lies on this curve, and it is tempting to associate this with a S_G-S_B change.* However, Smith *et al.* [2,3] considered that the phase formed on heating at this transition was an F phase. The transition from this phase to the C phase occurs at 48°.

Most complex of all, yielding five smectic phases is 50.6 which is discussed later. Since 50.4 shows no F phase or G phase, the curve from the S_G-S_F transition for 50.6 is drawn with a steep slope as m decreases. The fact that the S_F-S_B temperature for 50.6 lies on the dashed curve is of no significance.

A very involved smectic polymorphism therefore occurs in this series, and it will be interesting to complete the series and establish the behaviour of the 50.2, 50.3, and 50.5 members with certainty.

---

* as observed by Richter - see footnote to Table 5.

(b) *Systems studied in more detail*

Systems examined in greater depth by X-ray methods and in some cases by neutron scattering methods were 40.2, 40.8, 50.6, 50.7, 60.4, 70.1, 70.5, 70.7, and 90.4. Thermal data not already given for these compounds are:

40.2	K	41	G	51	N	66	Isotropic								
70.5	K	23	G	58.2	B	64.5	C	68.4	A	79.8	N	83.4	Isotropic		
70.7	K	33	G	55	B	69	C	72	A	83.7	N	84	Isotropic		

General comments - The N, A, and C phases were as expected; orientational distribution functions were obtained [20] for the A phase of 40.8. S_A-S_C transitions occurred either by layer tilting, with a displacement along the long molecular axes, or by tilting of fixed layers. S_C tilt angles were small ($\lesssim 10^o$).

Smectic B and G phases - A general result is that the molecular lengths in their most extended conformations are identical with the c parameter for the G phase and with the layer spacing for the B phase. Hence, the molecules retain their extended conformations and the simple layer structures are correct. Layer thicknesses for A (and C) phases are smaller in accordance with the long molecular axial fluctuations in essentially liquid-like layers. Cell dimensions for compounds examined by us and Doucet and Levelut [13] are in agreement.

(i) Smectic G phases (40.2, 50.6, 50.7, 60.4, 70.5, 70.7, 90.4) - Within the experimental resolution, packing within the layers normal to $\vec{c}$ (Fig. 6) is truly hexagonal, and the packing of the layers is of the AAA type. The G phases are unquestionably 3-dimensional; the diffraction spots give no broadening, and this shows the structures to have true long range order. However, only low order reflections are given, plus intense diffuse scattering, the intensity of the 00ℓ reflections diminishing very rapidly with ℓ and usually not more than three orders being observed. Thus, there is considerable disorder; the layer distributions are not *sharp*, and are very different from that of a crystal. Presumably this is associated with orientational disorder of the molecules about their long axes, but a displacement disorder is also required to explain a rapid decrease in intensity with h, k. Evidence for this has been obtained by neutron experiments.

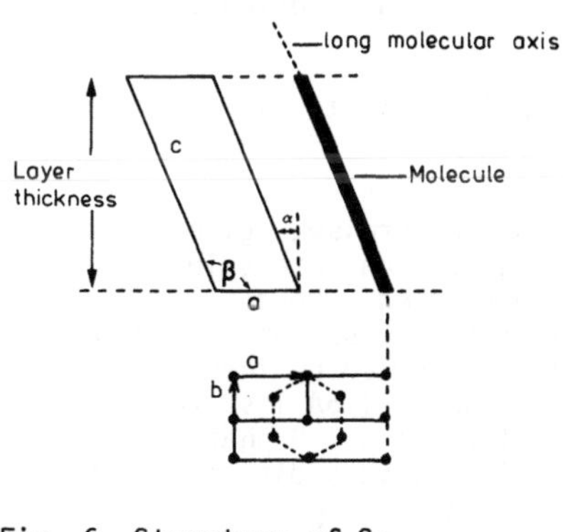

Fig. 6 Structure of S_G (Herringbone)

(ii) Smectic B phases (40.8, 50.6, 50.7, 60.4, 70.1, 70.5, and 70.7) - These were positively confirmed as having hexagonal symmetry (single domain samples from a single crystal; incident beam orthogonal to the layers). Generally, only two orders of reflection were observed, meaning that layer correlations were relatively weak, and that a picture of rigid sharp layers is not correct.

With well aligned samples, clear inter-layer correlations occur, and from intensity contour maps of the diffraction patterns (Fig. 7), evidence for different layer stackings, and changes of layer stacking with temperature were obtained - cf. results by Moncton and Pindak [21] from a very high resolution study of 40.8. Three stacking types are possible: AAA... (monolayer); ABA...(bilayer); ABCA...(trilayer). The monolayer type near to the S_B-S_G transition for 60.4 is the only one observed; the bilayer packing is the most common; examples of trilayer packing also occurred, eg, for 70.1. The most interesting aspect is that the packing arrangement often

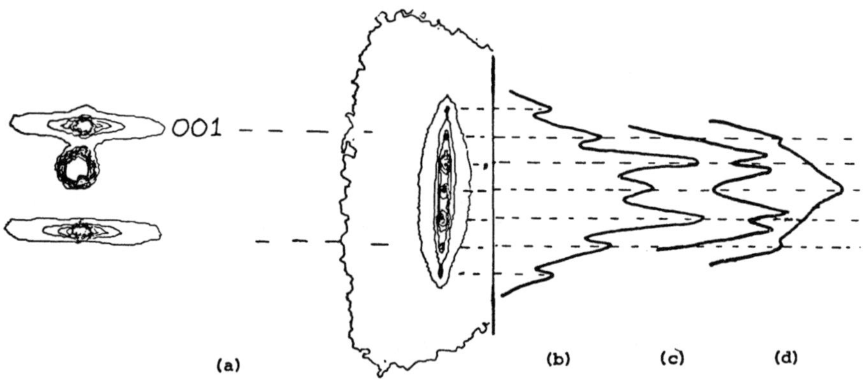

Fig. 7 Intensity contour map of the X-ray diffraction from a S_B phase with ABA...interlayer packing, (a) plus experimental intensity profiles showing (b) (40.8) ABA... (c) (50.7) ABCA... and, (d) (60.4) AA...stacking

changes with temperature. No enthalpy effects have been detected for changes such as those for 50.7 -

bilayer → trilayer → bilayer (close to S_B-S_G transition)

(decreasing temperature)

For compounds giving a S_B-S_G transition, but not otherwise, eg, not for 40.8 and 70.1, the S_B diffraction peaks for even, but non-zero values of ℓ develop satellites as the S_B-S_G transition is approached. Doucet and Levelut [13] observed these in powder diagrams, but did not discuss them. Here we note that we did *not* observe the extra reflection reported by these workers and explained by them in terms of a layer spacing about 1.5Å less than normal. We therefore have no evidence for a mixture of two B phase types. Close to the S_B-S_A transition, the (00ℓ) reflections give a disc of diffuse scattering from which, at temperature *above* $T_{G\text{-}B}$, eg, $T = T_{G\text{-}B} + 7K$ for 50.7, satellite peaks emerge (see Fig. 8). These indicate a transverse modulation of the structure parallel to the layers with a wavelength about 17 times the (100) spacing of the hexagonal net. At $T_{B\text{-}G}$, new peaks appear at a slight displacement from the satellites, and the central peak goes. The transition seems to be first order, but the satellites persist into the G phase for 1 or 2^0, suggesting a coexistence of the two phases over about 3K around the transition.

These results show that throughout the B and G phase ranges, the direction of the molecular long axes remains unchanged (along the $\vec{c}$ axis of the unit cell). At high temperatures, there are pronounced fluctuations involving molecular displacements along $\vec{c}$. As the temperature falls, these become periodic and the amplitude increases. The S_B-S_G transition occurs when the molecules in adjacent (100) planes are displaced by about 2Å relative to each other; at the same time, the bilayer structure of the B phase disappears. Since only one satellite is seen, the modulations would seem to involve essentially sinusoidal undulations of the layers; their occurrence is connected with the weak (though long range) inter-layer correlations, and presumably these modulations eventually trigger-off the S_B-S_G transition. Neutron experiments made on 50.7 were reported fully by Leadbetter *et al.* [19]. The rotational correlation times are 6×10^{-10} and 5.5×10^{-10} s rad^{-1} respectively for the B and G phases. In addition, the diffusive motion parallel to $\vec{c}$ gives $\langle z^2 \rangle^{\frac{1}{2}}$ = 1.35Å for B and 1.09Å for G. The correlation times are the same within experimental error (±10%) as those for rotation, implying that the two motions are strongly coupled.

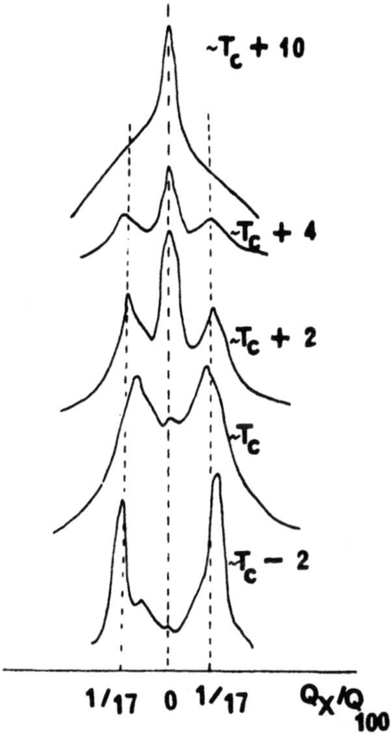

Fig. 8 X-ray intensity profiles [23] for 50.7

(iii) 50.6 and the Smectic F phase - As already noted, an F phase has been found for 90.4. Another F phase is now known to occur for 50.6 for which the phase sequence is as shown below. This sequence was confirmed by DSC and the enthalpies (in kJ mole^{-1}) associated with the transitions are given in brackets below the temperatures. Microscopic textures for the different phases were typical - see ref. [19], although the arced or chequered

K	36	G	38.1	F	42.4	B	50	C	51.8	A	60.3	N	72.8 Isotropic
	(15.2)		(0.39)		(0.15)		(2.5)		(v small)		(0.42)		(0.66)

pattern of the fans of the F phase was less well defined than normal. Transitions between all phases were clear by optical microscopy. Miscibility studies were made to confirm the phase assignments; the F phase was continuously miscible across the diagram of state for the binary system TBPA/50.6 - cf. the same situation for TBDA/90.4. A miscibility diagram for 50.6/90.4 was also constructed. Again this confirmed continuous miscibility of the two F phases (Fig. 9), although an unusual feature did arise as mentioned later.

For the X-ray studies, well aligned samples were used. At temperatures above 45°, the B phase has a bilayer (ABAB...) stacking. This is shown as usual by microdensitometer traces taken along the <00ℓ> (or $\vec{c}^*$) direction for the bar of scattering corresponding to the lowest order reciprocal lattice points (100, 1$\bar{1}$0 etc) of the hexagonal lattice. This diffuse scattering is much stronger than that found by Moncton and Pindak [21] using a free film of

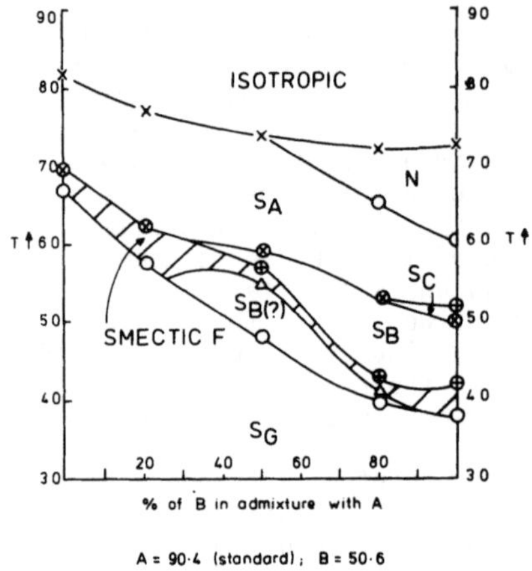

Fig. 9 Miscibility diagram for 50.6/90.4

the B phase of 40.8, and perhaps samples prepared by cooling in a field have a greater stacking disorder than that in free films.

Below 45°, the stacking changes to ABCA..., and about 1° above the S_B-S_F transition, this changes back to ABAB... . The disorder is however *greater* than that in the bilayer stacking at higher temperatures. Evidence for undulation motions in the B phase was again obtained. The diffraction pattern for the phase below the B phase clearly shows this to be F. The net width of the diffraction peak for F is about three times narrower than for the C phase. The profile is Lorentzian and its width gives a correlation length of about seven molecules; there is no correlation of molecular position between layers. The S_B-S_F transition is presumed to occur when the amplitude of the transverse modulations in the B phase become critical and a stable tilted structure is formed. The S_B-S_F transition therefore occurs by a movement of molecules along c, the direction of the molecular long axes remaining unchanged, resulting in a tilting of the layers in the F phase (tilt angle = 24°).

At about 38°, the G phase forms from the F phase with regeneration of long range order in the hexagonal packing. The G phase is monoclinic, with a tilted, hexagonal packing of the molecular long axes. However, the strong 100 and $1\bar{1}0$ reflections in the G phase are associated with strong diffuse scattering, similar to the F phase, and this suggests considerable regions of disorder. The direction of the long axes does not change at the S_F-S_G transition; the G tilt angle is 26°.

As first pointed out by Leadbetter *et al.* [22], and noted in more extensive work by Doucet and Levelut, the F phase has long range order of tilt direction (as for C phases), long range bond orientational order, but relatively short range positional order. The phase therefore corresponds to weakly coupled 2-D layers. Yet in 50.6, this phase exists on the temperature scale intermediate between a B and a G phase, both of which have long range, 3-D order. Why the F phase, lacking in 3-D order, appears between two phases which possess it is not known. Certainly however, compared with other nO.m's, a decoupling of the layers between the B and G phases occurs. Note that the stacking in G is AAA... and in B is ABA... .

This divergence from the normally expected sequence of increasing order with decreasing temperature emerges in another context involving 50.6. Whilst the miscibility diagram (Fig. 9) confirms the F character of the two components (90.4 and 50.6), the injection of the additional smectic phase between the F and G phases in the mid-region of the diagram is not only interesting, but unusual, because the additional phase seems to be B in character. It should be stressed however that we rely *at present* solely on the microscopic texture of the additional phase for its assignment to the B category. This must be confirmed by other means. However, if the injected phase is indeed B, then the 2-D smectic F appears on the temperature scale intermediate between 3-D smectic B phases, ie, at the 50% composition, the phase sequence on cooling would appear to be:

$$A \rightarrow B \rightarrow F \rightarrow B \rightarrow G$$

The injection of B properties in mixtures of F materials has been noted before for mixtures of 90.4 and 90.SF - see phase diagram in ref. [18].

CONCLUSIONS

1. The occurrence of F phases in the nO.m series is confirmed in two cases (90.4 and 50.6).
2. The occurrence of S_B-S_G transitions, as observed first by Doucet and Levelut [13] for several nO.m's, has been substantiated in other cases.
3. The mechanism of the S_B-S_G transition has been elucidated and appears to be triggered-off by an undulation mode in the B phase.
4. Different layer stackings occur in the B phases of the nO.m's, and changes between monolayer, bilayer, and trilayer stacking occur with temperature change.
5. No supporting evidence was obtained for the report that the B phases of certain nO.m's may consist of a mixture of two B phases with different layer spacings.
6. The smectic polymorphism of the nO.m's is highly sensitive to change in terminal alkyl chain length.
7. The compound 50.6 gives six liquid crystal phases which appear in the sequence N, A, C, B, F, G with falling temperature.
8. Compound 50.6 is of great interest in providing a case in which a 2-D smectic F is positioned on the temperature scale between two 3-D smectic phases (B and G).
9. These studies illustrate the value of combined studies by optical microscopy, miscibility, and X-ray methods for the investigation of smectic systems. As more is learned through X-ray and neutron scattering studies about the various polymorphic smectic phases, it is perhaps tempting to seek a classification of the phases based on structural factors, to replace the older system of letters (A to K), based on miscibility criteria, which has developed historically, and consequently in no very rational way. However, if a symbolism is to be developed and embrace *all* the structural characteristics, there is a danger that this will be too complex or cumbersome for ready adoption. Conversely, over-simplistic structural classifications are of little real help, and whilst new facets of smectic structure are coming to light so frequently, we feel strongly that it would be a pity if such a scheme were adopted and then had to be abandoned or seriously changed in the light of advancing structural knowledge. We think that it would be wise to wait for two or three years, after which time a rational and useful scheme may be developed on a secure basis of fact.

REFERENCES

[1] H Kelker and B Scheurle, *Angew Chem Internat Ed*, *8*, 884 (1969); H Kelker, R Hatz, and W Bartsch, *Angew Chem Internat Ed*, *9*, 962 (1970).

[2] GW Smith, Z Gardlund, and RJ Curtis, *General Motors Corporation Research Publication*, *GMR-1285*, November 1972; *Mol Cryst Liq Cryst*, *19*, 327 (1973).

[3] GW Smith and ZG Gardlund, *General Motors Corporation Research Publication*, *GMR-1354*, March 1973; *J Chem Phys*, *59*, 3214 (1973).

[4] JB Flannery (Jr) and WJ Haas, *J Phys Chem*, *74*, 3611 (1970).

[5] DL Fishel and YY Hsu, *J Chem Soc (D)*, 1157 (1971).

[6] LE Knaak, HM Rosenberg, and MP Servé, *Mol Cryst Liq Cryst*, *17*, 171 (1972).

[7] HJ Dietrich and EL Stieger, *Mol Cryst Liq Cryst*, *16*, 263 (1972).

[8] K Murase, *Bull Chem Soc (Japan)*, *45*, 1772 (1972).

[9] D Demus, S Diele, M Klapperstück, V Link, and H Zaschke, *Mol Cryst Liq Cryst*, *15*, 161 (1971).

[10] A de Vries and DL Fishel, *Mol. Cryst Liq Cryst*, *16*, 311 (1972).

[11] JW Goodby and GW Gray, *J Phys (Paris)*, *40*, 363 (1979).

[12] H Sackmann, *J Phys (Paris)*, *40*, 5 (1979).

[13] J Doucet and A-M Levelut, *J Phys (Paris)*, *38*, 1163 (1977).

[14] WH de Jeu and JA de Poorter, *Phys Lett*, *61A*, 114 (1977).

[15] L Richter, D Demus, and H Sackmann, *J Phys (Paris)*, *37*, 41 (1976).

[16] J Billard, *Compt rend Acad Sci*, *280B*, 573 (1975).

[17] D Demus, private communication to J Doucet and A-M Levelut.

[18] JW Goodby and GW Gray, *Mol Cryst Liq Cryst Lett*, *56*, 43 (1979).

[19] AJ Leadbetter, MA Mazid, and RM Richardson - presented at the Liquid Crystal Conference, Bangalore, December 1979 - to be published.

[20] AJ Leadbetter and PG Wrighton, *J Phys (Paris)*, *40*, 234 (1979).

[21] DE Moncton and R Pindak, *Phys Rev Lett*, *43*, 701 (1979).

[22] AJ Leadbetter, JB Gaughan, BA Kelly, GW Gray, and JW Goodby, *J Phys (Paris)*, *40*, 178 (1979); JW Goodby, GW Gray, AJ Leadbetter, and MA Mazid, *J Phys (Paris)*, to be published.

[23] AJ Leadbetter, MA Mazid, BA Kelly, JW Goodby, and GW Gray, *Phys Rev Lett*, *43*, 632 (1979).

The System of Nonamphiphilic Smectic Liquid Crystals with Layer Structures

H. Sackmann

Sektion Chemie, Martin-Luther-Universität Halle-Wittenberg, DDR-402 Halle (Saale), German Democratic Republic

1. Introduction

Before 1970 the interest in the field of the polymorphism of nonamphiphilic liquid crystals was turned to the spread of the nematic phases and the smectic phase types A, B and C. Later on new substances with higher polymorphism and new phases have been found and also intensive investigations on structures took place. This article gives a summarizing report on the latest development in this field, especially on investigations of smectic phases by miscibility (section 2). A comparison of these results with investigations on structure will be done in section 3.

It is expedient to exclude here phases with cubic and discotic structures and to concentrate in this report on phases with layer structures. Only here an extensive material on high polymorphism in both the fields (miscibility and structure) exists.

2. Miscibility and Phase Types

This report can be given by the observation of the polymorphism in some homologous series with a high degree of smectic polymorphism. In such series the existence of different phases is sure from the beginning.

2.1 The System of Phases in the Homologous Series of TBAA, AAOP and AOBAA [1] [2] [3] [4] [5]

At first the discussion is concentrated on the following three homologous series:

terephthalylidene-bis-[4-n-alkylanilines] (TBAA)

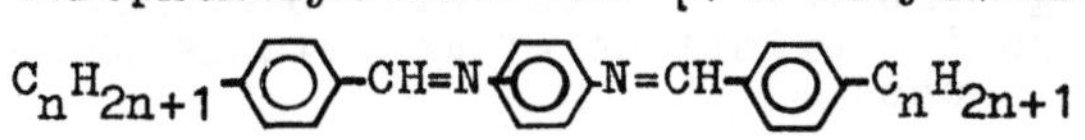

2-[4-n-alkylphenyl] -5- [4-n-alkoxyphenyl] -pyrimidines (AAOP)

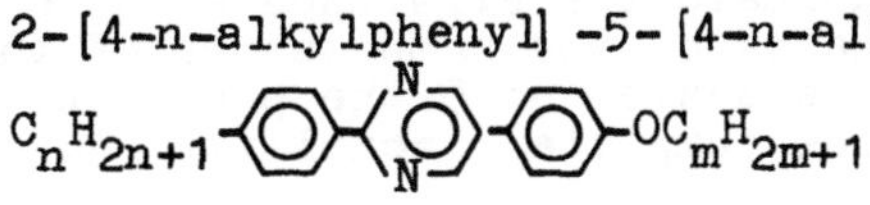

N– [4-n-alkoxybenzylidene] -4-n-alkylanilines (AOBAA)

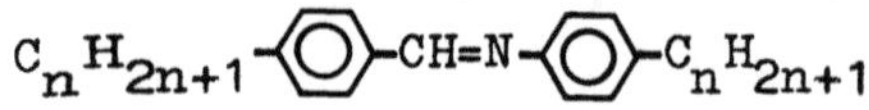

The relations of miscibility found between the members of the three series are listed in Fig.1 together with relations to other substances which are called reference substances. These reference substances are listed in Table 1 and its polymorphism can be seen in Fig.1.

Table 1 Reference substances

REF 1	ethyl-4-[4-ethoxybenzylideneamino] cinnamate This is the standard substance with a smectic B phase [6] [7].
REF 2	propyl-4-[4-n-octyloxybenzylideneamino] cinnamate Classification of the phases [8].
REF 3	n-pentyl-4-[4-n-dodecyloxybenzylideneamino] cinnamate Classification [9] has been changed by new miscibility investigations (see Fig.1) [1].
REF 4	n-pentyl-4-[4-n-decyloxybenzylideneamino] cinnamate Classification [9] has been changed by new miscibility investigations (see Fig.1) [1].
REF 5	4-n-propyloxy-4'-ethanoyl-biphenyl Classification [10].
REF 6	4-[4-n-propylmercaptobenzylideneamino] -azobenzene Classification [11].
REF 7	4,4'-bis-[nonyloxyazoxy] benzene Classification [12].
REF 8	4-[4-n-nonyloxybenzylideneamino] azobenzene Classification [7].

The series in Fig.1 are named by triangles (TBAA), circles (AAOP) and rectangles (AOBAA). The members of the series are signed by the number of C-atoms in the side chains. The lines characterize the binary combinations investigated. The numbers within the lines give the number of the binary system. The two linked substances are the compounds of the system. The letters below the substance give the phases found in the substances in the sequence of decreasing temperature from left to right. The letters in parentheses mean the existence of metastable phases with respect to the solid phase.

At the same time the lines indicate the existence of a complete miscibility between the phases of the same letter of the two compounds. For instance in the system no 5 the phases A, C, F and G of the two substances show complete miscibility. As can

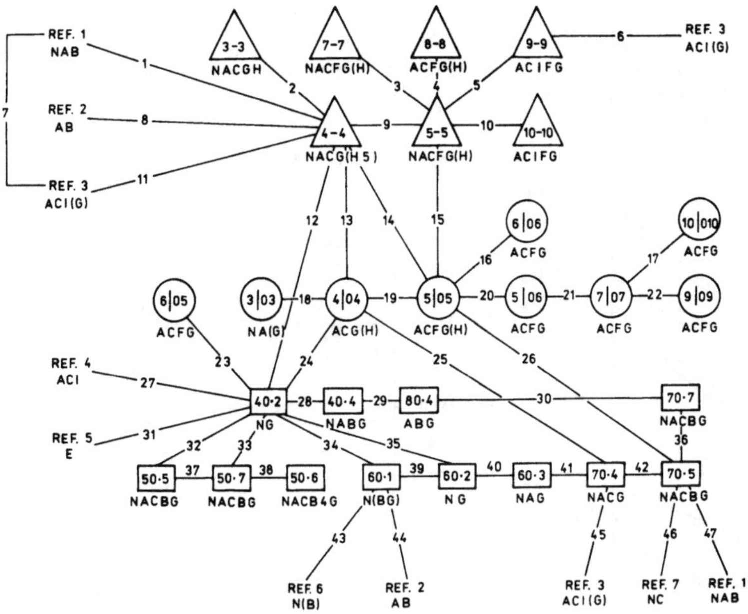

Fig.1 System of miscibility of the series TBBA, AAOP and AOBAA [1]

be seen some substances are used in several binary systems. In these cases there are several "ways" of characterizing the phases by miscibility testing the consistency of the system.

2.2 Code of Phases. Selected Examples

The denomination of the phases starts with standard substances. From the standard substances the phases of other substances have been denominated by miscibility found in the binary systems listed.

The compound 5/05 of the AAOP series (PPOP) serves as a standard substance. This is the compound on which DEMUS et al. have described two new phases F and G in 1971 [13]. It possesses the sequence of phases A C F G, and as we now known [1] a further phase (H), as a low temperature modification. The parenthesis means the existence of a metastable phase with respect to the solid phase in the supercooled region.

The substance REF 1 is the standard substance with a B phase.

Starting from these two substances we code the phases A, B, C, F, G and H of the three homologous series by miscibility in the binary systems. The use of other reference substances (Table 1) supports the coding in some cases. The denomination of the phases A, B, C, F and G in the system in Fig.1 therefore goes back to known phases and is necessarily fixed.

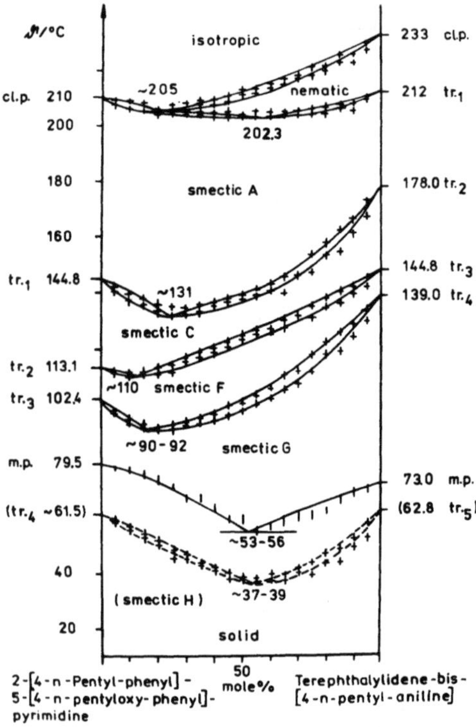

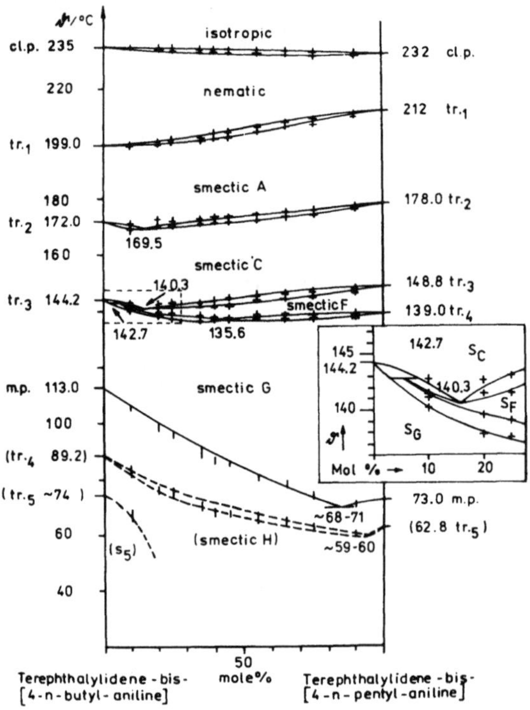

Fig.2 Miscibility in system no 15 [1]

Fig.3 Miscibility in system no 9 [1]

The relations between the phases of the AAOP and TBAA series are shown as examples in Fig.2 and Fig.3 (systems no 15 and no 9 in Fig.1). The co-miscibility between all smectic phases shown in Fig.2 transfers the code A C F G H from the phases of the standard substances PPOP to the phases of the 5-5 compound (TBPA). Starting with this substance makes the coding of the phases of the 4-4 compound (TBBA) possible (Fig.3). This system has also been proved by SAKAGAMI et al. [14] and by GOODBY et al. [15], and all the three systems show the same principal results on miscibility. There is a wide mixed phase region F, starting from the 5-5 compound. But the region is limited near the pure compound TBBA. Therefore TBBA possesses the sequence N A C G(H 5). In this way the phases G and H, which DOUCET et al. [16] have investigated by X-rays have been coded.

In the relation of the AOBAA series the compound 40.2 plays a special role. The smectic phase was first investigated with X-rays by DE VRIES and FISHL [17]. As can be seen in Fig.1 the use of this substance in numerous binary systems always proves the existence of the polymorphism N G.

Fig.4 Miscibility in system no 26 [1]

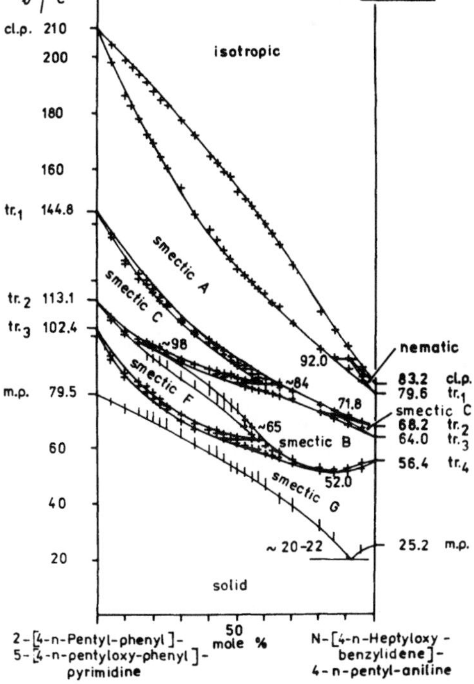

The existence of a sequence B G is realized in the substance 70.5. The existence of a B phase has been proved by miscibility with standard substance REF 1 (system no 47). As can be seen in Fig.4 the miscibility with the standard substance PPOP (system no 26) proves the existence of a G phase in this compound. In this diagram at concentrations of about 50 to 60 % PPOP the existence of a sequence of mixed phases A C B F G is shown.

Meanwhile this sequence B F G has been found in the compound 50.6. The nonidentified phase 4 has been proved to be an F phase by GRAY et al. [18].

2.3 I Phases

In the binary system no 5 with the members 5-5 and 9-9 (TBNA) of the TBAA series a phase in TBNA appears, which does not belong to the C, F and G phases and lies in the temperature sequence between the C and F phases. There is such a phase in the 10-10 compound, too (system no 10). This phase also appears in the substance REF 3 according to the miscibility in the binary system no 6. The miscibility between REF 3 and REF 1 (standard substance with a B phase) can be seen in Fig.5 (system no 7). The B phase region and a mixed phase region starting from REF 3 are separated by a small field of a phase transition.

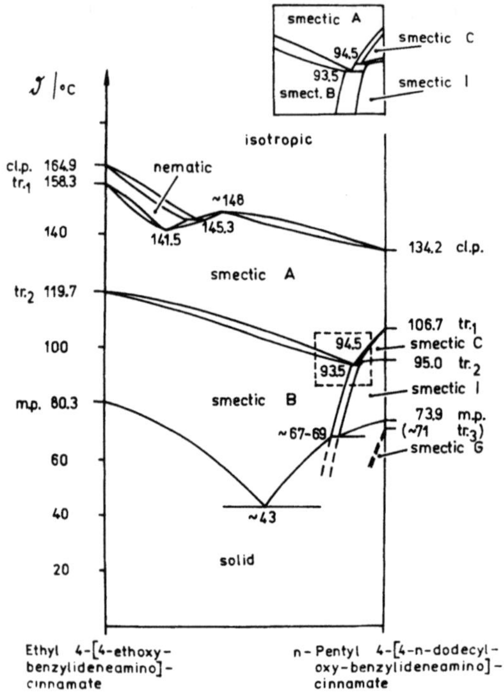

Fig.5 Miscibility in system no 7 [1] [19]

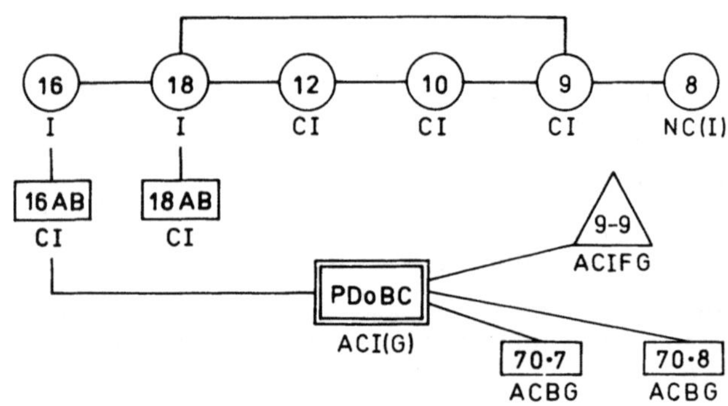

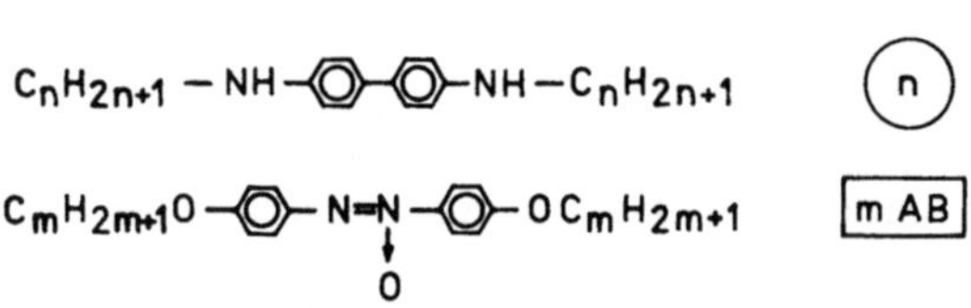

Fig.6 System of miscibility with I phases [19] [20]

These results are a strong indication of the existence of a new phase type I because this phase cannot be a C, F, G or H phase according to its appearence in the temperature sequence and it shows no uninterrupted miscibility with a B phase.

The substance REF 3 (PDoBC) can now be used as a reference substance for I phases. Further I phases have been found by miscibility with this standard (Fig.6). As can be seen in some members of two homologous series (derivatives of bis-alkylamino-biphenyls and alkyloxy-azoxybenzenes) I phases appear in a very simple polymorphism C I or in monomorphism. The diagrams of the binary systems of PDoBC with the 70.7 and 70.8 compounds of the AOBAA series show an interrupted miscibility between B and I phases similar to Fig.5. But no substance is known which possesses a B and an I phase. Therefore at first no statement on the temperature sequence of the B and I phases is possible.

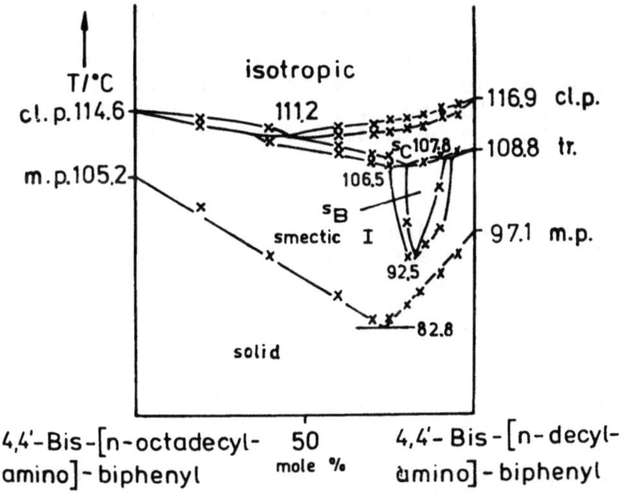

Fig.7 Miscibility between two members of the series of the bis-alkylamino-biphenyls [19] [20]

But in the diagram of the binary system of the two substances 18 and 10 in Fig.7 an intermediated mixed phase region B exists which is separated at lower temperatures from a mixed phase region I. This is an argument for the order of B phases as high temperature modifications with respect to I phases.

2.4 Summary and Supplementations

The results on miscibility in section 2 can be summarized in a common variant of polymorphism (1)

N A C B I F G H (1)

The temperature sequences of all variants of polymorphism described result from this variant (1) by omitting of one or several phases.

The miscibility investigations start from the standard substance POPP. The code of the phases A, C, F and G is transfered to the homologues of the series AAOP, TBAA and AOBAA. By the use of the reference substance REF 1 the code B has to be introduced in the AOBAA series. The phases which follow the G phases with decreasing temperature in the PPOP and TBAA series belong to a new phase type H. These phases H are linked by uninterrupted miscibility. Phases which show no uninterrupted miscibility with B phases and lie in the temperature sequence between C and F phases represent a further new phase type I. It has been found at first in the TBAA series. In mixed phases I phases are low temperature modifications with respect to B phases.

E phases have not been found in the substances reported here. E phases in polymorphism are only known together with A and B phases as low temperature modification in other substances (e.g.[21]). Therefore, they cannot be ordered in the temperature sequence (1) with respect to the I, F, G and H phases. The sequence (1) ...B I F G H has to be modified according to

$$. . . B \quad E \qquad (1\ a)$$

Intensive investigations independently done at the same time by GOODBY and GRAY [22][23] confirm and complete these results on the existence of the phases F, G and H and their temperature sequence. These authors have investigated homologous series of the alkylphenylesters and the alkoxy-phenylesters of 4'-n-alkyloxyphenyl-4-carboxylic acids and 4'-n-alkylbiphenyl-4-carboxylic acids. All the variants of polymorphism found can be derived from (1) (With respect to differences in the use of the letter symbols of the phases G and H see [24]).

Moreover BARRALL et al. [25] have investigated the substance bis-(4'-n-heptyloxybenzylidene)-1,4-phenylene-diamine with hexamorphic smectic polymorphism N C F J G H K. The existence of a phase between the temperature regions of the F and G phases refers to a new phase type J. A phase, which exists below the low temperature modification H indicates a further new phase type K. Comparing investigations on miscibility especially with substances forming sequences I F G have to be made.

3. Phase Types and Structure

The great number of smectic phases which are characterized by miscibility measurements faces an increasing number of phases, which are characterized by structure investigations. These investigations possess different methodical and experimental levels. In order to compare the phase types with structure a first summary can be given, when some simple but essential characteristics of the structures are used, which are available in extensive materials.

The smectic layer structures, as is known, are subdivided in two groups with disordered and ordered arrangements of the molecules within the layers. A second characteristic concerns the tilt of the molecular axes. In both groups normal and tilted structures are known i.e. structures with the average molecular axes parallel or tilted to the layer normal.

In both the ordered normal and ordered tilted structures hexagonally packed arrangements of the axes are known announcing a relatively high rotational motion along the molecular axes. Other structures are known with a more strongly hindered rotational motion. That leads to herring-bone packings in the normal and tilted groups. Further differentiations in structures are connected with the degree of rotational disorder, and especially with the local order in the layers and the correlation between the layers.

Since the X-ray investigations by LEVELUT and LAMBERT [26] in B phases examples have been known with hexagonally packed normal and tilted structures (pseudo-hexagonal structure), often shortly distinguished as B normal and B tilted structures. The tilted structure with a hexagonal arrangement has been found in the 4-4 compound (TBBA) of the TBAA series [16]. This compound possesses the phase sequence N A C G H 5 in the phase type designation according to the results in Fig.1. These new miscibility investigations proved the existence of a G phase at the former place of a B phase. Therefore, the problem of the existence of two B phases with different structures, which are miscible, goes back to the existence of a phase transition which could not be found in earlier miscibility investigations. B phases possess a hexagonal layer structure as it is found in the B phase of the standard substance REF 1 [26]. Hexagonally tilted structures do not belong to this phase type.

According to the X-ray investigations by DIELE et al. [27] [28], in the homologous series of the AAOP and TBAA the G phases have the structure of the G phase of TBBA i.e. a pseudo-hexagonal structure with local herring-bone order. In the F phases of these series a tilted structure with pseudo-hexagonal symmetry is also found. This is in agreement with the structure investigation in an oriented F phase of the 5-5 compound (TBPA) by LEADBETTER et al.[29] who found a locally c-centered monoclinic cell with regular hexagonal packing of the molecules.

In the I phases the existence of an ordered tilted structure can be found [28]. In the I phase of the substance REF 3 a more detailed investigation proves the existence of a hexagonally tilted structure with special orientations of the central part and the side chains of the molecules [30]. Therefore a trace of tilted ordered smectic phases exists in the sequence H G F I of the molecules with increasing temperature, starting with a herring-bone like structure in the H phases, which seems to be only a local one in the G phases. There is also a similar trace of normal ordered structures in the sequence E B. A herring-bone like arrangement of the molecules can be found in E phases and a hexagonal one in B phases.

These results give a first frame of the relations between phase types and strcuture in connection with the sequence rule (1). But it needs further investigations to differentiate more exactly between the structures in G, F and I phases. This emphasizes the valuable confrontation of the phase type system with a structure system.

4. Conclusion

This paper gives a survey on the order of smectic liquid crystalline phases by miscibility measurements. The phases, which are related by uninterrupted miscibility are coded by letter symbols. The phases with the same symbol belong to one phase type. The name "phase type" characterizes the thermodynamical treatment, a method of which is the investigation of the miscibility of phases. It is shown that this original intention of the Halle Liquid Crystal Group proved useful also under the expansion to substances with higher smectic polymorphism. The sequences (1) and (1 a) summarize the latest position of this order of smectic phases. These sequences contain the number of phase types derived from miscibility measurements as well as the realization of the phases in the temperature sequence in real cases.

The symbols of phases by letters have been used as symbols of structures at the same time. This occurs, above all, in the absence of a system of structures of liquid crystals. Nevertheless this proceeding proves to be successful, if the connection with the thermodynamical origin of these symbols has been observed.

The sequences (1) and (1 a) summarize smectic phase types, which possess layer structures. The system of these structures can be described in structures "disordered" in the layers (A, C) and "ordered" in the layers (B, E, I, F, G, H), in "normal" and "tilted" structures (A, B, E) and (C, I, F, G, H).

The relations between structure and phase type finally mean a way going from structure to phase transitions and from phase transitions to miscibility, each of them is a problem and they are all connected.

References

1 L.Richter Dissertation Halle (Saale) 1979

2 A.Biering, D.Demus, L.Richter, H.Sackmann, A.Wiegeleben and H.Zaschke Mol.Cryst.Liqu.Cryst. in print

3 L.Richter, D.Demus and H.Sackmann Mol.Cryst.Liqu.Cryst. in preparation

4 H.Sackmann J.de Phys.Colloq. C 3 40 C3-5 (1979)

5 H.Sackmann Reports III. Liquid Crystal Conference Budapest (1979) in print

6 H.Arnold and H.Sackmann Z.Elektrochemie, Ber.Bunsenges. f.physik.Chemie 63 1171 (1959)

7 D.Demus and H.Sackmann Z.physik.Chemie (Leipzig) 238 215 (1968)

8 D.Demus, G.Kunicke, G.Pelzl, B.Röhling and H.Sackmann Z.physik.Chemie (Leipzig) 254 373 (1973)

9 D.Demus, H.Sackmann, G.Kunicke, G.Pelzl and R.Salffner Z.Naturforsch. 23a 76 (1968)

10 D.Demus, L.Richter, C.E.Rürup, H.Sackmann and H.Schubert J.de Phys.Colloq. 36 C1-349 (1975)

11 D.Demus, M.Klapperstück, R.Rurainski and D.Marzotko Z.physik.Chemie (Leipzig) 246 385 (1971)

12 H.Arnold and H.Sackmann Z.physik.Chemie (Leipzig) 213 262 (1960)

13 D.Demus, S.Diele, M.KLapperstück, V.Linke and H.Zaschke Mol.Cryst.Liqu.Cryst. 15 161 (1971)

14 S.Sakagami, A.Takase and M.Nakamizo Mol.Cryst.Liqu.Cryst. 36 262 (1976)

15 J.W.Goodby, G.W.Gray and A.Mosley Mol.Cryst.Liqu.Cryst. 41 (Letters) 183 (1978)

16 J.Doucet, A.M.Levelut and M.Lambert Phys.Rev.Lett. 32 301 (1976)

17 A.de Vries and D.L.Fishl Mol.Cryst.Liqu.Cryst. 16 311 (1972)

18 G.W. Gray, J.W. Goodby, A.J. Leadbetter, M.A. Mazid: these proceedings, p.3

19 N.K.Sharma Dissertation Halle (Saale) 1979

20 N.K.Sharma, W.Weißflog, L.Richter, S.Diele, B.Walther, H.Sackmann and D.Demus Reports III. Liquid Crystal Conference Budapest (1979) in print

21 A.Biering, D.Demus, G.W.Gray and H.Sackmann Mol.Cryst. Liqu.Cryst. 28 275 (1974)

22 J.W.Goodby and G.W.Gray J.de Phys.Colloq. C 3 40 C3-27 (1979)

23 J.W.Goodby and G.W.Gray J.de Phys.Colloq. C 3 40 C3-363 (1979)

24 D. Demus, J.W. Goodby, G.W. Gray, and H. Sackmann: these proceedings, p. 31

25 E.M.Barrall, J.W.Goodby and G.W.Gray Mol.Cryst.Liqu.Cryst. 49 (Letters) 219 (1979)

26 A.M.Levelut and M.Lambert C.R.Séan.Acad.Sci. 272 1018 (1972)

27 S.Diele, D.Demus, A.Echtermeyer, U.Preukschas and H.Sackmann Acta Physica Polonica A 55 125 (1979)

28 S.Diele, H.Hartung, P.Ebeling, D.Vetters, H.Krüger and D.Demus Reports III. Liquid Crystal Conference Budapest (1979) in print

29 A.J.Leadbetter, J.P.Gaughan, B.Kelly, G.W.Gray and J.W. Goodby J.de Phys.Colloq. C 3 40 C3-178 (1978)

30 S.Diele, D.Demus and H.Sackmann Mol.Cryst.Liqu.Cryst. (Letters) in print

Recommendation for the Use of the Code Letters G and H for Smectic Phases

D. Demus[1], J.W. Goodby[2], G.W. Gray[2], and H. Sackmann[1]

1 Sektion Chemie, Martin-Luther-Universität, Halle-Wittenberg, DDR-402 Halle (Saale), German Democratic Republic

2 Department of Chemistry, The University, Hull, HU6 7RX, England

In the course of investigations on smectic polymorphism and of the coding of smectic phases by letters, an inverse denomination of the phases G and H came into use. Examples of this inverse use are demonstrated by the substances:

2-[4'-n-Pentylphenyl]-5-[4''-n-pentyloxyphenyl]pyrimidine (PPOP),
Terephthalylidene-bis-[4-n-butylaniline] (TBBA),
and
N-[4-n-butyloxybenzylidene]-4'-ethylaniline (BBEA).

Until now, the phases given by these substances with increasing temperature have been coded according to either (a) or (b)

	PPOP		TBBA		BBEA
(a)	H G F C A	(a)	H G C A N	(a)	G N
(b)	G H F C A	(b)	G H C A N	(b)	H N

Therefore all those phases which may be related to the G and H phases of those substances by uninterrupted miscibility were inversely coded either according to (a) or to (b). This arises because of the transfer of the code letters of phases from one substance to another by the criterion of uninterrupted miscibility; these two inverse codings for all known G and H phases exist in several publications in the literature.

In order to avoid this confusing situation, the authors have agreed to use in future the sequence (a) H G, following historical precedence.

In 1971 the *first* G phase was described for the compound PPOP using the sequence G F C A [1]. At this time, besides these smectic phases, only smectic phases B, D, and E were known. Later on phases were found which followed the G phase at lower temperatures. This was the case not only for TBBA and several of its homologues, but also for homologues in several other series. These phases were mutually related by uninterrupted miscibility, and the use of a further code letter became necessary; consequently this should be H.

Therefore for the above substances, only the following code sequence will be used by the authors in the future

PPOP	H G F C A
TBBA	H G C A N
BBEA	G N

As further examples, the codes of phases in two homologous series are shown in Figures 1 and 2.

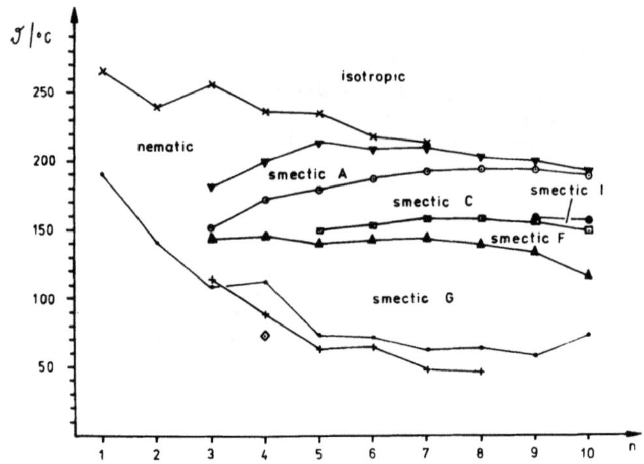

Fig.1 Phases and phase transitions for the homologous series of terephthalylidene-bis-[4-n-alkylanilines] [2]

n = number of C atoms in the alkyl chains
· = melting point
+ = transition smectic H → smectic G
◇ = transition smectic 5 → smectic H

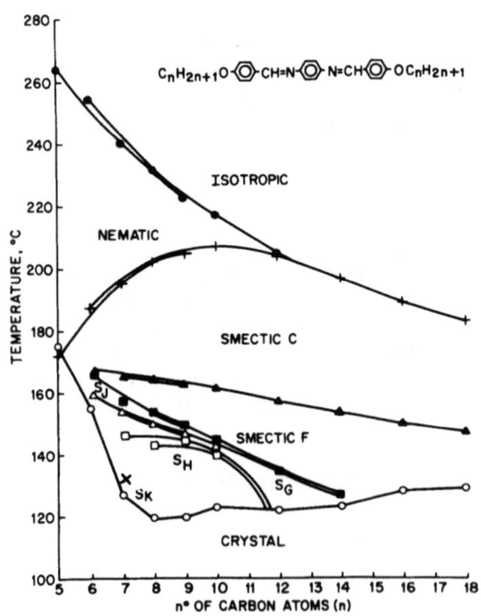

Fig.2 Phases and phase transitions for the homologous series of bis-[4'n-alkoxybenzylidene[-1,4-phenylenediamines [3]

If both phases G and H exist in one substance therefore, the G phase is always the higher temperature modification with respect to the H phase.

The authors strongly recommend this agreement for general use in the denomination of smectic phases G and H. They are preparing a list of all known substances possessing G and H phases.

References

[1] D Demus, S Diele, M Klapperstück, V Link, and H Zaschke, *Mol Cryst Liq Cryst*, *15*, 161 (1971).

[2] L Richter, D Demus, and H Sackmann, *Mol Cryst Liq Cryst* (in preparation).

[3] EM Barrall II, JW Goodby, and GW Gray, *Mol Cryst Liq Cryst Lett*, *49*, 319 (1979); JW Goodby and GW Gray, these proceedings

Magnetic Resonance of Chiral and Achiral Smectics

R. Blinc, M. Luzar, J. Seliger, M. Vilfan, and V. Rutar

J. Stefan Institute, E. Kardelj University of Ljubljana,
61000 Ljubljana, Yugoslavia

Abstract

A review is presented of the results of the Ljubljana group in studying the problem of local orientational ordering in the smectic phases of chiral DOBAMBC, chiral TBACA and achiral TBBA and IBPBAC as well as $(C_{10}H_{21}NH_3)_2CdCl_4$, the hydrocarbon part of which represents a smectic lipid bilayer exhibiting two structural phase transitions in analogy to biomembranes. A Landau theory describing the melting of bilayer membranes in terms of order parameters used in the theory of liquid crystals has been constructed to explain the obtained results.

1. Introduction

The main contribution of magnetic resonance to the physics of liquid crystals has been the determination of the local orientational order as a function of the atomic position in both thermotropic and lyotropic liquid crystalline molecules. One of the unsolved problems in this field concerns the nature of the orientational ordering of the short molecular axes in the various smectic phases and, in particular, the difference in this ordering between chiral and achiral biaxial smectics. According to one point of view, the tilting of the molecules with respect to the smectic planes is connected with the biasing of the rotational motion of the molecules around their long axes [1] in both chiral and achiral compounds. Alternative models have been however formulated where the molecular tilt takes place while the molecules are still freely rotating [2]. The discovery of ferroelectricity [3] in the tilted smectic phases of chiral liquid crystals has further enhanced the interest in this problem. Still another unsolved problem is the relation between the local orientational order and the nature of the two phase transitions [4] in lipid bilayer membranes.

Here we shall review the results of the Ljubljana group in studying:

i) the nature of the local orientational ordering in the tilted smectic (Sm) phases of achiral TBBA and chiral TBACA as compared to the untilted SmB phase of achiral IBPBAC,

ii) the nature of the local orientational order in the ferroelectric phase of chiral DOBAMBC as compared to achiral HOAB,

iii) the nature of the two phase transitions in $(C_{10}H_{21}NH_3)_2CdCl_4$, the hydrocarbon part of which represents a smectic lipid bilayer embedded into a crystalline matrix.

The main experimental techniques we used in this study are proton–nitrogen nuclear quadrupole double resonance [5] and proton enhanced ^{13}C nuclear magnetic resonance [6].

2. Chiral and Achiral Smectics

In the SmA phase the molecules are rotating "freely" around their long axes which are – on the time average – oriented perpendicularly to the smectic layers ($n_z \neq 0$, $n_x = n_y = 0$).

Achiral (centrosymmetric) SmA systems exhibit the point group symmetry $D_{\infty h}$ and may undergo a large variety of structural phase transitions induced by the various irreducible representations of $D_{\infty h}$. The irreducible representation E_{1u} induces the transition to the homogeneously ordered tilted SmC phase. The symmetry of the new phase is C_{2v} and the primary order parameter is the tilt angle, i.e. the quadratic combinations $\xi_1 = n_z n_x$ and $\xi_2 = n_z n_y$ of the components of the molecular director $\vec{n}$, describing the orientation of the long molecular axes with respect to the smectic plane normal.

Chiral (non–centrosymmetric) SmA systems belong to the point group D_∞. The two dimensional representation E_1 of D_∞ induces the transition to the chiral SmC* phase where the point symmetry of the layers is reduced to C_2. If the molecules have a permanent dipole moment perpendicular to their long axes, the new phase may be ferroelectric. The order parameters of the transition are $\xi_1 = n_z n_x$ and $\xi_2 = n_z n_y$ as well as the components of the inplane spontaneous polarization P_x and P_y. In contrast to the achiral SmC phase the chiral SmC* phase is not homogeneous but shows a helicoidal precession of the tilt (and the polarization) as one proceeds from one smectic layer to another along the z–axis.

The helicoidal distribution of the molecular tilt and the spontaneous polarization is the result of the presence of a Lifshitz term

$$\xi_1 \frac{\partial \xi_2}{\partial z} - \xi_2 \frac{\partial \xi_1}{\partial z} \tag{1a}$$

or

$$P_x \frac{\partial P_y}{\partial z} - P_y \frac{\partial P_x}{\partial z} \tag{1b}$$

in the expansion of the free energy in terms of the order parameters. Such a term is allowed by symmetry in D_∞ but forbidden in $D_{\infty h}$. The periodicity of the helix will be, in the general case, incommensurate with the one dimensional translational periodicity of the SmA phase. In a strong enough magnetic field the helicoidally modulated SmC* phase may become unstable and may undergo a transition to the homogeneous SmC phase.

3. The Biaxial Smectic Phases of TBBA and TBACA and the Smectic B Phase of IBPBAC

The local orientational ordering of the central part of the achiral terephtal–bis–butyl–aniline (TBBA) molecule and of the chiral p–terephtal–bis–amino–methyl–butyl–cinnamate (TBACA) molecule have been studied via ^{14}N NQR. The sequence of phase transitions in TBBA is as follows [7]

$$\begin{array}{ccccccccccccc} \text{IX} & & \text{VIII} & & \text{V} & & \text{IV} & & \text{III} & & \text{II} & & \text{I} \\ \text{Cr} & \xrightleftharpoons{-33^{\circ}\text{C}} & \text{Cr} & \xrightarrow{113^{\circ}\text{C}} & \text{SmH} & \xrightleftharpoons{144^{\circ}\text{C}} & \text{SmC} & \xrightleftharpoons{172^{\circ}\text{C}} & \text{SmA} & \xrightleftharpoons{195^{\circ}\text{C}} & \text{N} & \xrightleftharpoons{235.5^{\circ}\text{C}} & \text{Is} \\ & 52^{\circ}\text{C} & \uparrow & 68^{\circ}\text{C} & \Updownarrow 84^{\circ}\text{C} & & & & & & & & \\ & & \text{VII} & \rightleftharpoons & \text{VI (SmG)} & & & & & & & & \end{array}$$

and in TBACA [8]

$$\text{Cr}_1 \xrightarrow{130^{\circ}\text{C}} \text{SmH}^*(\text{Cr}_2) \xrightarrow{149^{\circ}\text{C}} \text{SmC}^* \xrightarrow{180^{\circ}\text{C}} \text{SmA} \xrightarrow{242^{\circ}\text{C}} \text{N} \xrightarrow{288^{\circ}\text{C}} \text{Is.}$$

In addition we as well studied the orientational ordering in the untilted SmB phase of IBPBAC:

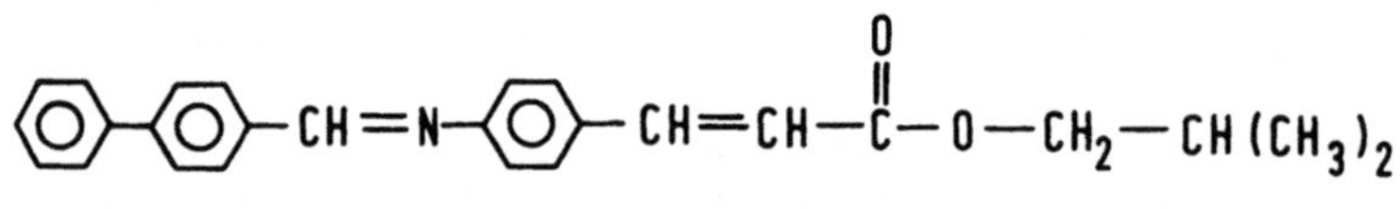

which exhibits the following sequence of phases [9]:

$$\text{Cr} \underset{\text{supercools}}{\xrightleftharpoons{86^{\circ}\text{C}}} \text{SmE} \xleftrightarrow{114^{\circ}\text{C}} \text{SmB} \xleftrightarrow{162^{\circ}\text{C}} \text{SmA} \xleftrightarrow{206^{\circ}\text{C}} \text{N} \xleftrightarrow{214^{\circ}\text{C}} \text{Is}.$$

The SmE phase is the untilted analogue of the SmVI phase with a distorted hexagonal molecular arrangement within each layer.

The possible biasing of the rotation of the C–N=CH–C groups around the long molecular axis is checked via a determination of the asymmetry parameter

$$\eta = |V_{XX} - V_{YY}|/V_{ZZ} \tag{2}$$

of the electric field gradient (EFG) tensor at the ^{14}N sites. For free uniaxial rotation, the largest principal axis V_{ZZ} will point along the long molecular axis and the EFG tensor will be axially symmetric: $V_{XX} = V_{YY}$, $\eta = 0$. Any anisotropy in the rotation around the long molecular axis will destroy the axial symmetry of the ^{14}N EFG tensor resulting in a finite η and the appearance of three ^{14}N pure NQR lines

$$\nu_{1,2} = (3/4)(eQV_{ZZ}/h)(1 \pm \eta/3)\,, \qquad \nu_3 = \nu_1 - \nu_2 \tag{3}$$

instead of a single one for $\eta = 0$.

If the 5 independent EFG tensor components in the molecular frame are known [5] and if a model for the molecular motion is assumed, one can calculate all the components of the time averaged EFG tensor and compare the resulting eQV_{ZZ}/h and η with the experiment. For uniaxial rotation we transform [5] the EFG tensor $V_{\alpha\beta}$ from the molecular frame x_o, y_o, z_o (Table 1) to a frame x, y, z rotating around the $z \parallel z_o$ axis where $x \perp y \perp z$ and $\varphi = \varphi(t)$ is the angle between x and x_o. The anisotropic fluctuations of the long molecular axis are taken into account [5] by still another transformation to a x', y', z' frame where z' is normal to the smectic planes, x' is the projection of the long molecular axis on the smectic plane and $y' \perp x' \perp z'$. The tilt angle $\theta(t)$ is the angle between the long molecular axis and z', whereas $\phi = \phi(t)$ is the angle between the projections of the instantaneous and average directions of the long molecular axis on the smectic planes.

Table 1 Rigid lattice ^{14}N quadrupole coupling tensor of TBBA [5] expressed in the x_o, y_o, z_o molecular frame. Here z_o is the long molecular axis, y_o is the normal to the C–N=CH–C plane and x_o is perpendicular to y_o and z_o.

$$eQV_{\alpha\beta}/h\,[\mathrm{MHz}] = \begin{pmatrix} -3.15\,, & \pm 0.15\,, & \pm 2.41 \\ \pm 0.15\,, & 1.95\,, & \pm 0.40 \\ \pm 2.41\,, & \pm 0.40\,, & 1.20 \end{pmatrix}$$

a) Bipolar biasing of uniaxial rotation: For bipolar biasing of the uniaxial rotation one has $\langle\cos 2\varphi\rangle \neq 0$, $\langle\cos\varphi\rangle = \langle\sin\varphi\rangle = \langle\sin 2\varphi\rangle = 0$. For $\langle\cos 2\varphi\rangle \leqslant 0.3$ we now find

$$\eta = r\langle\cos 2\varphi\rangle \tag{4a}$$
$$eQV_{ZZ}/h = eQV_{z_oz_o}/h \tag{4b}$$

where $r = 2[V^2_{x_oy_o} + 1/4(V_{x_ox_o} - V_{y_oy_o})^2]^{1/2}/V_{z_oz_o}$.
If, on the other hand, $|r\langle\cos 2\varphi\rangle| > 1$, we get

$$\eta = |(r\langle\cos 2\varphi\rangle - 3)/(r\langle\cos 2\varphi\rangle + 1)| \tag{5a}$$

$$eQV_{ZZ}/h = |-(1/2h)\,eQV_{z_oz_o}(1 + r\langle\cos 2\varphi\rangle)|\ . \tag{5b}$$

b) Polar biasing of uniaxial rotation: For small polar biasing of uniaxial rotation $\langle\cos\varphi\rangle \neq 0$, $\langle\cos 2\varphi\rangle = \langle\sin\varphi\rangle = \langle\sin 2\varphi\rangle = 0$ one gets in the limit $|\langle\cos\varphi\rangle| \ll 1$:

$$\eta = K\,.\,\langle\cos\varphi\rangle^2 \tag{6a}$$

$$eQV_{ZZ}/h = (eQV_{z_oz_o}/h)(1 + \eta) \tag{6b}$$

where

$$K = (2/3)(V^2_{x_oz_o} + V^2_{y_oz_o})/V^2_{z_oz_o}\ . \tag{6c}$$

c) Anisotropic fluctuations in the orientation of the axis of rotation: Anisotropic fluctuations in the direction of the long molecular axis as well result in a non–zero value of η though $\langle\cos 2\varphi\rangle = \langle\cos\varphi\rangle = 0$:

$$\eta = 3/2\,[\langle\sin\theta\,\cos\theta\,\cos\phi\rangle^2 - \langle\cos^2\theta\rangle\langle\sin^2\theta\,\cos^2\phi\rangle + \langle\sin^2\theta\,\sin^2\phi\rangle] \tag{7a}$$

$$eQV_{ZZ}/h = (eQV_{z_oz_o}/h)\,[1 + 3/2(\langle\sin\theta\,\cos\theta\,\cos\phi\rangle^2 - \langle\cos^2\theta\rangle\langle\sin^2\theta\,\cos^2\phi\rangle - \langle\sin^2\theta\,\sin^2\phi\rangle)]\ , \tag{7b}$$

but the temperature dependence of η is very different from the above two cases. The temperature dependence of η is not monotonous but exhibits a maximum as a result of the presence of competing terms.

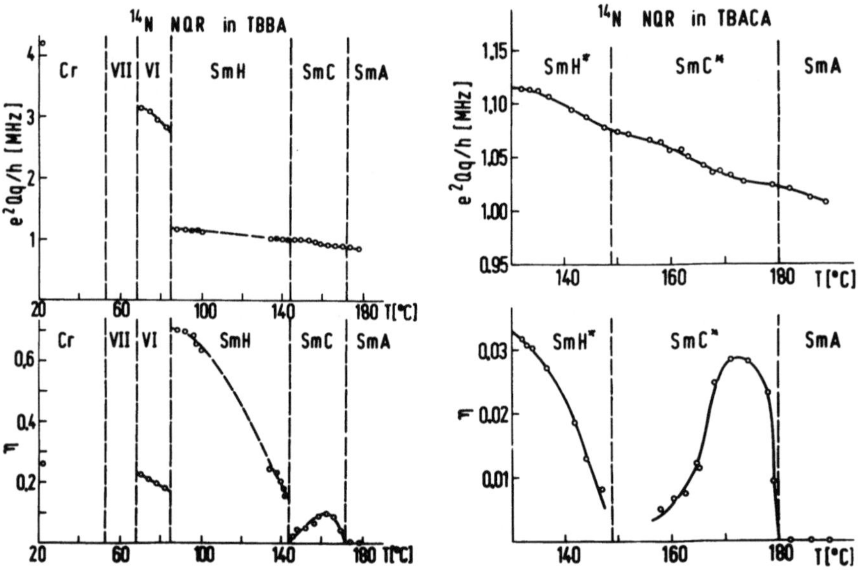

Fig.1 Temperature dependence of the ^{14}N quadrupole coupling constant and asymmetry parameter in TBBA and TBACA

The temperature dependences of e^2qQ/h (= eQV_{ZZ}/h) and η for TBBA and TBACA are shown in Fig.1. The results show [5] that in the biaxial smectic phases of achiral TBBA the bipolar order parameter $\langle\cos 2\varphi\rangle$ is different from zero whereas in chiral TBACA it is the polar order parameter $\langle\cos\varphi\rangle$ which is different from zero. The occurrence of a maximum in η in the SmC phase of TBBA and the SmC* phase of TBACA demonstrates that the finite value of η in these phases is mainly produced by anisotropic fluctuations in the direction of the long molecular axes whereas the orientational ordering of the central part of the molecules is relatively small. In the SmH and SmVI phases, on the other hand, anisotropic fluctuations are relatively unimportant and the temperature dependence of η is dominated by the orientational ordering.

The temperature dependence of the polar order parameter $\langle\cos\varphi\rangle$ in the tilted phases of chiral TBACA and of the bipolar order parameter $\langle\cos 2\varphi\rangle$ in the tilted phases of achiral TBBA is shown in Fig. 2.

The situation is quite different in the untilted SmB phase of IBPBAC. Here the ^{14}N eQV_{ZZ}/h is comparable to the values found in TBBA (1.11 MHz at 137°C and 1.09 MHz at 146°C) whereas the value of η is much smaller ($<$ 0.015 at 137°C and $\sim$ 0 at 146°C).

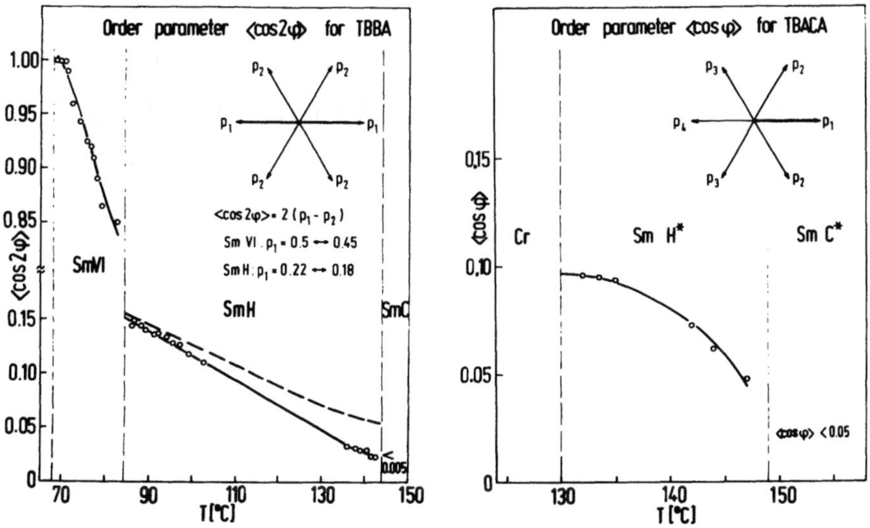

Fig.2 Temperature dependence of the bipolar orientational order parameter in TBBA and of the polar orientational order parameter in TBACA (broken line represents calculated $<\cos 2\varphi>$ if the model assumes fluctuations in the direction of the long axis as well)

This demonstrates that the rotation around the long molecular axis is practically "free" in the SmB phase, i.e. the molecules seem to reorient between 6 equivalent potential wells. The mean square fluctuations in the direction of the long molecular axis $(<\delta\theta^2>)^{1/2}$ are estimated to $\sim 12^\circ$ at 140°C.

The above results show that there is a close relation between the molecular tilt and the biasing of the molecular rotation around the long molecular axis in the two–dimensionally ordered chiral and achiral SmG and SmH phases though this relation may be unsignificant in the one dimensionally ordered SmC phase. The biasing is "bipolar" in achiral and "polar" in chiral smectics.

4. The Ferroelectric Smectic C* Phase of DOBAMBC

It is well known [3] that the SmC* liquid crystalline phase of chiral DOBAMBC (p–decyloxybenzilidene p'–amino 2–methyl–butyl–cinnamate) is ferroelectric. The pitch of the helix is of the order of several microns. Fig.3 contains a schematic presentation of the DOBAMBC molecule with numbers denoting different carbon nuclei whereas the temperature dependence of all resolved ^{13}C resonance lines of DOBAMBC is shown in Fig.4.

In the isotropic phase only the isotropic average of the chemical shielding tensor is observed: $\sigma_i = 1/3\, \mathrm{Tr}\, \underline{\sigma}$.

In the SmA phase the molecules are aligned along the direction of the magnetic field.

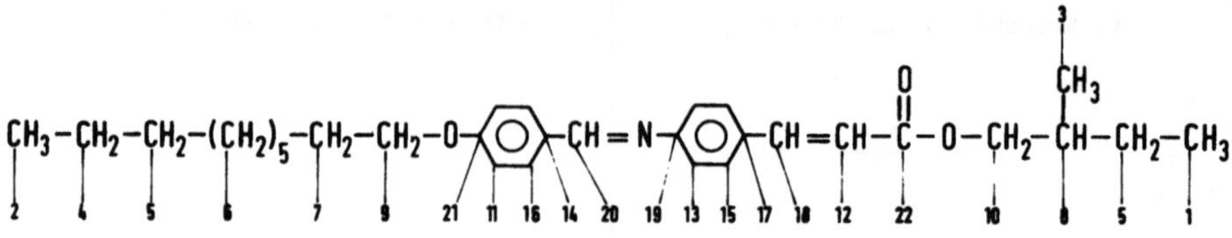

Fig.3 Schematic presentation of the DOBAMBC molecule with numbers denoting different carbon nuclei

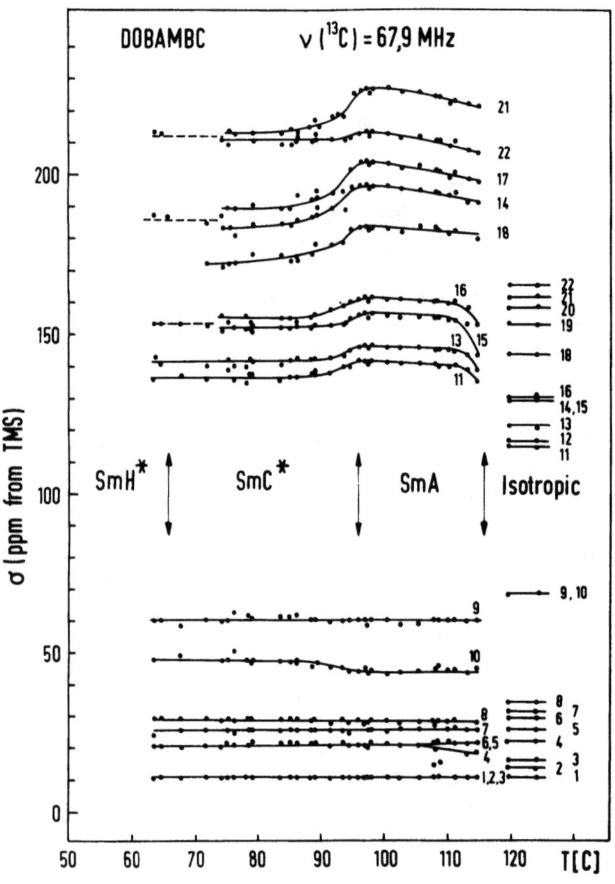

Fig.4 Temperature dependence of the ^{13}C chemical shifts (relative to TMS) in chiral DOBAMBC

Since for the aromatic ^{13}C nuclei the heaviest shielding occurs perpendicular to the aromatic plane, whereas for the aliphatic groups the most shielded element of $\underline{\sigma}$ lies parallel to the para axis we expect at the isotropic–SmA transition a downfield shift of the aromatic lines and an upfield shift of the aliphatic lines. This is exactly what has been indeed observed (Fig.4). The observed change in the chemical shift on going from the isotropic to the SmA phase is given by:

$$\Delta\sigma = \sigma - \sigma_i = \frac{2}{3} S (\overline{\sigma}_{\parallel} - \overline{\sigma}_{\perp}) , \tag{8a}$$

where S is the nematic order parameter, $\overline{\sigma}_{\parallel}$ and $\overline{\sigma}_{\perp}$ are the components of the chemical shift tensor averaged over molecular rotation around the long axes. The observed S values vary from $S \approx 0.7$ near the SmA–isotropic transition up to $S = 0.8$ at the SmA – SmC* transition. These values refer to the aromatic carbons whereas the S values for the aliphatic carbons are significantly smaller.

On going from the SmA to the SmC* phase a steady increase in S and a corresponding increase in the aromatic ^{13}C chemical shifts is expected. This effect is indeed seen in achiral HOAB but not in chiral DOBAMBC, where a significant decrease of the ^{13}C chemical shifts of the ring and C = O carbons is observed. This effect can be understood if one assumes that in chiral systems the helicoidal axis and the normals to smectic layers remain parallel to the direction of the external magnetic field, whereas the long molecular axes are tilted with respect to the field direction. This is opposite to the behaviour of achiral systems where the molecular long axes are oriented parallel to the field, whereas the layer normals are tilted. Since our experiment gives us only the component of $\underline{\sigma}$ along the magnetic field direction, the observed chemical shifts in the smectic C* phase of DOBAMBC are given by:

$$\sigma = \sigma_i + \frac{2}{3} S (\overline{\sigma}_{\parallel} - \overline{\sigma}_{\perp})(\frac{3}{2}\cos^2\theta_o - \frac{1}{2}) , \tag{8b}$$

where θ_o is the average molecular tilt angle and the fluctuations in the direction of the long molecular axes are assumed to be isotropic.

The above model explains the anomalous temperature behaviour of the ring and C=0 carbon chemical shifts at the SmA → SmC* transition in terms of the molecular tilt θ_o. If this explanation is correct the temperature dependence of

$$\frac{\sigma - \sigma_i}{\frac{3}{2}\cos^2\theta_o - \frac{1}{2}} = \frac{2}{3}(\overline{\sigma}_{\parallel} - \overline{\sigma}_{\perp}) S \tag{8c}$$

should reflect the temperature dependence of S which is known to increase monotonously with decreasing temperature. This is nearly so for the ring but not for the C=0 carbons (Fig.5). This discrepancy seems to be the result of a small polar biasing of the rotation of the C=0 group, which is the main ferroelectric dipole in the structure. In such a case expression (8b) has to be extended to

$$\sigma - \sigma_i = \frac{2}{3} S (\overline{\sigma}_{\parallel} - \overline{\sigma}_{\perp})(\frac{3}{2}\cos^2\theta_o - \frac{1}{2}) - \sigma_{\xi\zeta} \sin(2\theta_o) \langle\cos\varphi\rangle \tag{8d}$$

where the effect of fluctuations in the direction of the long molecular axis on the second term of expression (8d) has been neglected, $\zeta \| \vec{n}$ and $\sigma_{\xi\zeta}$ can be estimated from the literature [10] as amounting to ~ 100 ppm. By ascribing the deviation of the experimental data from the straight line predicted by (8c) to the "biasing" term $\sigma_{\xi\zeta} \sin(2\theta_o) \langle\cos\varphi\rangle$ we find that $\langle\cos\varphi\rangle$ amounts to $\sim 5 \times 10^{-2}$. The temperature dependence (Fig.5) of this "biasing" shift Δ is given by:

$$\Delta \propto \sin(2\theta_o) \langle\cos\varphi\rangle \propto (T_c - T)^{0.7} \quad . \tag{9}$$

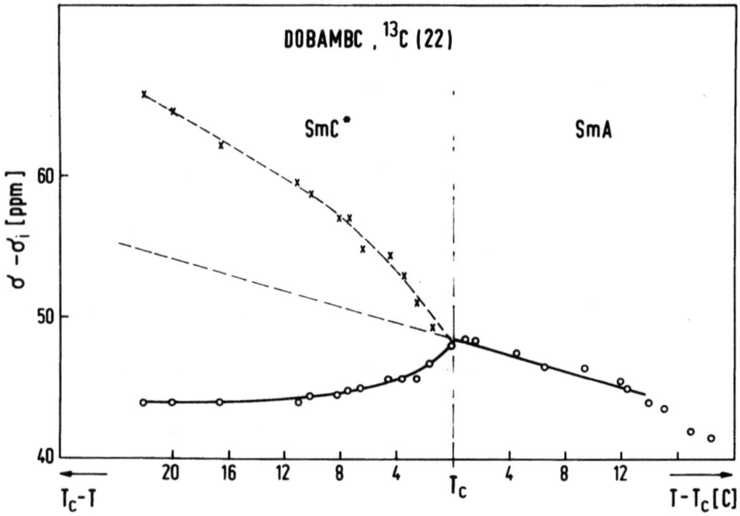

Fig.5 Temperature dependence of $\sigma - \sigma_i$ (ooo) for the C=O carbon at the SmA–SmC* transition in DOBAMBC. Crosses represent the temperature dependence of $x = (\sigma - \sigma_i)/(\frac{3}{2}\cos^2\theta_o - 1)$ and θ_o designates the molecular tilt angle. The deviation Δ of x from the straight line–extrapolated from the SmA phase – measures the effect of the polar orientational ordering: $\Delta \propto \sin(2\theta_o)\langle\cos\varphi\rangle$ in SmC* phase

The above result agrees rather well with the results of the direct measurements of the temperature dependence of the tilt angle

$$\theta_o \propto (T_c - T)^{0.31} \tag{10a}$$

and the polarization [11]

$$P_s \propto (T_c - T)^{0.37} \quad . \tag{10b}$$

The magnitude of $\langle\cos\varphi\rangle \cong 5 \times 10^{-2}$ as well agrees rather well with the one estimated from polarization measurements.

5. $(C_{10}H_{21}NH_3)_2CdCl_4$: a Lipid Bilayer Embedded in a Crystalline Matrix

The hydrocarbon part of $(C_{10}H_{21}NH_3)_2CdCl_4$ represents a smectic lipid bilayer exhibiting two structural phase transitions at T_{c1} = 35°C and T_{c2} = 39°C in analogy to biomembranes.

The projection of the structure of $(C_{10}H_{21}NH_3)_2CdCl_4$ (designated as C10Cd) on the b–c plane at room temperature is shown in Fig.6. The structure consists of $CdCl_4^{2-}$ layers sandwiched between well ordered alkylammonium chains which are tilted by 40° with respect to the normal to the layer. The ammonium end of each chain is linked to the layer by N–H ... Cl hydrogen bonds and each chain is coordinated by six others. The entropy change at the lower phase transition T_{c1} – (0.9 ± 0.3) R per mole chains – can be explained by an order–disorder transition of rigid chains between two equivalent sites. The entropy change at the higher phase transition T_{c2}, on the other hand, corresponds to 0.8 R per R–C–C–R bond and can be only explained by a "melting" of the chains, or what is equivalent, by rapid chain isomerization via kink diffusion.

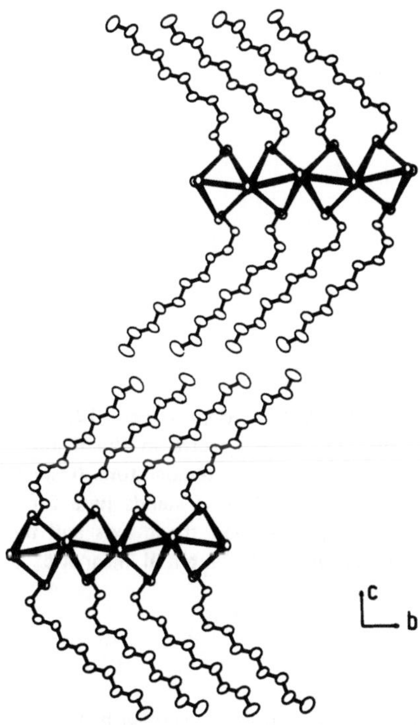

Fig.6 Projection of the structure of C10Cd on the b–c plane at room temperature

^{14}N quadrupole resonance data and proton magnetic resonance second moment data[12] show that on going from the low temperature to the intermediate phase the polar $-NH_3$ groups and the alkyl chains start flipping by 90° around their chain axes so that neighbouring chains move in opposite directions like a two–dimensional array of connected gears.

The ^{13}C spectra [12] of the low temperature and the intermediate temperature phases are nearly identical but there is a sharp change in the ^{13}C spectrum on going into the high temperature phase (Fig.7). The angular dependence of the $-CH_2$ group spectra in the low temperature phase is characteristic of two sets of rigid $-(CH_2)_n-$ chains which are tilted by $\pm$ 40° with respect to the layer normal. In the intermediate temperature phase the chains are still tilted by $\pm$ 40° and almost rigid. The bulk of the data can be described by an axially symmetric ^{13}C chemical shift tensor: $\sigma_\perp = \frac{1}{2}(\sigma_{11} + \sigma_{22}) = 43 \pm 3$ ppm, $\sigma_\parallel = \sigma_{33} = 15 \pm 5$ ppm with respect to TMS. The symmetry axis $\sigma_\parallel = \sigma_{33}$ is parallel to the chain direction. The axial symmetry of the ^{13}C tensor is due to the 90° flips around the chain axis which result in a rapid exchange of σ_{11} and σ_{22} but do not affect σ_{33}. The nematic order parameter S with respect to the preferred chain direction, $\theta_o = \pm 40°$, is here about one, $S \approx 0.95 - 0.99$. The high temperature phase data are completely different and show that all chains are equivalent and on the time average normal to the layers. The anisotropy of the chemical shift tensor is only partially removed by molecular motion and the nematic order is still significant:

$$S = \frac{(\sigma_\parallel - \sigma_\perp)_{obs}}{(\sigma_\parallel - \sigma_\perp)_{rigid}} = 0.55 \pm 0.10 \quad . \tag{11}$$

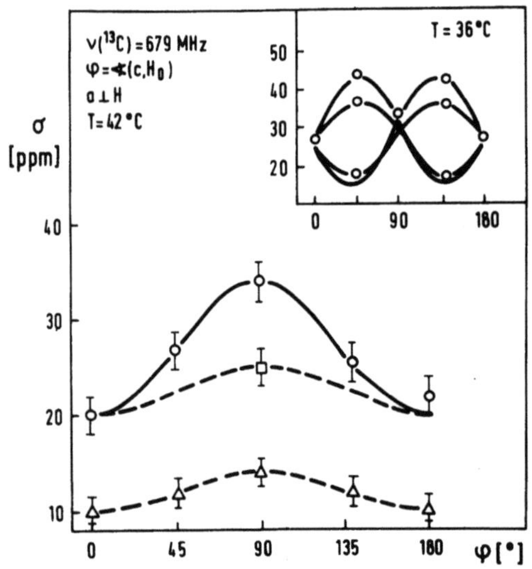

Fig.7 Angular dependence [12] of the ^{13}C NMR spectra in the high and intermediate temperature (insert) phases of C10Cd. φ stands here for the angle between the c axis and the direction of the external magnetic field.

The hydrocarbon part of C10Cd thus represents a smectic liquid crystal with a structure similar to the interior of the bilayer lipid membranes. Let us therefore try to describe [12] the two phase transitions in C10Cd by a Landau type expansion of the non-equilibrium free energy F in terms of order parameters θ_o, $S-S_c$ and p, used in the theory of smectic liquid crystals [12]:

$$F = \frac{1}{2}A\theta_o^2 + \frac{1}{4}B\theta_o^4 + \alpha(T-T_{c2})(S-S_c) - \frac{1}{2}c(S-S_c)^2 + \frac{1}{4}d\,(S-S_c)^4 + \frac{1}{2}ap^2 + \frac{1}{4}bp^4 - \frac{1}{2}A'\theta_o^2(S-S_c) - \frac{1}{2}a'p^2(S-S_c) \quad . \tag{12}$$

Here S stands for the average nematic order parameter and $A, B, \alpha, c, d, a, b, A'$ and a' are all positive constants. The SmC order parameter θ_0 gives the average tilt of the molecules with respect to the layer normal whereas p is the orientational order parameter for the 90° flipping of the chains and the terminal NH_3 groups between the two equilibrium orientations corresponding to $p = \pm 1$. S_c is the critical value of the average nematic order parameter S, i.e. the arithmetic average of S in the melted and rigid phase coexisting at the first order transition temperature T_{c2} (Fig.8). In deriving expression(12) we assumed that the melting of the chains is the important driving mechanism which induces the transitions in S as well as in θ_0 and p. It should be noted that S is – in contrast to the isotropic–nematic transition – here not a symmetry breaking order parameter and is different from zero both above and below T_{c2}. Because of that the linear term in the expansion of F in powers of $(S-S_c)$ is not identically equal to zero as in the case of nematic liquid crystals but has the same structure as the free energy at liquid–gas phase transitions.

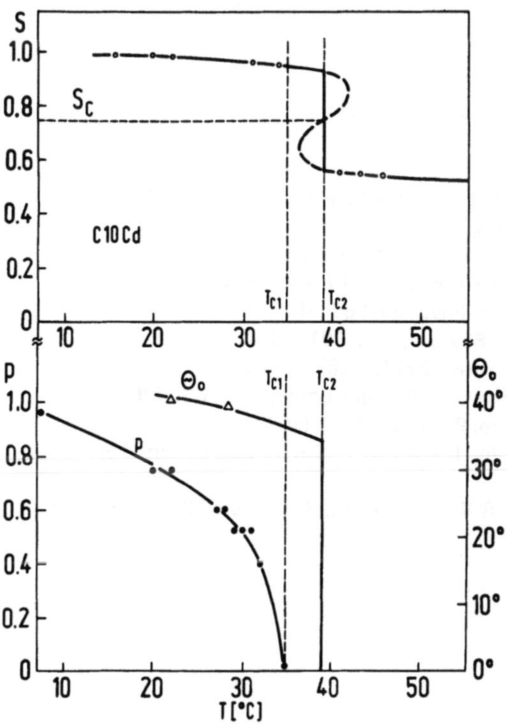

Fig.8 Comparison [12] between the calculated and the observed temperature dependences of the order parameters S, p and θ_0 for C10Cd

Minimizing the total free energy with respect to p, θ_0 and S we obtain the following stable solutions:

$$T > T_{c2}: S < S_c,\ \theta_o = 0,\ p = 0, \qquad (13a)$$

$$T_{c1} < T < T_{c2}: S > S_c,\ \theta_o \neq 0,\ p = 0\ , \qquad (13b)$$

$$T < T_{c1}: S \gg S_c,\ \theta_o \neq 0,\ p \neq 0\ . \qquad (13c)$$

The high temperature transition at T_{c2} corresponds to a partial melting of the chains which simultaneously destroys the tilting of the molecules whereas the low temperature transition at T_{c1} corresponds to an orientational transition and a disordering of the polar "heads". The temperature dependence of the order parameters S, θ_o and p is shown in Fig.8 together with the experimental values determined in this study.

Acknowledgment

The authors thank Professor Leadbetter for the sample of IBPBAC.

References

1. R.J.Meyer and W.L.McMillan: Phys. Rev. **A9**, 889 (1974); R.J.Meyer, Phys. Rev. **A12**, 1066 (1975)
2. R.G.Priest: J. Chem. Phys. **65**, 408 (1976)
3. R.B.Meyer, L.Liebert, L.Strzelecki, P.Keller: J. Physique Lett. **36**, 69 (1975)
4. D.Chapman: Q. Rev. Biophys. **8**, 185 (1975)
5. J.Seliger, V.Žagar, and R.Blinc: Phys. Rev. **A17**, 1149 (1978); R.Blinc, J.Seliger, M.Vilfan, V.Žagar: J. Chem. Phys. **70**, 778 (1979)
6. A.Pines, M.G.Gibby, J.S.Waugh: J. Chem. Phys. **59**, 569 (1973)
7. J.Doucet, A.Levelut, M.Lambert: Phys. Rev. Lett. **32**, 301 (1974)
8. W.Z.Urbach, J.Billard: C.R. Acad. Sci. (Paris) **B274**, 1287 (1972)
9. A.J.Leadbetter, R.M.Richardson, J.C.Frost: J. Physique Colloq. **40**, C3–125 (1979)
10. A.Pines, J.J.Chang, R.G.Griffin: J. Chem. Phys. **61**, 1021 (1974)
11. B.I.Ostrovski, A.Z.Rabinovich, A.S.Sonin, B.A.Strukov, S.A.Taraskin: Ferroelectrics **20**, 189 (1978)
12. R.Blinc, M.I.Burgar, V.Rutar, B.Žekš, R.Kind, H.Arend, G.Chapuis: Phys. Rev. Lett. **43**, 1679 (1979)

Physical Properties of Plastic Crystals

R.M. Pick

Département de Recherches Physiques*, Laboratoire associâ au C.N.R.S. n°71
Université P. et M. Curie, 4 Place Jussieu, F-75005 Paris, France

Summary

Plastic crystals are the counter part of liquid crystals, i.e. they are molecular crystals in which, in a certain temperature range, the molecular centers of mass form a regular crystal, while some disorder exists in the molecular orientations.

Only molecules with rather simple geometrical shapes (pseudo spheric, pseudo planar, or pseudo linear) can form plastic crystals, which are, usually, crystals of high symmetry. |1| and |2| give excellent reviews of such properties.

From a macroscopic point of view, these crystals are characterized

- by some low elastic constants (see |3| for a review) at low frequency; these constants recover a normal value at higher frequency |4-5|, when the coupling between deformations and molecular reorientations do not exist any more;

- by a high static plasticity, which means that permanent deformations can be easily obtained under small stresses. The relationship between plasticity and vacancy diffusion has been demonstrated close to the melting temperature Tm. The high plasticity is not related to a low activation energy of the vacancies, but to the high value of their diffusion constant close to Tm |6|. Nevertheless the relative role of their mobility and/or of their concentration is not yet known.

The orientational probability of the molecules is generally far from isotropy |7-8| which shows that the molecules sit, most of the time, at the bottom of potential wells. The anisotropy is related to steric hindrances which also leads to short range orientational correlations |9|.

Very little is presently known on the quasi harmonic dynamical properties of these systems. In practice, only the lowest part of the acoustical branches can be properly determined |4-5-10| . This is presumably related to a strong disorder in the force constants rather than to the short lifetime of these excitations. Cf. |10|.

The crucial problem of the reorientational mechanism of the molecules is far from understood, and certainly depends on the residence time at the bottom of the potential wells. In the case of CN^- in NaCN, it has been shown

* Laboratoire associé au C.N.R.S. n°71

Ref. |11|, that the molecules are trapped in their potential wells by the steric hindrance, and reorient, quasi freely, when a local deformation removes this hindrance.

[1] *Disorder in crystals* by N.G. Parsonage and L.A.K. Staveley, Clarendon, Oxford (1978).
[2] W.J. Dunning *in The Plastically Crystalline State*, 1, J.N. Sherwood ed. J. Wiley (New York) (1979).
[3] R. Pethrick *in The Plastically Crystalline State*, 123, J.N. Sherwood ed. J. Wiley (New York) (1979).
[4] J.M. Rowe, J.J. Rush *et al*, Ferroelectrics 16, 107 (1977).
[5] M. More and R. Fouret, Proceedings of the Faraday Discussion 69, *to be published.*
[6] J. Sherwood *in The Plastically Crystalline State*, 39, J.N. Sherwood ed. J. Wiley (New York) (1979).
[7] J.M. Rowe, D.J. Hinks *et al*, J. Chem. Phys. 58, 2039 (1973).
[8] M. More, J. Lefebvre and R. Fouret, Acta Crystal B33, 3862 (1977).
[9] G. Coulon and M. Descamps, *to be published in* Jour. of Physics C.
[10] J.C. Damien *et al, in Neutron Inelastic Scattering*, 331, IAEA Vienna (1978).
[11] D. Fontaine and R.M. Pick, Journal de Physique, 40, 1105 (1979).

Two-Dimensional Order in the SmF Phase

J.J. Benattar[1], F. Moussa[1], M. Lambert[1], A.M. Levelut[2]

[1] Laboratoire L. Brillouin, CEN Saclay BP N°2,
F-91190 Gif-sur-Yvette, France

[2] Laboratoire de Physique des Solides associé au CNRS
Université Paris-Sud Bât.510, F-91405 Orsay, France

A previous study by means of X-ray diffraction on single domain samples of the SmF phase of TBPA allowed us to collect information concerning the nature of the order in this phase [1] : the smectic layers are weakly coupled and within the layers, there is always a pseudo hexagonal packing as in the SmH phase but which extends only to short distance with the persistence of a local "herring-bone" packing of the molecular sections.

Meanwhile, because of a large sample mosaic , it has not been possible to analyse with accuracy the intensity profile of the diffuse ring.

The principal motivation of the present work is to eliminate the drawback of the mosaic effects by using the Debye-Scherrer method.

Powder pattern of the SmF phase exhibit only, at large angles, a broad diffuse ring. We made a calculation of the intensity profile, assuming that the smectic layers are completely uncorrelated, in two cases :

- within the layers, there is a quasi-long range positional order corresponding to a X-ray structure factor around each reciprocal lattice vector $\vec{G}$: $|\vec{q}-\vec{G}|^{-2+\eta_G}$
- there is a short range positional order corresponding to a Lorentzian structure factor.

Our computation is only consistent with a Lorentzian law and a correlation length within the layer plane : $\xi_{//} \simeq 250$ Å .

Therefore, the SmF phase could constitute an example of the "hexatic phase" predicted by the theory of Halperin and Nelson [2] .

[1] J.J.Benattar, J.Doucet, M.Lambert and A.M.Levelut, Phys.Rev. A20, 2505 (1979).

[2] D.R.Nelson and B.I.Halperin, Phys.Rev. B19, 2457 (1979).
B.I.Halperin, to appear in proceedings of the Kyoto Summer Institute on two-dimensional systems (1979).

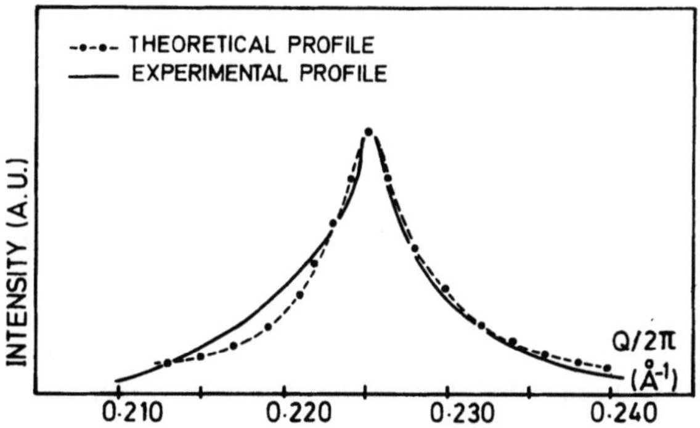

Molecular Conformational Changes in the Mesophases of TBBA *

A.J. Dianoux

Institut Laue Langevin, BP 156X, F-38042 Grenoble Cêdex, France

F. Volino

CNRS and Equipe de Physico-Chimie Molêculaire, S.P.S., D.R.F., CENG 85X, F-38041 Grenoble Cêdex, France

In a previous paper [1] we have presented results concerning the molecular orientational order in the mesophases of TBBA from a consistent analysis of neutron, ^{14}N NQR and DMR data. We consider here the problem of the relative temperature dependence of various DMR splittings, in the light of these results. Two extreme models have been considered (i) the most probable conformation changes with temperature [2,3] and (ii) the most probable conformation does not change with temperature, but more than one order parameter is needed to describe the molecular orientational order [3,4]. In the calculation, we introduce explicitly the internal motions, essentially rotations of the phenyl rings around their para-axis. In model (i), the molecule is assumed to rotate uniformly around its long axis Oz_o. This long axis is defined in the frame $Ox_my_mz_m$ attached to the most probable orientation of the central ring, with Oz_m along the para-axis and Ox_m in the plane of the ring, by its polar and azimuthal angles ε and ϕ, respectively. It is found that the relative temperature dependence of five splittings is satisfactorily explained, assuming a variation of ϕ, presumably due to variations of the mean dihedral angles between rigid molecular fragments, while ε is kept constant as suggested by the results of ref.[1]. On the contrary we show that explanation (ii) cannot be the main phenomenon since it predicts relative variations which are definitely too small.

References

1 Dianoux A.J. and Volino F., J. Physique 40 (1979) 181.
2 Charvolin J. and Deloche B., J. Physique Colloques 37 (1976) C3-69.
3 Bos P.J., Pirs J., Ukleja P., Doane J.W. and Neubert M.E., Mol. Cryst. Liq. Cryst. 40 (1977) 59.
4 Bos P.J. and Doane J.W., Phys. Rev. Lett. 40 (1978) 1030

* The corresponding publication has been submitted to J. Physique (1980)

Smectic Polymorphism of Some Bis-(4,4'-n-alkoxybenzylidene)-1,4-phenylenediamines up to 3 kbar by Differential Thermal Analysis

J. Herrmann, J. Quednau, and G.M. Schneider

Department of Chemistry, University of Bochum,
D-4630 Bochum, Federal Republic of Germany

Abstract

The p-T phase behaviour and other thermodynamic data of some homologues of the bis-(4,4'-n-alkoxybenzylidene)-1,4-phenylenediamines have been determined by high pressure differential thermal analysis (DTA) up to 3 kbar in the temperature range from 300 to 600 K [1].

The homologues with alkoxy chain lengths of 4, 5, 6, 7, 8, 12, 13 and 14 carbon atoms have been investigated [1, 2].

The substances show a very complex phase behaviour. At normal pressure they exhibit up to 7 mesomorphic phases [3], whereas with increasing pressure the number of smectic phases diminishes. Especially the low temperature smectic phases vanish at triple points where some of them are between smectic phases only.

The range of the nematic phase increases drastically with increasing pressure for homologues with short alkoxy chains. For homologues with longer alkoxy chains this effect is less pronounced but still holds for the dodecoxy homologue. The tridecoxy and tetradecoxy homologues do not exhibit a nematic phase at normal pressure but a pressure-induced nematic phase appears with increasing pressure at about 290 and 680 bar respectively.

References

1. W. Spratte and G.M. Schneider, Mol. Cryst. Liq. Cryst., 51, 101 (1979)

2. J. Herrmann, W. Spratte and G.M. Schneider, Proceedings of VII th AIRAPT International Conference, Le Creusot, France, 1979

3. E.M. Barrall II, J.W. Goodby, G.W. Gray, Mol. Cryst. Liq. Cryst. Lett., 49, 319 (1979)

Investigation of a Smectic H-Smectic C Phase Transition by X-Ray Diffraction

G. Albertini, B. Dubini, S. Melone, and M.G. Ponzi-Bossi

Istituto di Fisica Medica, Facoltà di Medicina e Chirurgia, Ancona, Italy

F. Rustichelli

European Communities J.C.R. Ispra, Varese, Italy, and
Dipartimento di Scienze Fisiche, Facoltà di Ingegneria, Ancona, Italy

Abstract

The smectic H-smectic C phase transition of TBBA was investigated using X-ray diffraction.

Kα radiation of Cu (λ=1,54 Å) was sent on a polycrystal sample in a powder diffractometer. Temperature control in the range of 125°C<T<150°C was provided by a suitable electronic device.

A progressive broadening of the diffraction peak (110) and a lowering of its height were observed during the heating of the sample.

This behaviour is to be related with a continuous fall of long range order within the smectic planes, until the short range order of smectic C phase is reached.

On the contrary, the long range order perpendicular to the smectic planes does not change during all the process, as it can be deduced by the fact that the corresponding Bragg peak does not suffer any broadening.

Moreover a plot of interplanar distance versus temperature shows a stronger change close to the transition temperature.

It is interesting to see that according to microcalorimetric measurements the investigated transition appears to be a first order transition.

A possible explanation of the observed pretransition phenomena is presented.

Submitted for publication in the proceeding of the Conference on Liquid Crystals - Bangalore (India), 3rd-8th/Dec./1979

fractometer equipped with a hot stage controlling the temperature to ± 0.1°C. The experimental details are similar to those reported in [5] and [6]. Ni filtered Cu Kα radiation (λ =1.54 Å) was used. The material was bought from Eastman Organics Chemicals and is expected to be of the same high purity as in previous experiments. Once fixed the temperature of the sample at a value inside the interval of existence of the mesophase, several diffraction patterns were recorded each one after the other without any change in the experimental conditions. The diffraction patterns appeared to be in general each one different from the other: this can be explained by the fact that the sample, which was obtained by heating a solid powder, is neither an "ideal polycrystal powder" nor "a single crystal" and that the texture in a plastic crystal can change as a function of time [7].

However some common features can be derived from the series of diffraction patterns. Few narrow Bragg peaks exist, whose number ranges from four to one and occasionally to zero. Moreover a strong diffuse scattering is observed below the Bragg peaks, taking the form of two broad peaks centered at $2\theta \sim 8.8$ and ~ 19 deg.

The intensity of this diffuse scattering can suffer slight changes from one diffraction pattern to the other one.

Figure 1b reports a typical diffraction pattern.

Figure 1a reports the diffraction pattern in the solid phase at a temperature of 50°C.

We expect that an "ideal polycrystalline powder" sample would produce a diffraction pattern including each one of the Bragg peaks observed in any scan and possibly more. By using the language of the set theory, and by defining B_i the set of all Bragg peaks appearing in the ith scan and with B_p the set of all Bragg peaks which would appear in the scan associated to an "ideal powder", one has:

$$\bigcup B_i \subseteq B_p \qquad (1)$$

where $\bigcup B_i$ is the union set of all the obtained sets B_i.

The validity of this expectation is confirmed by the results obtained by using a Laue flat camera. In fact some spots are observed, which belong to some Debye-Scherrer rings, instead of observing continuous rings, as one would obtain from an "ideal powder".

On the other hand two broad and isotropic diffuse rings are observed: they correspond to the diffuse scattering which was common to all the scans. The isotropy of these rings in the photos explains why a good reproducibility exists of the diffuse scattering in the goniometer scan, in spite of the time evolution of the sample texture.

Figure 2 reports in a schematic way the union set $\bigcup B_i$, above defined.

From Fig.1b, Fig.2 and from (1) it appears that a *three-dimensional long-range order* exists in the mesophase.

This result is in contradiction with the observation of only one Bragg peak reported in [1].

The discrepancy can be explained by assuming that also in the neutron diffraction experiment the sample was an imperfect "polycrystalline powder".

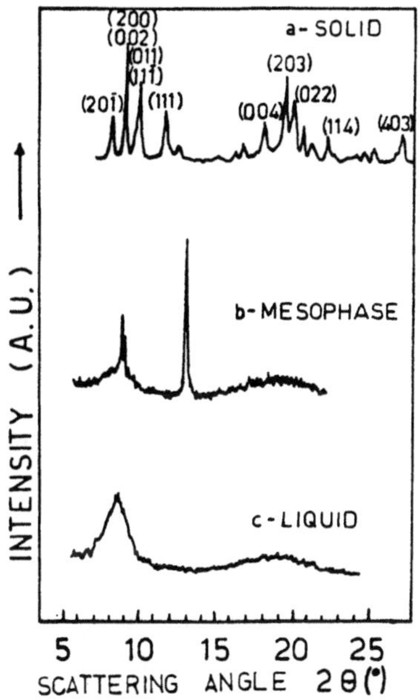

Fig. 1 a-c (a) X-ray diffraction pattern by OPCTS solid phase at a temperature of 50°C. (b) A typical X-ray diffraction pattern by OPCTS mesophase at temperature of 195°C. (c) X-ray diffraction pattern by the isotropic liquid phase of OPCTS at a temperature of 206°C

In fact in this case it is possible that only one single Bragg peak is recorded in the multidetector counter of D1B neutron diffractometer, which covers only a linear section of the Debye-Scherrer rings. In fact also in our experiment, as previously mentioned, only one Bragg peak was sometimes observed.

Then, by removing the assumption leading to the suggestion [2] that the mesophase is of smectic nature, namely that only one Bragg peak would exist, also the suggestion loses its validity. Therefore from the present experiment, from the optical isotropy and from the globular shape of the molecules, it is to be tought that the mesophase is a plastic crystal and in any case not a smectic phase similar to a smectic A phase. In fact the strong background scattering and the observation of a small number of reflections are characteristic features of X-ray diffraction patterns from plastic crystals and they are consequence of the high degree of disorder in plastic crystals associated to three-dimensional long-range positional order.

Unfortunately the difficulties associated to the Bragg peak observation do not allow to derive the lattice structure characteristics associated to the plastic crystal phase, and therefore a safe Bragg peak ordering.

This situation is not unusual when dealing with plastic crystal "powder" samples, as it was the case, for instance, of the β-CF_4 phase, for which the indexing of six Bragg reflections was not possible in spite of the relatively good accuracy of their positioning [9].

An experiment on "single crystal" seems now necessary in order to get the lattice characteristics of the OPCTS plastic crystal.

Finally Fig.1c reports the X-ray diffraction pattern for the isotropic liquid phase. In the liquid phase the diffuse part is similar to the one in

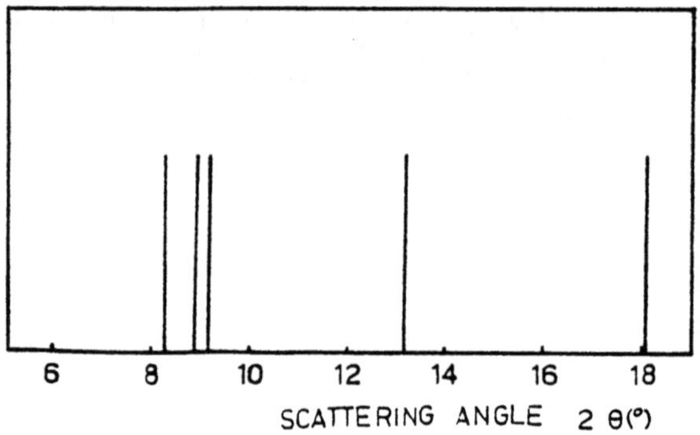

Fig.2 The union set $\bigcup B_i$ of the Bragg peak sets obtained in successive scans (see text) of OPCTS

the mesophase. Thus the rotational order should not have to change in the mesophase-liquid transition.

An apparent growth of the peak at ∿8.8 deg may be due to an increase of the diffused radiation as a consequence of the disappearing of long-range order.

Aknowledgments

Thanks are due to Dr F.VOLINO for suggesting investigation, to Prof A. LEADBETTER and dr A.FILHOL for useful discussion and to Mr L.FERRANTINI and Mr S.POLENTA for their technical assistance.

References

1 P.H.KEYES, W.B.DANIELS, J. Chem. Phys. 62, 5 (1975)
2 F.VOLINO, A.J.DIANOUX, Ann. Phys. 3, 151 (1978)
3 D.BRAGA, G.ZANOTTI, Acta Cryst. B36, (1980)
4 B.DUBINI, S.MELONE, M.G.PONZI BOSSI, F.RUSTICHELLI, submitted for publication in the proceeding of the 3.rd Liquid Crystal Conference of Socialist Countries - Budapest - August 1979
5 G.ALBERTINI, B.DUBINI, S.MELONE, M.G.PONZI BOSSI, F.RUSTICHELLI, G.TORQUATI J. Physique C3, 384 (1979)
6 G.ALBERTINI, B.DUBINI, S.MELONE, M.G.PONZI BOSSI, P.PULITI, F.RUSTICHELLI submitted to "Il Nuovo Cimento"
7 A.LEADBETTER - private comunication
8 P.A.WINSOR - Liquid Crystal & Plastic Crystals - Ellis Horwood Publisher Vol.1, 48 (1974)
9 S.C.GREER, L.MEYER J. Chem. Phys. 51, 10 (1969)

A Chiral Smectic F Phase

P. Keller[1], A. Zann[2], J.C. Dubois[2], and J. Billard[3]

[1] Laboratoire de Physique du Solide, Faculté des Sciences, F-91405 Orsay, France

[2] Laboratoire Central de Recherches, Thomson-CSF, F-91401 Orsay, France

[3] Laboratoire de Physique de la Matière Condensée, Collège de France, F-75231 Paris, France

Introduction

Today five twisted mesomorphic phases are known ; first the nematic phase N^*, or cholesteric ; then two smectic phases : the twisted smectic C, $S_C{}^*$ [1,2,3,4,5] and the twisted smectic H, $S_H{}^*$ [6,7,4] ; and now the two blue phases, BP I and BP II, which are twisted [8,9].

The smectic F phase is a tilted phase [10,11,12] as the S_C or S_H one, but no compounds exhibiting a chiral $S_F{}^*$ phase are already published ; S_F phases containing chiral compounds in their racemic mixtures were only described [13].

We describe here for the first time, a twisted smectic F phase, $S_F{}^*$. A new chiral compound (I), without S_F phase, induces a chiralization of the S_F phase of a non chiral compound (II). The compound (I) is the terephtalidene-bis-4-(3-methylpentylaniline) or TBMPA* :

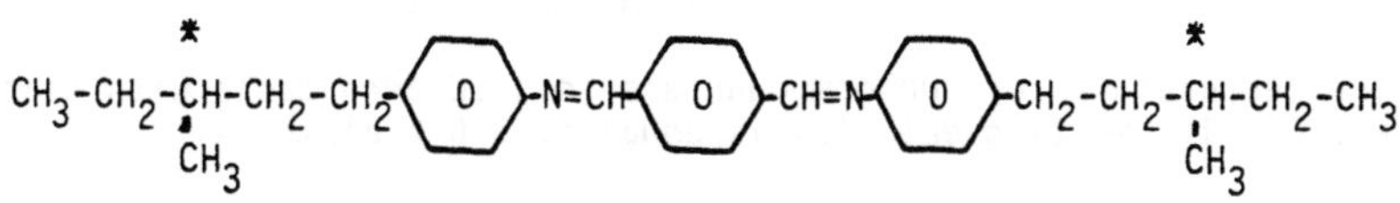

The compound (II) is the terephtalidene-bis-4 n-pentylaniline or TBPA :

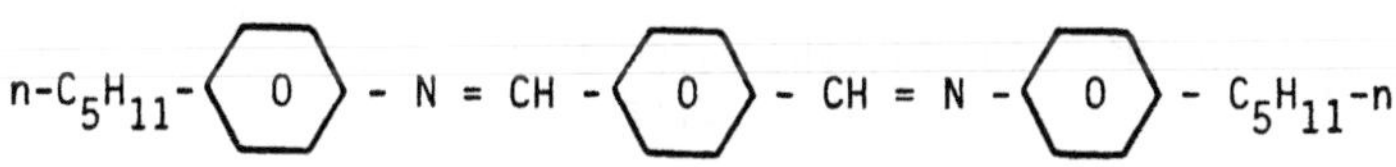

This compound was first prepared by S. SAKAGAMI et al. [14] and M.E. NEUBERT et al. [15].

Synthesis of TBMPA*

The last step of the preparation is the condensation of the 4(3-methylpentyl) aniline with the terephtaldehyde :

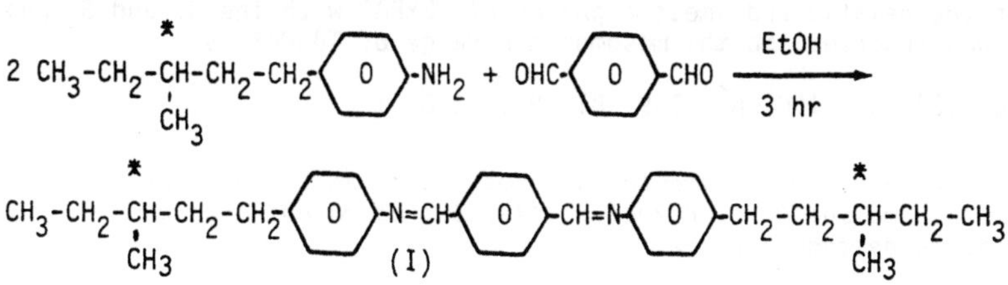

The 4-(3-methylpentyl) aniline is obtained, via nine steps, from the commercial S(-) amylic alcohol (α_D^{20} = - 4.76°).

The main intermediate compounds are successively :

- the 3-methylpentanoic acid : $CH_3-CH_2-\overset{*}{C}H(CH_3)-CH_2-COOH$, obtained by the method of VOGEL [16],

- the 3-methyl-pentyl-phenyl : 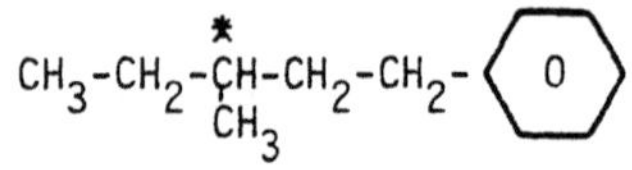 [17],

- the final 4-(3-methyl-pentyl) aniline prepared according to the method of J. VAN DER VEEN et al. [18].

The synthesized TBMPA* is purified by several recrystallizations from cold absolute ethanol.

Characterization and Identification of the Mesophases

The two synthesized compounds, TBPA and TBMPA*, were studied by means of microscopic observation (Leitz Orthoplan-Pol) with polarized light and programmable hot stage (Mettler FP 5). The temperatures and enthalpies of transition were determined by differential enthalpic analysis (Mettler TA 2000).

-The mesophases of TBPA were previously identified by J.W. GOODBY et al. [13, 19]. TBPA has the following succession of mesophases :

$$K\ 68\ S_H\ 141\ S_F\ 150\ S_C\ 178\ S_A\ 213\ N\ 232.5\ I$$

The transition temperatures of our compound are in good accordance with that of reference [19]. Its melting enthalpy is equal to 3.6 kcal.mol^{-1}.

- TBMPA* exhibits two smectic phases, a nematic phase and a blue phase, with the following transition temperatures :

K	91	S_2	133	S_1	160	N*	169	Bl. Ph.	170	I[(a)]
	2.6		*1.5*		*0.5*				*0.23*	

The transition enthalpies, in italics, are in kcal.mol^{-1}. A very beautiful blue phase is observed on the microscopic preparation, when the temperature increases or decreases. The nematic-blue phase transition is clearly visible on the thermogram, but the transition enthalpy is not measurable. Any second transition, blue phase I to blue phase II, which might exist, according to STEGEMEYER studies [8] is not detected. The smectic mesophases of TBMPA* were identified by means of the contact method [20]. The binary phase diagram under atmospheric pressure of TBPA and TBMPA* was plotted (figure 1).

Isomorphy of the nematic and smectic phases of TBMPA* with the S_H and S_C phases of TBPA was observed. So the mesomorphic range of TBMPA* is :

$$K\ 91\ S_H\ 133\ S_C\ 160\ N^*\ 169\ Bl.\ Ph.\ 170\ I$$

(a) The calorimetric DSC measurements show two other small peaks at 96.8°C (0.6 kcal mol^{-1}) and 104°C (0.16 kcal.mol^{-1}). But we cannot observe any texture change at these temperatures.

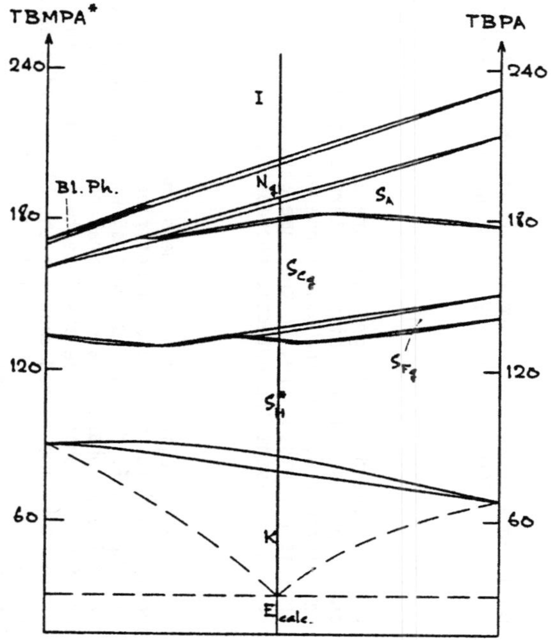

Fig. 1 : Binary phase diagram under atmospheric pressure of TBMPA* and TBPA.

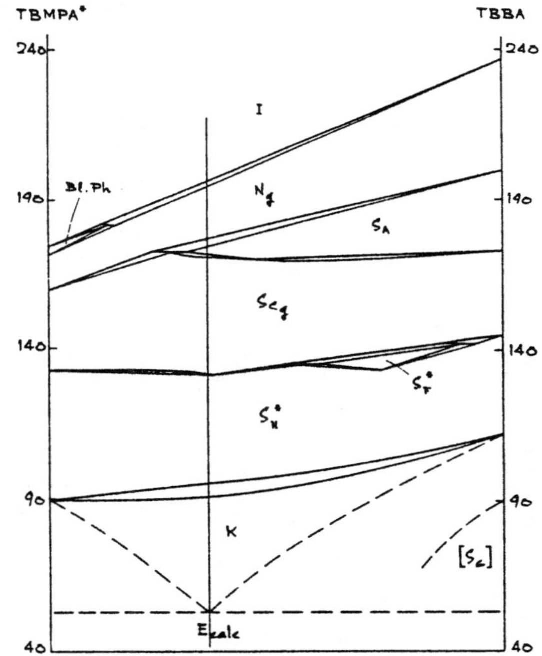

Fig. 2 : Binary phase diagram under atmospheric pressure of TBMPA* and TBBA.

To perform the binary diagram, we calculated and plotted the melting curves using the LE CHATELIER [21] and VAN LAAR [22] relations (figure 1, dashed lines) ; then we realized the eutectic mixture, $E_{calc.}$, which corresponds to 0.49 [TBMPA*] : 0.51 [TBPA] in molar fraction and would melt at 30°C. In fact, the experimental mesomorphic range of the previous mixture, determined by enthalpic analysis, is :

$$K \;\; 85 \;\; S_H \;\; 132.5 \;\; S_F \;\; 134.5 \;\; S_C \;\; 179 \;\; S_A \;\; 190.5 \;\; N^* \;\; 205 \;\; I$$

No eutectic drop is observed; the two compounds form a solid solution.

Some Properties of the Twisted Smectic Phases

The rotatory power in the various phases of the contact preparation between TBPA and TBMPA* has been studied, using polarized light and a rotating stage. The nematic phase exhibits an apparent rotatory power and, too, the smectic C and F phases. So TBMPA* induces a chiralization of these three phases of TBPA, and especially, that is new, in the S_F phase.

The relative and absolute senses of the twist in the S_F^*, S_C^* and N^* phases were determined. Firstly, no area without twist is observed in the preparation in the S_C^* and S_F^* phases. Hence, the sense of the twist is the same in the two each phase. Then, the absolute sense of the twist is determined by the observation, between silane-coated slides [23], of the isochromatic bands in homogeneous areas of the preparation. This method has been described by N. ISAERT et al. [24]. When the analyzer rotates in the sense of the twist, the isochromatic bands diverge from a critical area. Inversely, when the analyzer rotates in the inverse sense of the twist, the isochromatic bands converge to this area. The so-determined sense of the twist for the three studied phases is a sinister one :

$$S_{Fg}, \; S_{Cg}, \; N_g$$

Similar results were obtained with a contact preparation between TBBA (terephtalidene bis-butyl aniline) and TBMPA* (figure 2). TBBA has the following mesomorphic range :

$$K \;\; 114 \;\; S_H \;\; 144 \;\; S_C \;\; 172 \;\; S_A \;\; 199 \;\; N \;\; 236 \;\; I \qquad [25]$$

TBBA do not exhibit S_F phase. The binary diagram (figure 2) confirms the identification of the S_C and S_H phases, and exhibits an intermediate S_F phase. The N, S_C and S_F phases are twisted too, with the same sense of twist that in the diagram of figure 1, that is a sinister one. In the same way, the mixture forms a solid solution and does not present the eutectic decreasing.

Conclusion

A new chiral compound, the terephtalidene bis-4(-3-methylpentylaniline) or TBMPA*, has been synthesized and its mesophases identified ; this compound exhibits successively the S_H^*, S_C^*, N^* and blue phases, when the temperature increases. In contact with the S_F phase of a non chiral compound, TBBA, it induces a chiralization of such a phase. The absolute twist sense of the S_F^* phase was determined, using the same method than for the N^* and S_C^* phases. These three phases have the same sinister sense of twist. Consequently, the existence of a chiral smectic F phase, implies that there is some orientation correlation between the layers of this mesophase. This is in good agreement with the biaxiality of the achiral S_F phase [15] and with the weak correlation observed by X-ray structural investigation [12].

Bibliography

1 W. Helfrich, Chan S. Oh., *Mol. Cryst.* 14, 289 (1971).

2 W.Z. Urbach, J. Billard, *C.R.A.S. Paris,* 274-B, 1287 (1972).

3 P. Keller, L. Liebert, L. Strzelecki, *J. Phys. Colloq.* 37, C3-27 (1976).

4 P. Keller, S. Suge, L. Liebert, L. Strzelecki, *C.R.A.S. Paris,* 282-C, 639 (1976).

5 J.P. Berthault, P. Keller, *Bull. Soc. Chim. Fr.*, 135 (1976).

6 R.B. Meyer, L. Liebert, L. Strzelecki, P. Keller, *J. Phys. Lett.*, 36 L, 69 (1975).

7 D. Coates, G.W. Gray, *Mol. Cryst. Lett.*, 34, 1 (1976).

8 H. Stegemeyer, K. Bergmann, these proceedings, p. 161

9 R.M. Hornreich, S. Shtrikman, these proceedings, p. 185

10 J.J. Benattar, J. Doucet, N. Lambert, A.M. Levelut, *Phys. Rev.*, A 20, 2505 (1979).

11 S. Diele, D. Demus, A. Echtermeyer, U. Preukschas, H Sackman, *Acta Phys. Polon.*, A 55, 125 (1979).

12 A.J. Leadbetter, J.P. Gaughan, B. Kelly, G.W. Gray, J. Goodby, *J. de Phys.* 40, C-3, 178 (1979).

13 J. Goodby, G.W. Gray, *J. de Phys.* 40, C-3, 27 (1979).

14 S. Sakagami, A. Takase, M. Nakamizo, *Mol. Cryst. Liq. Cryst.*,36, 261 (1976).

15 M.E. Neubert, J.L. Maurer, *Mol. Cryst. Liq. Cryst.,* 43, 313 (1977).

16 A. Vogel, "Elementary Practical Organic Chemistry", part.1 p. 205, 2nd ed. (Langmans, Green and Co., London 1966).

17 A. Vogel, "Elementary Practical Organic Chemistry", part.1 p. 732, 2nd ed. (Langmans, Green and Co., London 1966).

18 J. Van der Veen, W.H. De Jeu, A.H. Grobben, J. Boven, *Mol. Cryst. Liq. Cryst.,* 17, 291 (1972).

19 J.W. Goodby, G.W. Gray, A. Mosley, *Mol. Cryst. Lett.*, 41, 183 (1978).

20 L. Kofler, A. Kofler, "Thermomikromethoden" (Verlag Chemie, Weinheim 1954).

21 H. Le Chatelier, *C.R.A.S. Paris,* 100, 50, (1885).

22 J.J. Van Laar, *Z. f. Phys. Chem.,* 63, 216 (1908) and 64, 257 (1908).

23 F.J. Kahn, G.N. Taylor, H. Schonhorn, *Proc. IEEE,* 61, 823 (1973).

24 N. Isaert, B. Soulestin, J. Malthête, *Mol. Cryst. Liq. Cryst.,* 37, 321 (1976).

25 J. Doucet, These Doctorat es Sciences Physiques. Univ. Paris-Sud (1978).

Coherent Neutron Scattering Study of the SmV → SmVI Transition in TBBA*

A.M. Levelut[1], F. Moussa[2], M. Lambert[1,2], B. Dorner[3]

[1] Laboratoire de Physique des Solides associé au CNRS, Université de Paris-Sud, Bâtiment 510, F-91405 Orsay, France

[2] Laboratoire Léon Brillouin, C.E.S. Saclay, B.P. N°2, F-91190 Gif-sur-Yvette

[3] Institut Laue Langevin, B.P. 156, F-38042 Grenoble-Cedex, France

Abstract

We have shown [1] by X-rays studies that an herring-bone local order exists in the hexagonal SmB_A phase or in the pseudo-hexagonal SmB_C (SmH) phase. When the studied compounds undergo a $SmB_C \rightarrow SmE_C$ transition, only four of the twelve diffuse spots characteristic of the herring-bone order become Bragg spots of the SmE_C phase. We have studied the dependence of the neutron scattered intensity, near the two kinds of diffuse spots of a single domain of the SmB_C phase of fully deuterated TBBA versus energy, scattering vector and temperature. We could not measure any energy broadening of the scattered intensity on a diffuse spot (the resolution of the spectrometer was of .16 mev). We followed the width and the intensity of the diffuse spots in the reciprocal space and from these measurements one can deduce the correlation length for the herring-bone order and the anisotropy of the rotation of TBBA molecules around their long axis in the SmB_C phase. We compare our results with those deduced from incoherent neutron quasi elastic scattering [2] and NQR experiments [3].

[1] A.M. Levelut, Journ. de Phys., 37, C_3 51 (1976).

[2] A.J. Dianoux, H. Hervet, F. Volino, J. de Phys., 38, 809 (1977).

[3] R. Blinc, J. Seliger, M. Vilfan, V. Zagar, J. Chem. Phys., 70, 778 (1979).

---

* Soumis au Journal de Physique.

Part II

A and C Smectic Phases and Structures

High Resolution X-Ray Scattering from Smectic A, B, and C Phases

J.D. Litster

Center for Materials Science and Engineering and Department of Physics, Massachusetts Institute of Technology, Cambridge, MA 02139, USA

In my lecture this morning I am going to summarize the results of a number of high resolution x-ray scattering measurements on smectic phases. These have been carried out at M.I.T. and at Risø National Laboratory. My collaborators have been Prof. R.J. Birgeneau at M.I.T. and Dr. Jens Als-Nielsen at Risø as well as graduate students and staff members at both institutions. The names of people involved in the various experiments will be found in references to work which is either published or in press in various scientific journals. In my lecture I will discuss the scientific questions addressed and the results we obtained in these experiments.

1. X-Ray Scattering in the Smectic A Phase

In these experiments [1] we addressed the question of whether the smectic A (SmA) phase has true long range order. The SmA phase is formed by the establishment of a one-dimensional density wave in the orientationally ordered nematic phase. The wave vector of this density wave lies along the nematic director (taken as the z axis) and has magnitude $q_0 = 2\pi/d$, where d is the smectic "layer" spacing. Thus the density in an incompressible SmA phase can be written

$$\rho = \rho_0 \left\{1 + \mathrm{Re}\left(\psi e^{iq_0 z}\right)\right\} \qquad (1).$$

Here $\psi = |\psi|\, e^{iq_0 u}$ is a complex order parameter for the SmA phase. Many years ago Peierls [2] and Landau [3] gave simple arguments that a one dimensional density wave in a three dimensional liquid could not have long range order. It is worthwhile to repeat those arguments here; we first need to understand the unusual elastic properties of the SmA phase.

When the phase of the complex order parameter (see above) is written as $q_0 u$, u corresponds to displacements along z of the smectic layers from their equilibrium positions. The Goldstone mode [4] that results from SmA ordering involves fluctuations in u. The dispersion relation for this mode is given [5]

$$\hbar\omega = B\, k_z^2 + K(k_x^2 + k_y^2)^2 \qquad (2),$$

and has this unusual form because it requires energy to compress the smectic layers (given by the elastic constant B) but they slide freely on one another. Thus when $k_z = 0$, the restoring force for layer fluctuations comes only from the *curvature* elastic constant K for the layers; K is identical to the nematic phase splay elastic constant. The anisotropic form of (2) has a profound effect on the SmA phase properties. If we calculate the mean square fluctuations in u, the result is

$$\langle u^2(\vec{r})\rangle = \frac{kT}{(2\pi)^3}\int \frac{d^3k}{Bk_z^2 + K(k_x^2 + k_y^2)^2} \tag{3}$$

where the range of integration for a sample of linear dimension L is $(2\pi/L) \lesssim |\vec{k}| \lesssim q_0$. The result is

$$\langle u^2(\vec{r})\rangle \simeq \frac{kT}{4\pi(BK)^{1/2}} \ln (q_0 L) \tag{4},$$

and was used by Landau and Peierls to argue that fluctuations in the smectic layers would destroy long range order as $L \to \infty$. Thus the SmA phase has the same logarithmic singularity from long wavelength Goldstone modes that is found for solids in two dimensions and one might expect the same "topological" long range order predicted for two dimensional solids by Kosterlitz and Thouless [6]. The actual situation is a bit more complicated, as I shall explain presently.

Let us return to a discussion of the experimental situation. If the SmA phase had long range order, one could expect to see by x-ray scattering a Bragg peak at $\vec{q} = q_0\hat{z}$. The x-ray scattering intensity in the vicinity of this Bragg peak is given by the Fourier transform of the correlation function

$$G(\vec{r}) = \langle e^{iq_0[u(\vec{r})-u(o)]}\rangle \tag{5}.$$

This correlation function in a three dimensional solid consists mostly of a constant term with a very small term that goes to zero approximately as $(1/r)$. Its Fourier transform then gives a $\delta(\vec{q})$ Bragg peak and weak thermal diffuse scattering whose intensity varies as $(1/q^2)$, where $\vec{q}$ is the momentum transfer relative to the Bragg peak position. The same correlation function evaluated for liquid decays rapidly (exponentially) to zero, since the liquid has no long range order. It is also quite straightforward to calculate $G(r)$ using the harmonic approximation in a two dimensional solid; the result is that $G(r)$ goes to zero as $r \to \infty$ (no long range order) but rather slowly as $r^{-\eta}$, where η is small ($<1/4$). This algebraic decay of correlations is characteristic of topological long range order. For SmA phase $G(r)$ has been calculated by Caillé [7], and has the form

$$G(\vec{r}) \sim \frac{1}{(x^2 + y^2)^{\eta}}\, e^{-\eta E_1\left[\frac{x^2+y^2}{4\lambda z}\right]} \tag{6}$$

where $\eta = kTq_0^2\lambda/8\pi K$ (typically ~ 0.1), $\lambda = (K/B)^{1/2}$, and E_1 is the exponential integral. From the properties of E_1, it is readily shown that $G(r)$ also decays algebraically with distance

$$G(\vec{r}) \sim \frac{1}{z^{\eta}} \qquad (z^2 \gg x^2+y^2) \tag{7a}$$

$$\sim \frac{1}{(x^2+y^2)^{\eta}} \qquad (z^2 \ll x^2+y^2) \tag{7b}.$$

Taking the Fourier transform of (6), one finds there is no Bragg peak in the SmA density wave scattering, but a power law singularity in the scattering cross section

$$S(\vec{q}) \sim q_z^{-2+\eta} \qquad q_z \gg q_\perp = (q_x^2+q_y^2)^{1/2} \tag{8a}$$

$$\sim q_\perp^{-4+2\eta} \qquad q_z \ll q_\perp \tag{8b}$$

To observe the algebraic decay of the correlation function G(r), it is necessary to determine that the x-ray scattering is given by (8) rather than by a usual Bragg peak. We have done this in experiments on 8OCB (octyloxy-cyanobiphenyl) and the experimental difficulties and results are discussed at length in [1]. They arise chiefly from the fact that the observed scattering is a convolution of $S(\vec{q})$ with the spectrometer resolution function. Even with perfect crystal collimators the resolution has a finite width (the Darwin width) due to the fact that only a finite volume of the crystal is illuminated by the x-ray beam. Even more troublesome is the exponential fall off of the beam intensity in the crystal; this adds q^{-2} tails to the resolution function and makes deconvolution from (8) practically impossible. A solution to the problem came through the use of channel-cut crystals [8] with three Bragg reflections to give q^{-6} tails.

With channel-cut monochromator and analyzer crystals it was possible to show that the scattering from the SmA density wave was not that expected from a Bragg peak. It was also possible to convolve the theoretical scattering, the Fourier Transform of (6), with the measured resolution function. This involves two parameters: λ and η. We had previously measured λ by light scattering [9] and used η as an adjustable parameter to fit the data. The results are shown for two different temperatures in Fig.1. From λ and η one can determine the splay constant K; we found $K \sim 7\times10^{-7}$ dynes which agrees well with the measured value [10].

These measurements represent the first experimental observation of the algebraic decay of correlations in any system. There are some differences between the SmA liquid crystal and 2d solids however. First, it is possible to transform to a system in which the SmA phase does have long range order [11] so that the algebraic decay we have observed may not be associated with the lower marginal dimensionality. Second, there are prominent pretransitional phenomena at the SmA to nematic transition in contrast to the melting of 2d "solids". Finally, it has been calculated [12] that $\eta \lesssim 0.25$ in 2d solids, while we find $\eta \gtrsim 0.38$ in the SmA phase. More theoretical progress is required before we can understand all of these results.

2. The Nature of the Smectic B Phase

The well-ordered smectic phases (B, E, F, H) are still poorly understood. A variety of models have been proposed beginning with the stacking of 2d harmonic solids by de Gennes and Sarma [13]. Since then important progress has occurred in theories of 2d melting. The relevance of dislocation mediated melting, proposed by Kosterlitz and Thouless [6] to liquid crystals was first realized by Huberman *et al.* [14]. Recently the important effect of lattice symmetry on 2d melting has been shown by Young [15] and Halperin and Nelson [16]. Birgeneau and Litster have proposed [17] a SmB model combining features of references [16] and [13]. In this model a SmA crystal may transform on cooling

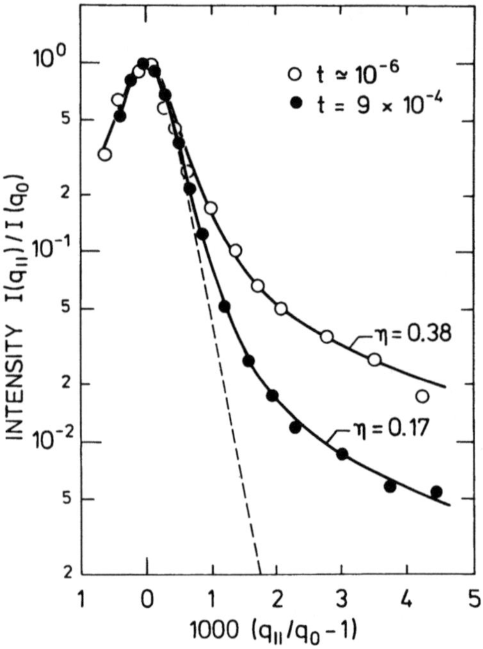

Fig.1 X-ray scattering from the SmA density wave in 8OCB. The dashed curve is the resolution function, which would be seen if there were a Bragg peak. The solid curves are fits to (6) for reduced temperatures shown on the figure

into either a 3d solid with positional long range order or into a 3d "stacked hexatic" phase with 3d bond orientational long range order and positional short range order in the smectic planes; only this latter phase would be a true liquid crystal.

We have carried out a high resolution x-ray study of butoxybenzylidene octyloxyaniline (40.8) to examine the SmB phase and the SmA-SmB transition [18]. Concurrently a study of freely suspended films of 40.8 was carried out by Moncton and Pindak [19]. Our experiments showed the SmA-SmB transition to be strongly first order and that the "smectic B" phase of this material has 3d crystalline long range order. In particular we observed a hexagonal close packed ABAB structure, in agreement with [19], and that the positional correlations extended at least 1.4 μm in the smectic planes and the between plane correlations extended at least 0.14 μm. The resolution limited in plane Bragg peaks were accompanied by extraordinarily strong diffuse scattering which could be represented as the sum of anisotropic phonon-like scattering consistent with the elastic properties of the "SmB" phase and an additional 2d ridge (q_z independent) contribution whose origin we do not understand.

Thus the SmB phase of 40.8 appears not to be a true liquid crystal. It remains to be seen if there are any of the phases B, D, E, F, H which are liquid crystals and if they correspond to the model of [17].

3. The Smectic A to Smectic C Transition

There are a number of materials that have a second order SmA to SmC transition which should be much simpler than the SmA-N or SmA-SmB transitions, and in the simplest model [20] is isomorphous to the superfluid transition in helium.

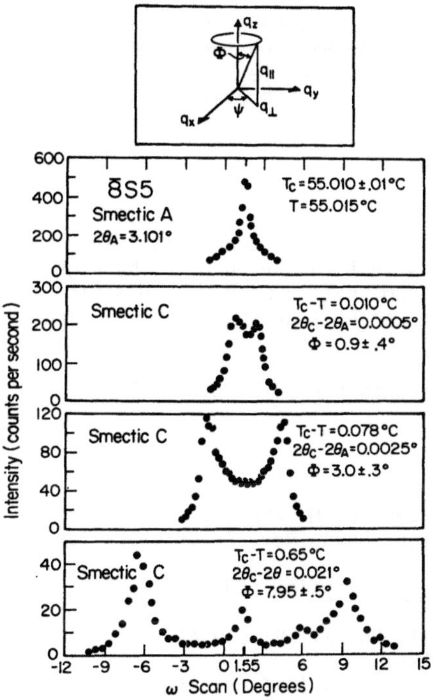

Fig.2 X-ray scattering near the SmC-SmA phase transition of $\bar{8}$S5 where the director orientation is held fixed by a magnetic field

One might therefore expect critical, rather than mean field, type of behavior near the phase transition. I wish to report our high resolution x-ray scattering study of the SmA-SmC transition in pentylphenylthiol octyloxybenzoate ($\bar{8}$S5) [21]. The experimental situation, reviewed in [21], is mixed with both critical and mean field behavior reported. We found that in sufficiently strong magnetic field the director could be held fixed in space so that both the layer tilt angle Φ and the smectic layer spacing could be measured simultaneously by x-ray scattering. Typical scans are shown in Fig.2.

Our experiments showed for $\bar{8}$S5 that the tilt angle Φ is the primary order parameter of the transition and the change in layer spacing is a secondary order parameter ($\sim\Phi^2$). We also found Φ vanished in the SmC phase as $(T_{CA}-T)^{\beta}$ with $\beta = 0.47 \pm 0.04$. This mean field behavior was observed over a reduced temperature range $5 \times 10^{-3} > 1 - T/T_{CA} > 3 \times 10^{-5}$. In order to understand this we may use the arguments of Ginsburg [22] that predict mean field behavior for

$$\frac{\Delta T_c}{T_c} > \frac{k^2}{32\pi^2\,(\Delta C)^2\,(\xi_{0||})^2\,(\xi_{0\perp})^4} \quad .$$

The heat capacity jump consistent with measurements of Schantz and Johnson [23] is $\Delta C = 10^6$ erg cm^{-3} K^{-1} which leads to mean field behavior for $\Delta T_c > 10^{-5}\,T_c$ if $\bar{\xi}_0 = \xi_{0||}^{1/3}\,\xi_{0\perp}^{2/3} > 13$ Å. Recent light scattering measurements

[24] indicate $\overline{\xi_0}$ to be about 18 Å in $\overline{8}$S5. Thus it appears that classical behavior is observed at the SmA-SmC transition in $\overline{8}$S5 for the same reason as in the superconducting transition - the material parameters cause the fluctuations to be small and one does not observe true asymptotic critical behavior until extremely close to T_{NA}.

4. Conclusion

I hope this presentation has given you some feeling for the insight into phase transition behavior in liquid crystals that can be obtained by high resolution x-ray scattering experiments in combination with quasielastic light scattering. More detailed discussion can be found in [1], [18], and [21].

This work was supported in part by NSF grants DMR-78-2355 and DMR-76-80895.

References

1. J. Als-Nielsen, J.D. Litster, R.J. Birgeneau, M. Kaplan, C.R. Safinya, A. Lindegaard-Andersen, and S. Mathiesen, Phys. Rev. B (in press for June 1980 issue).
2. L.D. Landau and E.M. Lifshitz, Statistical Physics (Addison-Wesley, Reading, Mass., 1969) p. 403.
3. R.E. Peierls, Helv. Phys. Acta 7, Suppl. No. 11, 81 (1934).
4. J. Goldstone, Nuovo Cimento 19, 145 (1961).
5. P.G. de Gennes, The Physics of Liquid Crystals (Oxford Univ. Press, Oxford, 1974)
6. J.M. Kosterlitz and D.J. Thouless, J. Phys. C6, 118 (1973).
7. A. Caillé, Compt. rendus Acad. Sci. Paris 274B, 891 (1972).
8. U. Bonse and M. Hart, Appl. Phys. Lett. 7, 238 (1965).
9. J.D. Litster, J. Als-Nielsen, R.J. Birgeneau, S.S. Dana, D. Davidov, F. Garcia-Golding, M. Kaplan, C.R. Safinya, and R. Schaetzing, J. de Phys. (Paris) 40, C3-339 (1979).
10. P.P. Karat and N.V. Madhusudana, Mol. Cryst. Liq. Cryst. 47, 21 (1978). These experiments were incorrectly analyzed; the correct analysis gives $K_1 = 6.8 \times 10^{-7}$ dynes for 8OCB.
11. B.I. Halperin and T.C. Lubensky, Sol. St. Comm. 14, 997 (1974).
12. D.R. Nelson and J.M. Kosterlitz, Phys. Rev. Lett. 39, 1201 (1977).
13. P.G. de Gennes and G. Sarma, Phys. Letters A38, 219 (1972).
14. B.A. Huberman, D.M. Lublin, and S. Doniach, Sol. St. Comm. 17, 485 (1975).
15. A.P. Young, Phys. Rev. B19, 1855 (1979).
16. B.I. Halperin and D.R. Nelson, Phys. Rev. Letters 41, 121 (1978).
17. R.J. Birgeneau and J.D. Litster, J. de Physique Lettres 39, 399 (1978).
18. P.S. Pershan, G. Aeppli, J.D. Litster, and R.J. Birgeneau, submitted to J. de Physique.
19. D.C. Moncton and R. Pindak, Phys. Rev. Lett. 43, 701 (1979).
20. P.G. de Gennes, Mol. Cryst. Liq. Cryst. 21, 49 (1973).
21. C.R. Safinya, M. Kaplan, J. Als-Nielsen, R.J. Birgeneau, D. Davidov, J.D. Litster, D.L. Johnson, and M.E. Neubert, Phys. Rev. B9, 4149 (1980).
22. V.L. Ginsburg, Sov. Phys. Sol. St. 2, 1824 (1960).
23. C.A. Schantz and D.L. Johnson, Phys. Rev. A17, 1504 (1978).
24. R. Schaetzing, private communication.

Dielectric Properties and Structure of the Smectic Phases

L. Benguigui

Solid State Institute, Technion-Israel Institute of Technology,
Haifa, Israel

Recent dielectric measurements (static dielectric constants and relaxation times) in the smectic phases have shown interesting connections with smectic structure. We present some examples.

1. The dielectric anisotropy has a definite different behavior in the SmA and SmB phases, than in the nematic phase in which the dielectric anisotropy is a consequence of the orientational ordering. This situation is believed to be due to the strong increase in the dipole-dipole correlations in a layered structure.

2. In the SmC, one can find a relation between the dielectric anisotropy and the tilt angle. This fact has been used as a new macroscopic method of measuring the tilt angle. This method has been used with success either on magnetic field oriented polydomain sample [1], or single domain samples [2].

3. In the ferroelectric chiral smectic C* phase, one can observe different phenomena, which are related to the helix pitch or to the tilt angle. Particular properties of this phase come from the fact that the measuring electric field cannot be parallel or perpendicular to the molecules, but only to the layers. Thus one can observed the molecular flip around the short axis whatever the field direction, and strong anomalies in both $\varepsilon_{\parallel}$ and $\varepsilon_{\perp}$ near the SmA-SmC* transition.

4. In TBBA, the very small dielectric anisotropy is puzzling since we shall expect a relatively important anisotropy due to the transverse dipole of the molecule, in the cis-configuration. One possible explanation could be that an important fraction of the molecules is in the trans-configuration with no dipole moment.

[1] L. Benguigui and D. Cabib, Phys. Stat. Sol. (a) 47, 71 (1978).
[2] A. Buka and L. Bata, Mol. Crys. Liq. Crys. Lett. 49, 159 (1979).

Molecular Packing Coefficients of the Homologous Series 4,4'-Di-n-alkyloxyazoxybenzene

N.C. Shivaprakash, P.K. Rajalakshmi, and J. Shashidhara Prasad
Department of Physics, University of Mysore, Mysore 570 006, India

Abstract

The molecular packing coefficients have been evaluated using the approach of KITAIGORODSKY for some members of the series 4,4'-di-n-alkyloxyazoxybenzene. The packing coefficient increases as alkyl chain increases contrary to what is observed for cholesteryl alkanoates. This shows that packing coefficient is intimately related with thermodynamic parameter.These features can be qualitatively explained by considering a mechanical rigid rod model. In principle, this can be used to explain the observed changes in the gradient of S factors in symmetric (4,4'-di-n-alkyloxyazoxybenzene) and asymmetric (p-(p-ethoxyphenylazo) phenyl alkanoates) molecules. The S factor of asymmetric molecules drops off more rapidly than that of symmetric molecules which is in conformity with the packing coefficient calculations.

It is very well established that molecular arrangements in the crystalline phase of liquid crystal forming substances lead to a better understanding of the existence and properties of liquid crystals and perhaps eventually to better the materials. In an earlier paper [1] it has been demonstrated how thermal stabilities in the mesogenic homologous series p-methoxy-XY-p'-alkyl tolanes could be explained beautifully by the help of the molecular packing coefficient without the actual conformation of the molecules in the crystalline state. Here we have extended the idea to the case of 4,4'-di-n-alkyloxyazoxybenzene. Detailed crystal structure analyses are available for two members of the series viz., para azoxyanisole (PAA) and 4,4'-di-n-heptyloxyazoxybenzene (HAB) which were studied by KRIGBAUM et al. [2] and LEADBETTER and MAZID [3] respectively. Prior to this, unit cell dimensions and space groups had been obtained for para azoxyphenetole (PAP) by CARLISLE and SMITH [4] and 4,4'-di-n-pentyloxyazoxybenzene (PAB) by RAJALAKSHMI et al. [5].

The molecular packing coefficients have been evaluated as in an earlier study [6] using the crystallographic data for para azoxyanisole [4], 4,4'-di-n-pentyloxyazoxybenzene [5] and 4,4'-di-n-heptyloxyazoxybenzene [3].

The intermolecular radii used for hydrogen, carbon, oxygen, nitrogen are respectively 1.17, 1.80, 1.52 and 1.58Å. Bond lengths have been taken from the work of KRIGBAUM et al. and LEADBETTER et al. The results are tabulated in Table 1.

Table 1 Unit cell volumes, number of molecules per unit cell (Z), crystal class, densities, transition temperatures, geometrical volumes of molecules and packing coefficients of para azoxyanisole (PAA), para azoxyphenetole (PAP), 4,4'-di-n-pentyloxyazoxy benzene (PAB) and 4,4'-di-n-heptyloxyazoxy benzene (HAB).

	Volume [Å^3]	Z	Crystal class	Densities (cal) [g/cc]	Transition temperatures [°C]	Geometrical volume [Å^3]	Packing coefficients
Para azoxyanisole	1291.67	4	Mono-clinic	1.328	119.5-136.5	243.37	0.754
Para azoxyphenetole	1482.49	4	Mono-clinic	1.283	136.9-167.5	277.51	0.749
4,4'-di-n-pentyloxy azoxy benzene	930.00	2	Tri-clinic	1.323	68.5-76.5-123.0	368.14	0.792
4,4'-di-n-heptyloxy azoxy benzene	1243.29	2	Tri-clinic	1.138	73.1-95.1-123.9	433.69	0.698

Figure 1 gives a plot of the packing coefficients versus the number of carbon atoms in the alkyl chain. As we can see from the graph, packing coefficient increases as we go from first member to fifth member of the series. We observe an odd/even effect in the variation of packing coefficient with alkyl chain for the first two members of the homologous series.

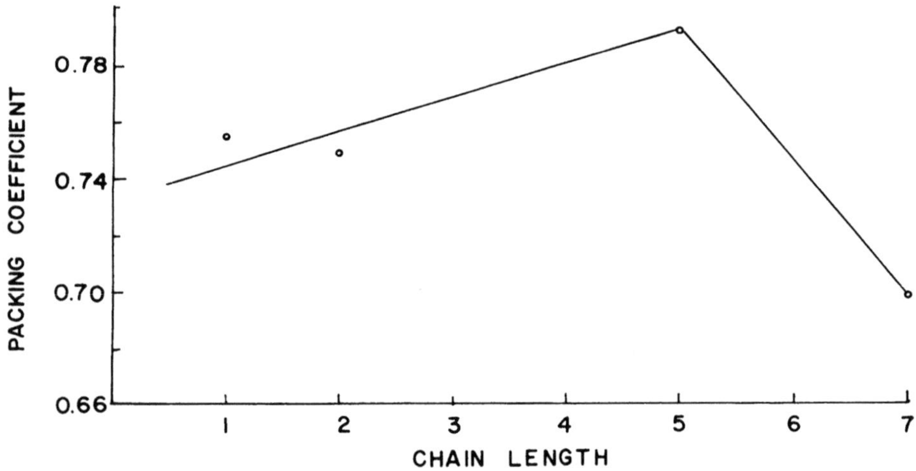

Fig. 1 The variation of the packing coefficient with number of carbon atom in the alkyl chain for the homologous series 4,4'-di-n-alkyloxyazoxybenzene

There is a change in the slope of the packing coefficients versus chain length at the fifth member of the series indicating the onset of a different molecular order and hence the onset of an additional or different mesogenic phase. This is in conformity with the observed anomalous smectic phase at the fifth member of the series [7]. Presumably it marks the appearance of a more efficiently packed antiparallel molecular assembly of alkyl chain in both crystalline and smectic phases. This study predicts that sixth member should also exhibit a smectic phase. This shows that packing coefficient is intimately related with thermodynamic parameter. In an earlier paper [6] we have studied the variation of packing coefficient with increasing carbon atom in the alkyl chain for the homologous series cholesteryl alkanoates. In the case of cholesteryl alkanoates as the alkyl chain increases the packing coefficient decreases, the thermal stability of which also decreases, whereas in the case of 4,4'-di-n-alkyloxyazoxybenzene series the packing coefficient increases as we go up the series. These features can be qualitatively explained by considering a mechanical rigid rod model. In the homologous series cholesteryl alkanoates the alkyl chain is extended on only one side of the cholesterol moiety, the molecule as a whole is asymmetric with respect to the cholesterol moiety whereas in the case of 4,4'-di-n-alkyloxyazoxybenzene series the side chain is extended symmetrically on both sides of the phenyl groups. A symmetric molecule will have greater tendency to be held in a particular orientation in spite of the fact that the molecular length is increased as compared to the addition of greater asymmetry into the system. Thus packing coefficient of symmetric molecules tends to increase in the beginning unlike in the case of asymmetric molecules where the tendency for the molecule is to get flipped off from the mean orientation. This in principle can be used to explain the observed changes in the gradient of S factors in symmetric [8] (4,4'-di-n-alkyloxyazoxybenzene) and asymmetric [9] molecules (p-(p-ethoxyphenylazo) phenyl alkanoates). The S factor of asymmetric molecules drops off more rapidly than that of symmetric molecules as seen in Figs. 2 and 3, for the two homologous series. This can be explained as the greater tendency of the symmetric molecules to retain their orientational order as compared to asymmetric molecules when the temperature is raised.

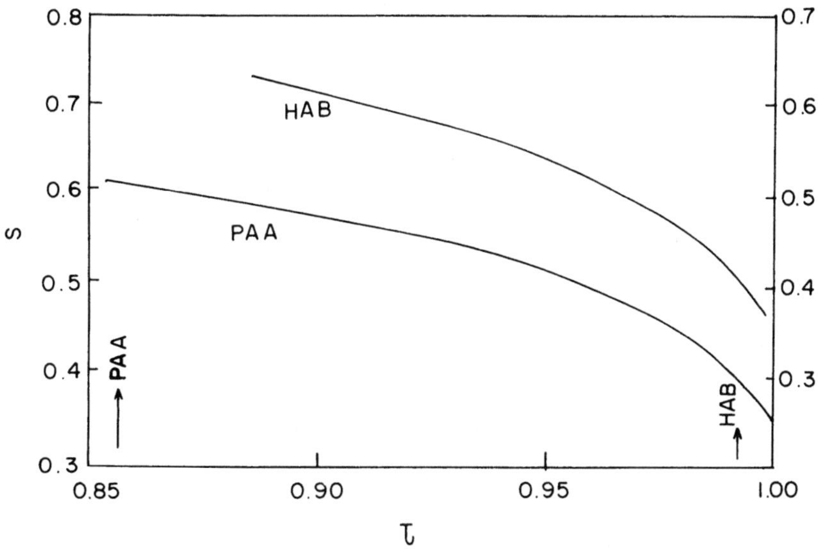

Fig. 2 Order parameter versus temperature for PAA and HAB

A detailed theoretical analysis of the rigid rod mechanical model is under progress.

One of us (JSP) would like to thank University Grants Commission, India for a Career Award. The award of research fellowships to (NCS) and (PKR) by CSIR India is gratefully acknowledged.

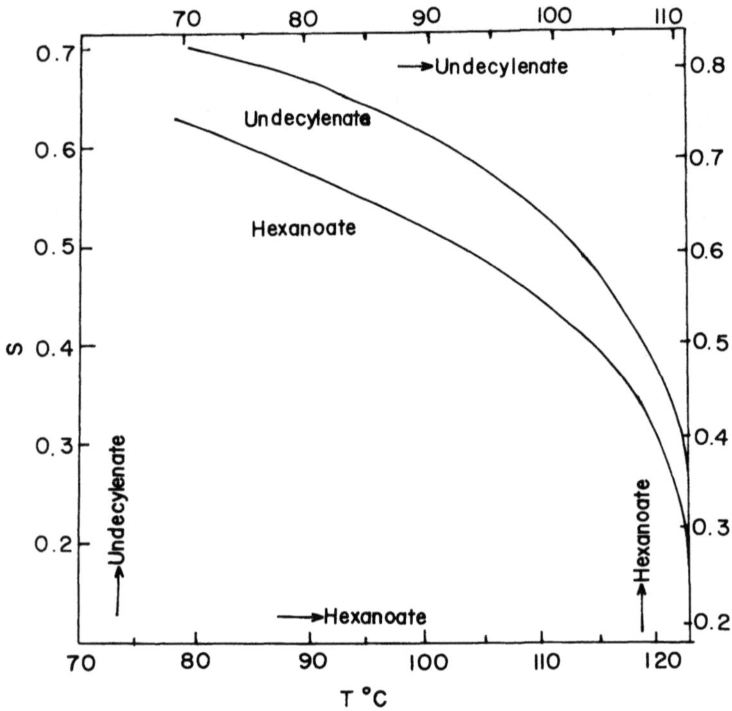

Fig. 3 Order parameter versus temperature for p-(p-ethoxyphenylazo) phenyl hexanoate and undecylenate

References

1. J. Shashidhara Prasad, Mol. Cryst. Liq. Cryst. *47*, 115 (1978).
2. W.R. Krigbaum, Y. Chatani and P.G. Barber, Acta Cryst. *B26*, 97 (1970).
3. A.J. Leadbetter and M.A. Mazid, Mol. Cryst. Liq. Cryst. *51*, 85 (1979).
4. C.H. Carlisle and C.H. Smith, Acta Cryst. *A25*, S47 (1969).
5. P.K. Rajalakshmi, N.C. Shivaprakash and J. Shashidhara Prasad, J. Appl. Cryst. *12*, Pt. 3, 316 (1979).
6. N.C. Shivaprakash, P.K. Rajalakshmi and J. Shashidhara Prasad, Mol. Cryst. Liq. Cryst. *51*, 317 (1979).
7. S.N. Prasad, S. Venugopalan and J. Billard, Mol. Cryst. Liq. Cryst. (Letters) *49*, 271 (1979).
8. W.H. de Jeu and W.A.P. Claassen, J. Che. Phy. *68* (1), 102 (1978).
9. C.L. Watkins and C.S. Johnson Jr, J. Phys. Chem. *75* (16), 2452 (1971).

Theoretical Conformational Study of Alkyl and Alkoxy Chains in MBBA, EBBA, and TBBA

J. Berges and H. Perrin

Laboratoire de Physique Moléculaire, Université P. et M. Curie, Tour 22, 4, place Jussieu, F-75230 Paris Cédex 05, France
and U.E.R. d'Etudes Médicales et Biologiques, Université R. Descartes, F-75230 Paris, France

The calculations were performed with a semi-empirical method (PCILO) and an empirical method (Buckingham potential). The results are quite similar and we present only the PCILO ones.

We have checked that the conformational study can be carried out separately in the core and in the chains. So we have studied the molecules of methoxy-benzene, ethoxy-benzene and butyl-benzene.

1. Methoxy-Benzene
The rotation around the Φ-O bond is quite free (0.5Kcal/mole). For the methyl-group the preferred positions are the usual ones (180,60,300)

2. Ethoxy-Benzene
We present on the Fig.1 the iso-energy curves for the rotations around the two first bonds (β_1,β_2). The results for the methyl group are the same as for the methoxy.y.

3. Butyl-Benzene
The iso-energy curves presented on the Fig. 2 concern the two first bonds (α_1,α_2) which determine the orientation of the chain with regard to the aromatic core.

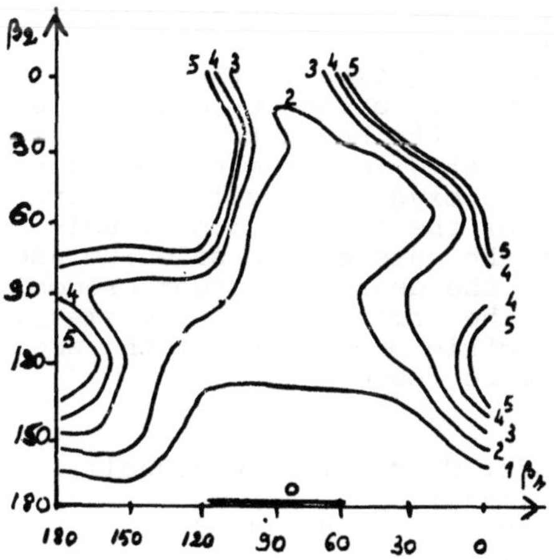

Fig.1

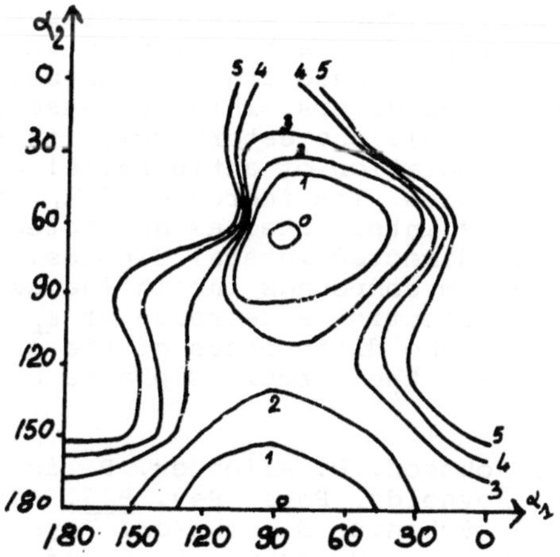

Fig.2

Refractive Indices and Dielectric Constants of the Nematic and Smectic Phases of O-, S-, Se-, and Te-4'-pentylphenyl-4-alkyloxychalcogenbenzoates

M. Bock, G. Heppke, B. Kohne, and K. Praefcke

Fachbereich für Synthetische und Analytische Chemie,
Technische Universität Berlin, D-1000 Berlin 12, Germany

The isoelectronic X-4'-pentylphenyl-4-heptyloxy- and octyloxy-chalcogenbenzoates ($\overline{7}\cdot X\cdot 5$; $\overline{8}\cdot X\cdot 5$; X= O, S, Se, Te) exhibit nematic and smectic A as well as smectic C phases showing some phase transitions of second order [1,2,3].

The refractive indices n_e and n_o were measured by means of a Leitz-Jelley-microrefractometer using a Mettler FP 2 hot stage. Corresponding to the molar refraction, the refractive index of the isotropic phase or the isotropic average of the refractive indices of the liquid crystalline phases respectively, increases with the atom number of the bridging chalcogen. Its temperature function depends on the density which can be estimated using the Lorenz-Lorentz equation. The birefringence itself increases with decreasing temperature showing marked discontinuities at the nematic-smectic A phase transitions, whereas slight decreases occur at the transitions nematic-smectic C ($\overline{7}\cdot S\cdot 5$) and smectic A-smectic C ($\overline{8}\cdot S\cdot 5$) showing a finite change of the tilt angle.

The order parameter calculated using the Haller extrapolation [4] deviates strongly from the values given by the Maier-Saupe theory.

The dielectric constants of magnetically oriented samples (thickness: 1 mm , B = 1 T) were investigated at a frequency of 1 kHz. The dielectric anisotropy increases with the atom number of the chalcogen. When lowering the temperature the dielectric anisotropy decreases strongly in the neighbourhood of the nematic-smectic phase transitions becoming negative in the case of the oxygen esters. This decrease is mainly caused by a decrease of $\varepsilon_\parallel$ of all the esters which is to be attributed to an increasing association of antiparallel orientated longitudinal molecular dipoles. The increase of $\varepsilon_\perp$ of the oxygen and sulfur esters is to be interpreted as an increase of the association of uniformly directed lateral dipoles. In the case of the selenium and tellurium compounds the influence of the degree of order predominates, effecting a decrease of $\varepsilon_\perp$ at the transition to the smectic A phase. No differences of the dielectric properties of the smectic A and the smectic C phases could be found.

1 D. Johnson, D. Allender, R. de Hoff, C. Maze, E. Oppenheim, R. Reynolds, Phys. Rev. B16, 470 (1977)
2 C.A. Schantz, D. Johnson, Phys. Rev. A17, 1504 (1978)
3 G. Heppke, B. Kohne, K. Praefcke, E.J. Richter, Israel J. Chem. 18, 199 (1979)
4 R.G. Horn, J. Physique 39, 105 (1978)

Crystal Structures of Smectogenic *p*-*n*-Alkoxybenzoicacids

R.F. Bryan and P. Hartley

Department of Chemistry, University of Virginia,
Charlottesville, VA 22901, USA

The isotypic crystal structures of three smectogenic *p*-n-alkoxybenzoic acids having eight, nine and ten carbon atoms in the alkyl chain are reported together with the crystal structure of a second modification of *p*-n-octoxybenzoic acid. The isotypic forms are identical in type to that previously reported for *p*-n-heptoxybenzoic acid (Bryan, Miller, & Shen, *Abstr. Amer. Crystallogr. Assoc.*, 5, 37 (1977)).

In each case the acids are present in the crystal as centrosymmetric hydrogen-bonded dimers, parallel to one another by crystallographic requirement. The molecules have a markedly non-linear and non-planar conformation arising from a *gauche*-relation of the ether oxygen and C(γ) about the C(α)-C(β) bond of the chain.

In the isotypic form the molecules pack in an imbricated stratified layer structure characterized by segregation of the aliphatic chains and aromatic cores into separate strata. In the chain strata the chains of molecules in adjacent layers are arranged in an antiparallel interdigitated fashion so that there are twice as many chains per stratum as cores. The chains adopt an extended conformation and there are slight differences in the mode of packing in the acids of even and odd chain lengths.

In the second crystal form of *p*-n-octoxybenzoic acid, the dimers are arranged in sheets in a head-to-tail fashion with chains in adjacent sheets in antiparallel juxtaposition.

Direct extrapolation of either crystal form to possible mesophase organization is not feasible since all undergo solid-solid transitions on heating before forming smectic phases. However, calculation of the projected molecular length of the dimers in the orientation derived by Blumstein and Patel (*Mol. Cryst. Liq. Cryst.*, 48, 151 (1978)) produces exact agreement with the observed smectic phase x-ray periodicities which they report, suggesting that the molecular conformation found in the crystal may persist into the smectic phase.

This work was supported by a grant, DMR78-19884, from the National Science Foundation, U.S.A. Full details of the x-ray analyses will appear in *Molecular Crystals and Liquid Crystals* as Parts VI and VII of the series: An X-Ray Study of the *p*-n-Alkoxybenzoic Acids.

The Nematic-Smectic C Phase Transition: A Renormalization Group Analysis *

D. Mukamel and R.M. Hornreich

Department of Electronics, Weizmann Institute of Science, Rehovot, Israel

Abstract

The phase transition from a nematic to smectic C liquid crystal is described by an infinite dimensional order-parameter. A LANDAU-GINZBURG-WILSON model appropriate for this transition is introduced and analyzed by renormalization group techniques. It is shown that for $d > 5$ dimensions the model exhibits a continuous transition with Gaussian-like critical behavior. However, for $d = 5 - \varepsilon$ ($\varepsilon > 0$), dimensions, the model does not possess a stable fixed point, indicating that a first order transition occurs. This is consistent with results derived previously by BRAZOVSKII and by SWIFT and LEITNER. A full account of this work appears in J. Phys. C13, 161 (1980).

* Supported in part by a grant from the U.S.-Israel Binational Science Foundation (BSF), Jerusalem, Israel.

Use of the Far Infrared Spectroscopy for the Determination of the Order Parameter of Nematic Compounds

D. Decoster, M. Bouamra

Université des Sciences et Techniques de Lille,
Centre Hyperfréquences et Semiconducteurs,
Laboratoire associé au C.N.R.S. n°287
F-59655 Villeneuve d'Ascq Cédex, France

Abstract

The far infrared spectra of some alkyl-cyano-biphenyl compounds and cyclohexane derivatives are presented in liquid, nematic and smectic A phases ; they are interpreted in terms of intra and intermolecular modes. Dichroism measurements performed with an optically pumped F.I.R. waveguide laser are also given. The conditions for deducing the order parameter $\overline{P_2}$ of nematic compounds from dichroism measurements in the F.I.R. range are studied with the aid of recent theoretical developments of absorption spectrometry based on orientational and vibrational correlation function formalism.

Introduction

F.I.R. spectra of some alkyl cyano biphenyl compounds and cyclohexane derivatives are given in solid, smectic, nematic and isotropic phases. Furthermore dichroism measurements on well oriented samples performed with a Far infrared waveguide laser are presented. Our present purpose is to study the conditions for deducing the order parameter $\overline{P_2}$ from F.I.R. measurements using the analysis of our experimental results and with the aid of recent theoretical developments.

1. Far Infrared Spectra

The measurements are performed in the 45 - 500 μm wavelength range with a grating spectrometer ; the resolution is about 3 - 4 cm^{-1}. We give (Fig. 1) the F.I.R. spectra of alkyl cyano biphenyl compounds and cyclohexane derivatives in liquid, nematic and smectic A phases (Table 1).

Table 1 Compounds studied

		C - SmA	C-N or SmA-N	N - I or C-I
CB_4	C_4H_9–⟨O⟩–⟨O⟩–CN	-	-	46,5°C
CB_5	C_5H_{11}–⟨O⟩–⟨O⟩–CN	-	22,5°C	35 °C
CB_7	C_7H_{15}–⟨O⟩–⟨O⟩–CN	-	28,5°C	42 °C
CB_8	C_8H_{17}–⟨O⟩–⟨O⟩–CN	21°C	32,5°C	40 °C
PCH_7	C_7H_{15}–⟨H⟩–⟨O⟩–CN	-	30 °C	57 °C
CCH_7	C_7H_{15}–⟨H⟩–⟨H⟩–CN	-	71 °C	83 °C

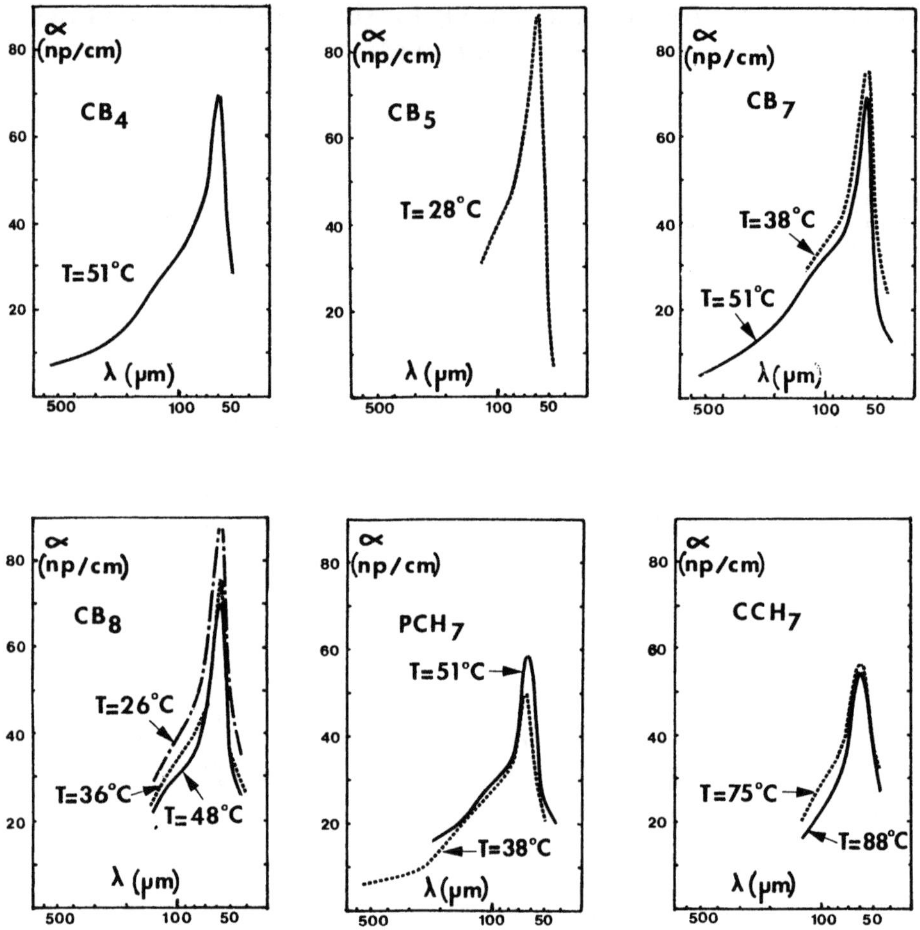

Fig. 1 Far infrared spectra of alkyl cyano biphenyl compounds and cyclohexane derivatives in liquid ———, nematic ------ and smectic A —.—.— phases

In agreement with previous works for CB_7 [1], all these spectra present a strong absorption in the 50 - 150 μm range. Note that in nematic and smectic A phases no attempt was made to fully orient the substances; nevertheless, we have observed [2] a variation of the intensity versus temperature in nematic and smectic A phases, while in the isotropic phase no variation could be observed ; these effects will be studied in section 2. For these substances, the dielectric relaxation time is very long [3]; thus the Debye absorption corrected by inertial effects is very weak (< 1 np/cm) [4] and can be neglected. In order to know the contribution of librational movements about the short axis of the whole molecule (Poley absorption), we have computed the integrated absorption in the isotropic phase

$A_c = \frac{1}{n}\left(\frac{n^2+2}{3}\right)^2 \frac{\Pi}{3c^2}\frac{2\mu^2}{I}$ deduced from Gordon sum rule [5] including the Polo-Wilson internal field correction [6] which leads to $A_c \simeq 270.10^{-21}$ cm for the CB_7 ($\mu \simeq 5$ Debye [3], $I \simeq 314.10^{-39}$ $g.cm^2$ [1] $n \simeq 1.61$ [3, 7, 8]). We have compared this result with the experimental area $A_{exp} = \frac{1}{N}\int_0^\infty \alpha(\omega)\, d\omega$ where N is the number of molecules in a unit volume ; taking the density $d \simeq 1$ [7, 8]; we have got $A_{exp} \simeq 4000.10^{-21}$ cm which is greater than the area corresponding to the theoretical prevision ; hence we could assume that the librational contribution is about 5 - 10 % of the total spectrum of CB_7 and the main contribution is of an intramolecular origin. According to EVANS [1] librational modes would appear in the 68 μm range. In order to confirm the importance of the intramolecular modes we have studied CB_7 in dilution (Fig. 2) and CCH_7 in solid phase (Fig. 3) ; in all cases a similar thin structure appeared. Since only the - C ≡ N molecular group is conserved in CB_7 and CCH_7 in these spectra (Fig. 2, 3) and since the modification of the alkyl chain length and the substitution of phenyl rings by saturated ones have no important effects on FIR spectra (Fig. 1), we could assume that the most important contribution would rise from vibrational torsional internal modes of the - C ≡ N molecular group. The spectrum of benzonitril in the liquid phase (Fig. 4) seems to confirm this assumption : the strong absorption peak of benzonitril is well known and assigned to a γ C ≡ N internal vibration [9] perpendicular to the molecular axis.

Fig. 2

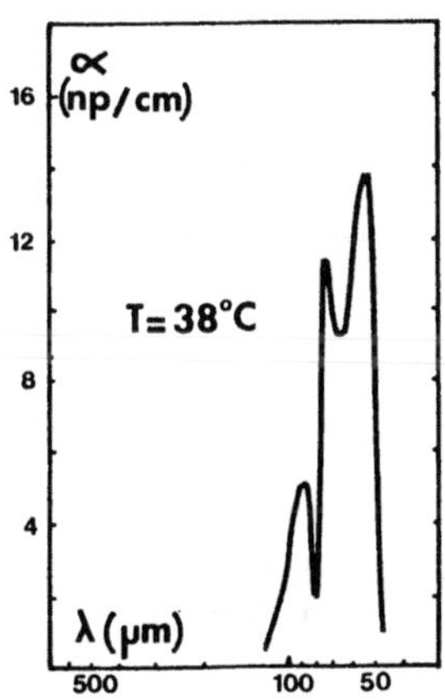

Fig. 3

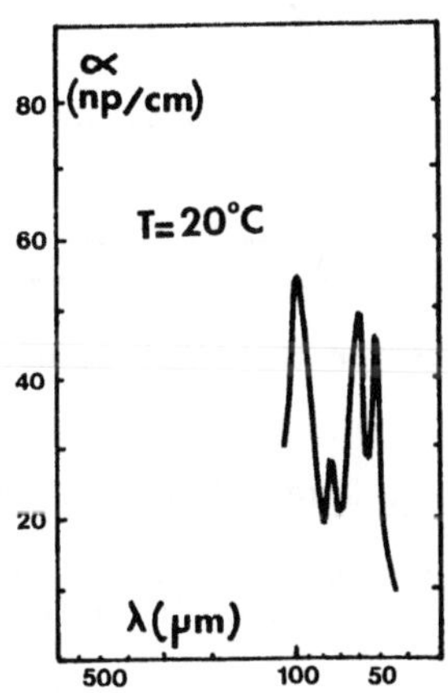

Fig. 4

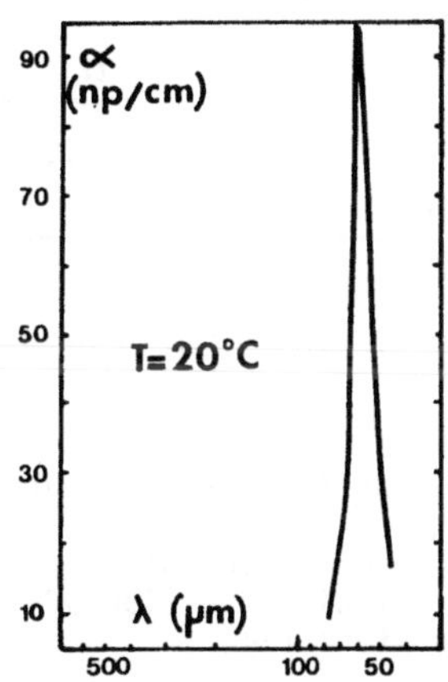

Fig. 2 Far infrared absorptions of CB_7 in dioxan (10 % mol.), corrected for solvent absorption

Fig. 3 Far infrared absorptions of CCH_7 in solid phase

Fig. 4 Far infrared absorptions of benzonitril in liquid phase

2. Dichroism Measurements

They are performed with an optically pumped F.I.R. waveguide laser for the CH_3OH 118.8 μm wavelength. The power available from this source (> 1 mW) would enable us to oriente the substances with a magnetic field (H > 2 KG). A polarizer placed between cell and source determines the axis of the FIR electric field. More details about our experimental set up are given in other papers [10, 11]. In the nematic phase (Fig. 5) two distinct values $\alpha_{/\!/}$ and $\alpha_{\perp}$ of the absorption can be observed, depending on the substance being oriented parallel or perpendicular to the F.I.R. electric field. In the isotropic phase only one value of the absorption can be observed.

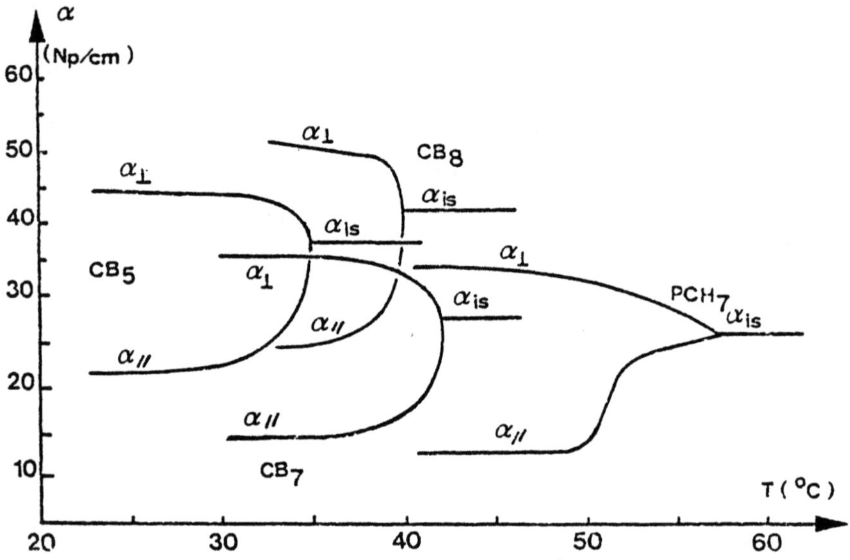

Fig. 5 Dichroism measurements for the 118.8 μm wavelength.

3. Far Infrared Absorption and Orientational Order

For dielectric materials, the electromagnetic absorption is related to the correlation function of the macroscopic dipole moment M(t) by a Fourier Transform [12] :

$$\alpha(\omega) = \frac{\omega\varepsilon''}{nc} = \frac{\omega}{kT\ V\ \varepsilon_V}\ \frac{1}{nc}\ \frac{Em(\omega)}{E(\omega)} \int_0^{\infty} < \vec{M}(o).\dot{\vec{M}}(t)> \sin \omega t\ dt$$

with $\vec{M}(t) = \sum_i \vec{\mu}^i(t)$ where $\vec{\mu}^i(t)$ is the microscopic moment of the i-th molecule. The FIR absorption is due to librational, torsional and vibrational modes ; hence $\vec{\mu}^i(t)$ can be put as follows [13, 14]

$$\vec{\mu}^i(t) = \vec{\mu}^i p(t) + \sum_m \Delta\mu_m^i\ \vec{\mu}_m^i(t)\ q_m^i(t)$$

$\vec{\mu}_p^i(t)$ represents the permanent dipole moment of the whole molecule (which contributes to librational modes) or of a molecular group (which contributes to torsional modes) ; $\Delta\mu_m^i \cdot \vec{u}_m^i = (\frac{\partial \vec{\mu}^i(t)}{\partial q_m})_{q_m^i = 0}$ is the component of the vector of the dipole moment along the normal coordinate q_m^i of the m-th vibrational mode. If vibrational, torsional and librational modes are not correlated to each other, these three contributions can be separated. As far as the contribution of vibrational modes is concerned, assuming : all vibrational modes not correlated, orientation of vibrations well defined, reorientation movements very slow, a same vibrational relaxation time in nematic and isotropic phases, and using a usual internal field correction in the FIR range [6], we get

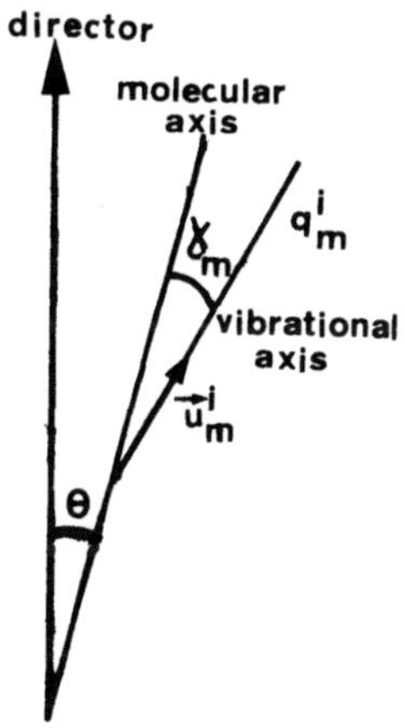

$$\frac{\alpha_{/\!/}}{\alpha_{\perp}} = \frac{n_{\perp}}{n_{/\!/}} \left(\frac{n_{/\!/}^2 + 2}{n_{\perp}^2 + 2}\right)^2 \frac{1 + 2S < 1 - \frac{3}{2} \sin^2\gamma_m >_m}{1 - S < 1 - \frac{3}{2} \sin^2\gamma_m >_m} \quad (1)$$

where S is the order parameter $\overline{P}_2 = \frac{1}{2} < 3\cos^2\Theta - 1>$

and $$<1 - \frac{3}{2} \sin \gamma_m >_m = \frac{\sum_m \Delta\mu_m^2 (1 - \frac{3}{2}\sin^2\gamma_m) \int_0^{\infty} <q_m(o).q_m(t)> \sin \omega t \, dt}{\sum_m \Delta\mu_m^2 \int_0^{\infty} < q_m(o).q_m(t) > \sin \omega t . dt}$$

The equation (1) is well known for a single mode [15, 16]. If the torsional modes (and weak librational contribution) at high frequencies correspond to small displacements, knowing that the permanent dipole moment lies essentially along the - C ≡ N molecular group [3], we can show that these contributions are similar to perpendicular vibration ones; hence equation (1) is valid with $\gamma_m = \frac{\Pi}{2}$.

4. Order Parameter

By using equation (1) with $\gamma_m = \frac{\Pi}{2}$ and with refractive indices obtained from literature [7, 8] we deduced order parameter from dichroism measurements for the 118.8 μm wavelength (Fig. 6). The values of $\overline{P}_2$ obtained by this method agree with some previous results [7, 8] but seem too low when we compare them to more recent results [17, 18, 19]. We could assume that this discrepancy lies first in experimental errors in particular in the vicinity of the clarification temperature, second by using an inaccurate value of γ_m which can be different from $\frac{\Pi}{2}$ for wavelengths too far from the 55-60 μm wavelength range; thus it would be interesting to perform dichroism measurements versus frequency in all the F.I.R. range. Moreover it might be necessary to study more carefully the orientation of the sample which could be partially due to window effects (these effects would explain the inaccurate values obtained for PCH_7 for temperatures closer to clarification temperature).

Conclusion

FIR measurements can lead directly to $\overline{P}_2$ if the quantity $<1 - \frac{3}{2} \sin^2\gamma_m>_m$ is determined with accuracy. In any case FIR dichroism measurements contain

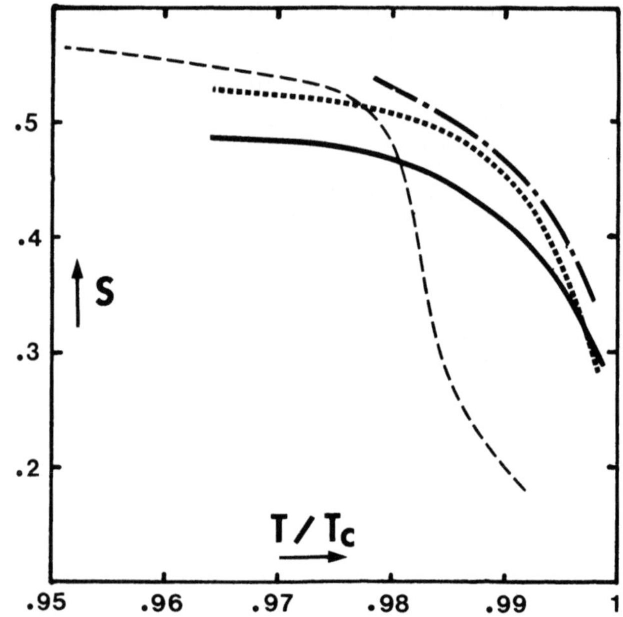

Fig. 6 Order parameter as a function of reduced temperature for : CB_5 ——— ; CB_7 .... ; CB_8 —·—·—· ; PCH_7 - - - - - -

orientational order information and it can be regarded as complementary to the other methods of determination of the order parameter of nematic compounds.

Acknowledgments

The authors would like to thank Pr. E. CONSTANT, Drs. M. CONSTANT and J.P. PARNEIX for several comments and fruitful discussions, and to thank Dr. W.H. DE JEU for helpful remarks.

References

[1]G.J. Evans, M. Evans, J.C.S. Faraday II 73, 285 (1977).
[2] M. BOUAMRA, Rapport de D.E.A. Lille (1979)
[3]D. Lippens, J.P. Parneix, A. Chapoton, J. Physique 38, 1465 (1977).
[4]Y. Leroy, Thesis Lille (1967).
[5]R.G. Gordon, J. Chem. Phys. 38, 1724 (1963).
[6]A. Gerschel, I. Darmon, C. Brot, Molec. Phys. 23, 317, (1972).
[7]P.P. Karat, N.V. Madhusudana, Mol. Cryst. Liq. Cryst. 36, 51 (1976).
[8]D.A. Dunmur, M.R. Manterfield, W.H. Miller, J.K. Dunleavy 45, 127 (1978).
[9]G. Varsanyi, Vibrational Spectra of Benzene derivatives.(Acedemic, N.Y. 1969)
[10]J. Depret, Thesis Lille (1979).
[11]J. Depret, D. Decoster, Coll. O.H.D. Lille (1979).
[12]R. Kubo, Lectures in theoretical Physics (Interscience Pub, New York, 1959) 1, p. 120.

[13] R. Fauquembergue, Thesis Lille (1977).
[14] M. Constant, Thesis Lille (1978).
[15] A. Saupe, W. Maier, Z. Naturforschg. 16a, 816 (1961).
[16] J.R. Fernandes, S. Venugopalan, J. Chem. Phys. 70, 519 (1979).
[17] P.L. Sherrell, D.A. Crellin, J. Phys. Coll C3, Suppl. n° 4, 40, C3-211 (1979)
[18] K. Miyano, J. Chem. Phys. 69, 4807 (1978).
[19] Hp. Schad, G. Baur, G. Meier, J. Chem. Phys. 71, 3174 (1979).

Investigation of Molecular Motions in Smectic A and C TBBA by NMR Relaxation Dispersion

Th. Mugele, V. Graf, W. Wölfel, and F. Noack

Physikalisches Institut der Universität,
Pfaffenwaldring 57, D-7000 Stuttgart 80, Federal Republic of Germany

We have studied the proton spin T_1 relaxation dispersion for the smectic A and C phase of TBBA using time dependent fast field-cycling techniques [1] in the Larmor frequency range from ν_p = 100 Hz to 10 MHz. Our investigation considerably extends recent works by BLINC et al. [2] performed with different NMR methods for frequencies greater than 140 kHz. The new experimental data are consistent with BLINC's results for Sm C but not for Sm A. The difference is that the essential T_1 dispersion observed with our technique for Sm A occurs at much lower frequencies, namely below 100 kHz, which means that the relaxation dispersion of both smectic phases looks rather similar. The $T_1(\nu_p)$ behaviour can be described quantitatively in terms of relaxation by order fluctuations (OF), self-diffusion (SD), and a third molecular mechanism with Debye-type power spectrum (possibly rotation or a second OF contribution), if one takes into account both the local field effect and the cut-off frequencies of the OF model [1,2]. Such an analysis shows that the OF relaxation rate is of similar magnitude as for high-temperature nematics, whereas the cut-off frequencies are considerably smaller. The diffusion constants of best curve fits differ somewhat from neutron scattering results, nevertheless it is also possible to obtain satisfactory model fits using data in accordance with the literature. Details of this relaxation analysis will be presented elsewhere [3].

1. W.WÖLFEL, Dissertation Universität Stuttgart, Minerva Publikation (1978); V.GRAF, Dissertation Universität Stuttgart (1979)
2. R.BLINC, M.VILFAN, M.LUZAR, J.SELIGER, and V.ZAGAR, J.Chem. Phys. 68,303 (1978), and references given there
3. TH.MUGELE, V.GRAF, W.WÖLFEL, and F.NOACK, submitted to Z. Naturforsch.

Study of Conformational Motions in the Aromatic Core of Schiff's Bases by Semi-Empirical Quantic or Empirical Methods. Influence of the Packing of Molecules on the Conformation

H. Perrin and J. Berges

Laboratoire de Physique Moléculaire, Université P. et M. Curie, Tour 22, 4, place Jussieu, F-75230 Paris Cédex 05, France
and U.E.R. d'Etudes Médicales et Biologiques, Université R. Descartes, F-75230 Paris, France

In order to explain the polymorphism of some mesogen compounds, we show the ability for the aromatic core of Schiff's bases to exhibit some deformations by isomeric rotations around the bonds.

I. Isolated Molecule of Benzylidene-Aniline.

The calculations are carried out by a semi-empirical method (PCILO) and an empirical method (Buckingham potential). Both methods give results in good agreement with the experimental results (figure below).

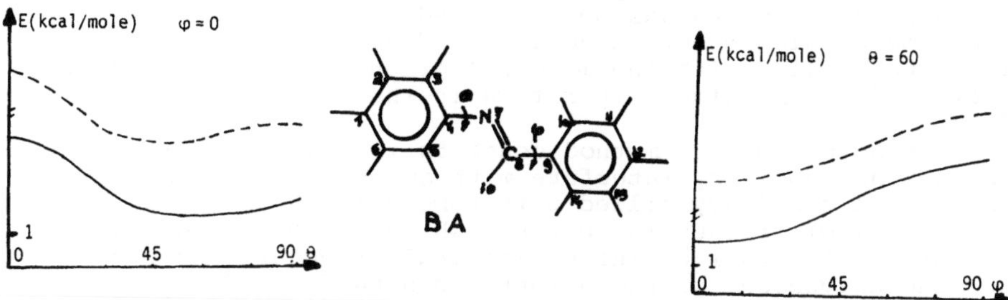

— PCILO method --- empirical method

II. Simulation of Packing.

The calculations are performed with the empirical method. The conformational abilities are limited but the intramolecular deformations cannot be neglected. These results are obtained for the calculated equilibrium distance d = 4.5 Å (see figure below).

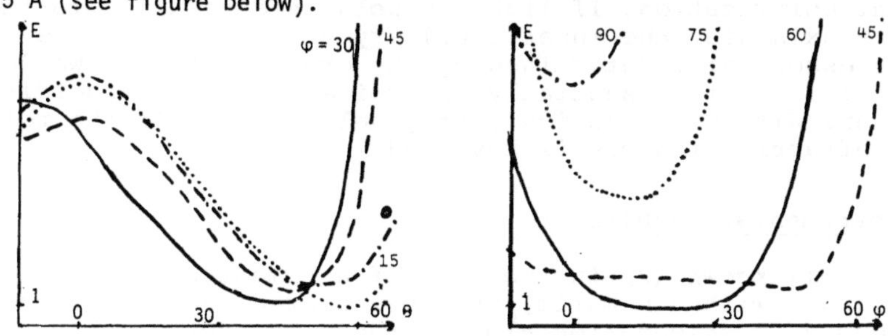

Homeotropic Alignment of Liquid Crystals in Cylindrical Geometry

F. Scudieri, M. Bertolotti, and A. Ferrari

Istituto di Fisica, Facoltà di Ingegneria, Università Roma, Italy
and Gruppo Nazionale Elettronica Quantistica Plasmi, of CNR, Roma, Italy

1. Introduction

Recently a new interferometric technique has been described which allows to study the refractive index distribution in cylindrical structures [1].

In the case of a capillary field with a liquid crystal it is possible to obtain information on the director orientation, and in some cases also the ratio K_B/K_S can be derived, where K_B and K_S are the elastic bend and splay constants respectively.

In the following COB (cyanooctyl biphenyl) liquid crystal in a cylindrical geometry has been studied.

This material has both a nematic and smectic phase. The capillary was filled with the material oriented homeotropically and studied by changing temperature in the range from 20 to 50° C.

The principle of the method consists in making the light crossing the capillary interfere with another beam crossing an identical capillary filled with some reference oil. In this way a system of fringes is obtained which is corrected for the curvature of the wall of the cylindrical structure and from which the alignement of the director can be derived.

It can be easily shown that in the case of a smectic filled capillary, observation with light linearly polarized orthogonal to the capillary axis gives information on the extraordinary index n_e, whilst the information on n_o is obtained with parallel polarization.

If the material is a nematic both ordinary and extraordinary refractive indices play always a role, irrespectively of the light polarization. If light is polarized parallel to the capillary axis near the inner capillary surface the fringe shift is essentially determined by the ordinary index, while near the axis it is essentially determined by the extraordinary index. With polarization orthogonal to the axis the role of the refractive indices is reversed.

2. Experimental results

As it is well known [2] in the case of a homeotropically filled capillary with a nematic, the molecular director changes continuously in orientation along a radial direction going from a disposition normal to the capillary axis near the inner wall to a disposition parallel to the capillary axis at

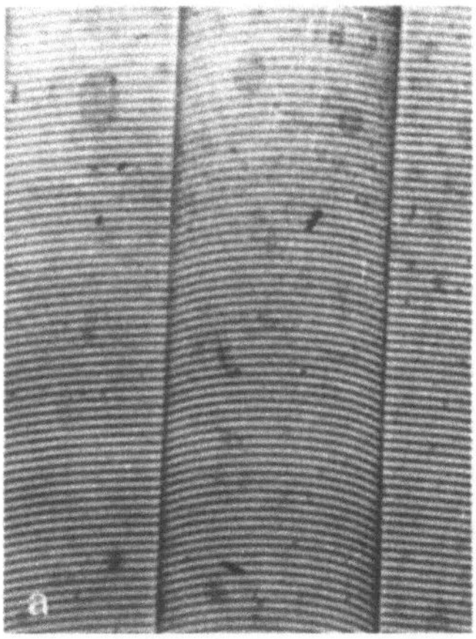

Fig. 1 *Fringe patterns of a smectic-A with light parallel to the axis (a) or orthogonal to it (b).*

its center; in the case of a smectic-A material the molecules are arranged in cylindrical layers with their director oriented radially. On the capillary axis a localized defect in the form of a disclination line is therefore formed.

The resulting fringe patterns are shown in Fig. 1 where a smectic A material is observed with light polarized parallel to the axis (Fig.1a) and perpendicular to it (Fig.1b). The contribution of the extraordinary index is present only in b, clearly separated by the nearly flat behavior of the ordinary index contribution in a. The small curvature change of the fringes out of the capillary is due to a temperature difference between the samples in the interferometer.

The director orientation in cylindrical structures has been calculated by Cladis and Kleman (3) and by a comparison of these calculations with the obtained fringes the ratio K_b/K_s

Fig. 2 *A disclination line along the axis of a smectic-A filled capillary.*

can be derived, if independently the refractive index values are known.

The method is also suitable to observe the presence of defects and their behavior as a function of temperature.

Fig.2 shows the localized axial defect, characteristic of the smectic-A phase which breaks off into small segments, which annihilate in a short time, at $T_{A \to N}$. Observation this time is made in white light.

The movement of a wall between homeotropic and parallel regions is shown in Fig.3, again made in white light.

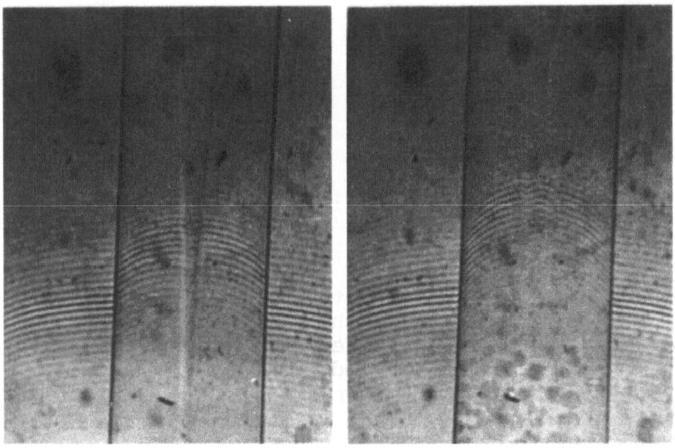

Fig.3 *Movement of a wall between homeotropic and parallel region. The wall is visible as the triangular shaped region which moves upwards in the right photograph.*

3. Conclusions

The present method is suitable to control the molecular director arrangement.

If the molecular arrangement is known, as in the smectic-A and in nematic phases where the single elastic constant approximation can be utilized, it is possible to determine n_o and n_e by the fringe shift. In the S_A phase the radially and axially polarized light components are completely separable and the two fringe systems can be independently observed. In the N phase the refractive indices can be obtained by analyzing the minimum and maximum fringe shifts (see ref.1).

If the refractive indices are independently determined it is possible to perform accurate measurements of the molecular arrangement in N phase.

In particular in the case where the material exhibits S_A and N phases it is possible, by fitting the observed molecular arrangement with the theoretical calculation, to known the ratio K_b/K_s. This method applied to static deformation in cylindrical geometries, is practically useful near $T_{A \to N}$ where K_b diverges.

Moreover the method allows observation of the dynamics of localized defects as here illustrated with a few simple examples.

REFERENCES

1. F.Scudieri-Appl.Opt.18, 1455, 1979

2. P.G.de Gennes - The Physics of Liquid Crystals,Oxford, Univ.Press, Oxford, England, 1975

3. P.E.Cladis and M.Kleman, J.Phys.(Paris) 33, 591, 1972

Part III

Defects, Elasticity, and Rheology of Smectics

Curvature Defects in Smectic A and Canonic Liquid Crystals

M. Klêman

Laboratoire de Physique des Solides, Bât.510, Université Paris-Sud, F-91405 Orsay, France

1. Introductory remarks

Smectic A and canonic liquid crystals have this common character (that they share with 3 d crystals but not nematics) to possess quantized symmetry translations. In the case of smectics, there is one such translation, whose magnitude is equal to the layer thickness at equilibrium. In the case of canonics, there are two such translations, the two lengths of the unit cell in a plane perpendicular to the rods (we have here adopted FRANK's proposal to call canonic the phases made of rods), a good example being the case of the hexagonal discotic phases recently discovered [1].

Dislocations of translation should therefore be common defects in these phases, and are indeed well-known in smectics [2]. The same topological objects should exist in canonics, and no doubt little time will pass before they are observed and analyzed. Other defects involving other kinds of symmetry breaking (dislocations break locally translation symmetries, disclinations break rotations; also topological singular points and non singular topologically stable configurations, if any) might exist. They all are classified by the methods of algebraïc topology [3], as it is well-known to-day.

Dislocations create long range strain fields whose description depends on an exact knowledge of elastic constants; this is different with each material. However they are cases in which the long-range distortion is material independent, for example when dislocations of the same sign gather in special clusters, or in some situations which imply only curvature of the lattice. In such cases geometrical considerations play a crucial role, and it is the purpose of this paper to review some of them, with special emphasis on the case of curvature defects.

1.1 Curvature defects ; a definition. We consider here deformations which do not distort elastically the medium : local densities and local angles are conserved. This situation exists in 3 d crystals (see 1.2 below) but is a priori much more relevant to the physics of liquid crystals. For example the layers of a Sm A phase can take any shape at constant density and thickness, because they are 2 d liquids; the texture we look for will be made of a stacking of such layers. Similarly the rods of a Ca phase can bent at constant density, and the texture we look for made of parallel bent rods, preserving equispacing. The geometrical problem is to classify these textures and describe their specific (curvature) defects. The case of smectics is known : the solution consists in the so-called cofocal domains; the case of canonics has been explored very recently by BOULIGAND [4], and established on firm geometrical grounds by the present author [5]. The solution consists in "developable domains".

1.2 Why curvature implies geometry mostly. The concept of curvature deformation was first introduced by NYE [6] for the case of solid crystals. Let us analyze it by pointing the difference with strain deformation.

In a 3 d crystal the reticular directions of the Bravais lattice define a local trihedron of vectors, which is deformable in two ways : by changing the angles between the coordinate vectors, and by allowing their length to vary. This is the usual description of the deformation of a crystal in terms of the strain tensor e_{ij}, but here we have insisted on a local definition : if we consider the components of this tensor in the axes of the local trihedron, the diagonal components $e_{11}, \ldots$ correspond to the variation in length of the coordinate vectors, and the off-diagonal $e_{12}, \ldots$ to the angular variations between them. Now, if we look to 2 neighboring trihedra, the strain imposed on one of them imposes some small rotation ω_i on the second, and limits its possibilities of strain. This is expressed by the relation (which implies first derivatives of e_{ij})

$$K_{ji} = \partial\omega_i/\partial x_j = e_{ik\ell}\, \partial e_{\ell j}/\partial x_k \qquad (1)$$

K_{ij}, which is called the tensor of contortion, is here the gradient of the gradient of the rotation vector density. But this is not the only possibility. Assume now that a contortion tensor is given, which is not gradient; it defines a variable local rotation

$$d\omega_i = \sum_j K_{ji}\, dx_j \qquad (2)$$

for each trihedron. The physical meaning of this deformation is easily visualized in the 2 d case : atomic layers suffer a process which is called single glide (Fig.1), and dislocations have to appear in order to maintain the system at zero strain ($e_{ij} = o$). In practice, the dislocations have a finite Burgers'vector b (equal to the lattice parameter), and it appears a number of dislocations per unit area equal to 1/bR, where R is the radius of curvature. In the limit $b \to o$, this number becomes infinite and the strain e_{ij} is strictly zero. The process decreases e_{ij} as much as possible when $b \neq o$. $\alpha = 1/R$ is called a dislocation density and is here equal to the contortion $K = 1/R$. It is noteworthy that the dislocations are all of the same sign; let us stress once more that their global effect is geometrical rather than strain. Also notice that $K = 1/R$ becomes infinite somewhere on each layer, on a surface which is a focal surface for the normals to the layers. This central region features a disclination, whose total strength can reach S = 1 (for an angle of 2π).

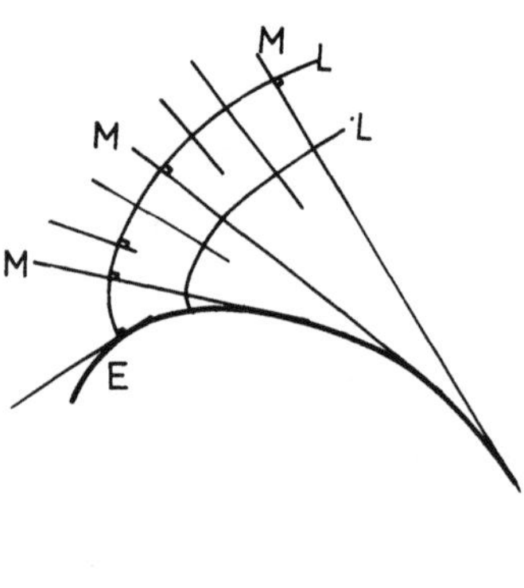

Fig.1 (L) are the layers (or reticular planes) which suffer curvature; dislocations are present at the end of some (M) layers. E is the singular focal surface

This two dimensional analysis can be extended to 3 d [7] ; the general formula (for small curvatures) is

$$K_{ji} = \varepsilon_{ik\ell} e_{\ell j,k} - \alpha_{ij} + \frac{1}{2} (\alpha_{11} + \alpha_{22} + \alpha_{33}) \delta_{ij} \tag{3}$$

where K_{ji} is the 3 d extension of $K = 1/R$ and α_{ij} the extension of α. α_{ij} is the density of dislocations parallel to the i direction and whose Burgers' vector is along the j direction. Two sources of lattice curvature appear clearly in (3) ; one is of elastic origin and does not interest us; the second one is a geometric effect due to dislocations and can be called NYE's curvature.

1.3 Two types of Nye's curvature in a liquid crystal. In the first type, the dislocations which are used in building the curvature are of infinitesimal strength; the elastic energy is vanishing totally. The dislocations in question have Burgers' vector in the layers (for smectics), or along the rods (for canonics). This is another way of stating that bending of layers or rods can be relaxed by fluid motion which brings back the system to a constant density. In the second type we use dislocations of finite strength ; this is the exact analog of the 3 d case, and, as in solid crystals one expects that smooth densities of dislocations of finite strength are unstable with respect to polygonization, i.e. the formation of tilt (or twist, or mixed) boundaries in which dislocations of the same sign gather, and through which the layers rotate abruptly by an angle proportional to the dislocation content of the wall (Fig.2) [8].

Fig.2 Tilt boundary in a system of layers (or reticular planes)

Such boundaries have been studied in smectics [2] and observed in canonics [9]. We will meet them occasionally in this paper.

2. Curvature defects in smectics

The subject of cofocal domains is so well-known (or should be so well-known, at least) since the classical work of G. FRIEDEL and F. GRANDJEAN [10] that we shall restrict ourselves to quoting the main theorems which come to hand here and to a few opened remarks.

The normals to a set of parallel layers are straight lines; they form what is called a congruence of straight normals (CSN in short) depending on two parameters. The only gradient of the director $\vec{n}$ which is different from zero is div $\vec{n}$. This congruence envelops two focal surfaces F_1 and F_2

(cf. a similar problem in geometrical optics) on which the layers curvature becomes infinite, and by consequence the energy density too. The energy is decreased if F_1 and F_2 are degenerated into lines. In that case we use :

Dupin's theorem : one of the focal line (F_1, say) is an ellipse and the other one an hyperbola whose foci are the apices of F_1 and which is located in a plane perpendicular to F_1. Conversely the foci of F_1 are the apices of F_2. The layers are curved into Dupin cyclides [11].

Simple examples are when the layers curve into parallel cylinders, or parallel tori (illustrations in [12]). ROSENBLATT et al. [13] have stressed the importance of cofocal domains built on cofocal parabolae; KLEMAN [14] has calculated the energies.

Small variations to perfect cofocal domains, made of a CSN, can be analyzed using the concepts of congruence of normals (CN in short; the molecules no longer align along straight lines, the layers are no longer parallel; $\vec{n}_\wedge \text{curl}\ \vec{n} \neq o$; $\vec{n}.\text{curl}\ \vec{n} = o$) and of congruence of straight lines (CSL in short; the notion of layers disappear; $\vec{n}.\text{curl}\vec{n} \neq o$; $\vec{n}_\wedge \text{curl}\vec{n} = o$). In both cases, energetical considerations lead us to expect F_1 and F_2 still to be lines, but there is no geometrical obstruction to them being surfaces. Mixed situations can prevail.

A CN is elastically distorted, but this can be relaxed by the appearance of a number of quantized edge dislocations (all of the same sign, Fig.3a). This situation exists in cholesterics [15], where it is observed that dislocations frequently polygonize. The variation in shape of F_1 and F_2 with respect to conics allows for an easier space filling than with perfect cofocal domains [16].

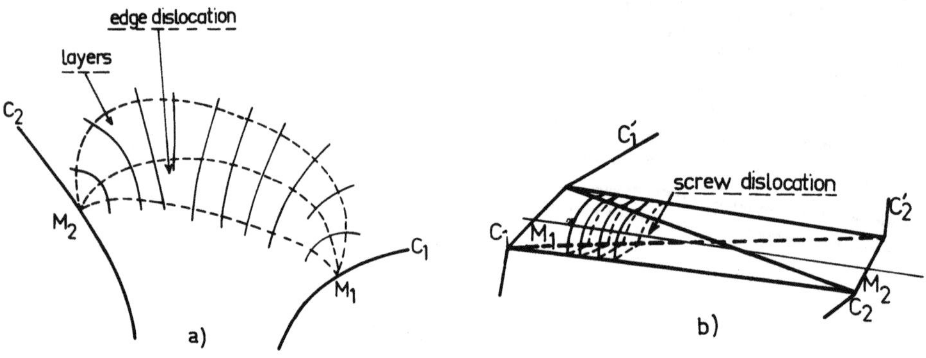

Fig.3 a) edge dislocations in a focal domain ; C_1 and C_2 are elements of focal lines
b) a typical situation which must show screw dislocations

It is possible to reintroduce the physical notion of layers in a CSL : the quantity $\vec{n}.\text{curl}\vec{n}$ can be analyzed as a density of quantized screw dislocations imposed on a set of layers, all of the same sign (Fig.3b). However they will not necessarily polygonize, because the layers in their close vicinity are curved into minimal surfaces (hence the curvature energy vanishes) and because their elastic energy is so small [17].

There is a theorem which is common to CSL and CSN, but does not apply to CN.

Darboux'theorem states that, given a congruence of straight lines D, two planes tangent to the focal surfaces F_1 and F_2 along any line D are orthogonal.

G. FRIEDEL noticed that the ellipse and hyperbola pertaining to the same cofocal domain and observed by optical microscopy always cut in projection at right angles and concluded that the molecular alignments obey Darboux' theorem. This reasoning is certainly a landmark in the physics of defects since it brought FRIEDEL to the concept of parallel smectic layers (he also knew DUPIN's theorem) without using any X-ray analysis.

We have just been describing the (small) variations to cofocal conics in terms of (quantized) dislocation densities (in the sense of NYE). Similarly one could use (unquantized) dislocation densities in order to describe the curvature in a perfect cofocal domain, by introducing a suitable contortion tensor K_{ij}. However this procedure is not straightforward here because the curvatures are large, and the above theory valid only for small curvatures (for an extension, see [18]). Let us remark that our "empirical" description of defects paves the way towards an understanding of the more abstract gauge theories, where dislocation densities are given the status of gauge field densities, as first proposed by DZYALOSHINSKII and VOLOVIK [19]. The interest of this description is stressed in [20].

Our final remark concerns situations in which the variation with respect to perfect cofocal domains is large; we have here in mind C. WILLIAMS'observations of double helices in focal relationship [21]. Here the analysis would require the introduction of the notion of virtual focal surface. We expect to publish this analysis in another paper [22].

3. Curvature defects in canonics

3.1 General considerations. Call $\vec{t}$ ($\vec{r}$) the unit vector along the physical rod, $\vec{r}$ being the position. We assume that rods do not break in the distorted state. This implies, by virtue of the conservation of rod flux

$$\operatorname{div} \vec{t} = 0 \qquad (4)$$

Since we are considering a situation of pure curvature, the rods are parallel in the distorted state. We can visualize this property of parallelism on surface Σ's generated by a one parameter subset of rods. Such surfaces can be safely defined in the continuous limit, because the distance between rods is small compared to their length. On any of those surfaces the orthogonal trajectories of a rod are geodesic lines, since these trajectories cut orthogonally a set of parallel rods (this result refers to the property of frontality of geodesic lines; for a simple and illustrative review of this geometrical concept as well as of others appearing here and elsewhere in this paper, see [22]).

Two different surfaces Σ_i and Σ_j intersect along a rod and make a constant angle on this intersection; this angle is equal to the angle in the undistorted state, as required in a situation of pure curvature. Here we use :

Joachimstahl's theorem, stating that if two surfaces cut at a constant angle along their intersection, then the intersection is a line of curvature on both surfaces, or is on none.

If the intersection is not a line of curvature, it is not possible to avoid a certain quantity of twist $q = \vec{t}.\mathrm{curl}\ \vec{t}$, which is necessarily attended by elastic distortions, as a detailed analysis can show. But this is what one wants to avoid. Therefore the only solution is when the rod is along a line of curvature on any Σ; this implies that the orthogonal trajectories of the rods are at the same time geodesic lines and lines of curvature. Therefore they are planar curves, because being lines of curvature their geodesic torsion must vanish, and being geodesic lines their geodesic torsion is equal to their natural torsion [11,22]. Hence the rods are perpendicular to a family of planes Σ in a situation of pure curvature (a result devised by BOULIGAND), which means also that each surface Σ is generated by a planar curve invariable in shape (the geodesic line) moving in such a way that the velocities of all its points are normal to the planes Π which contains it. Σ surfaces deserve to be called MONGE's surfaces, by the name of the famous geometer who studied them in detail for the first time [23].

Fig.4 represents two infinitesimally close Π planes; D is their line of intersection, which envelops a space curve L when the plane Π moves. The infinitesimal motion of Π can be divided in a motion of pure rotation about D, viz. $d\phi$, and a motion about a normal to Π passing through the point of contact of D with L, viz. $d\psi$.

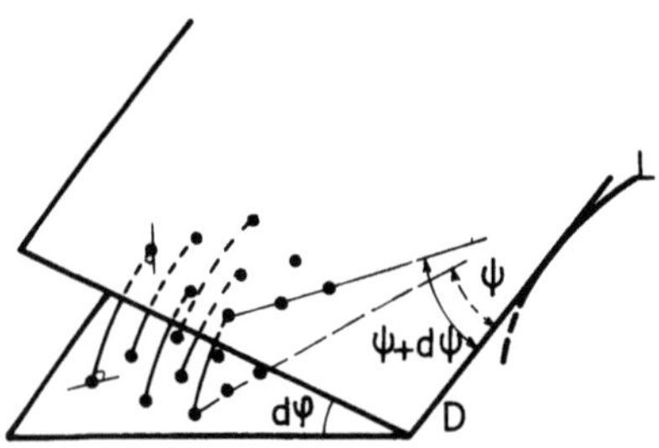

Fig.4 Disposition of rods between two infinitesimally close Π planes; on each plane there is a perfect hexagonal lattice. See text for other details

If one considers any rod, its center of curvature is on D, and its osculating plane perpendicular to D. The curvature is $1/R = |\vec{t}.\mathrm{curl}\ \vec{t}| = |d\phi/ds|$ where s is the curvilinear abscissa on the rod. In what concerns $d\psi$, it is related to the curvature of L at its point of contact M with D, i.e. $1/\rho = |d\psi/d\sigma|$, where σ is the curvilinear abscissa of L. The osculating plane of L in M is Π. Properties of reciprocity exist between L and any rod, viz.

$$\frac{\rho}{\tau} = -\frac{T}{R} \qquad (5)$$

where T and τ are the torsions of the rod and of L, oriented by the motion of the Π planes. The curvature 1/R of a rod becomes infinite on D, whose locus is consequently the curvature defect of the system. D describes a developable surface (Fig.5) which limits the domain of existence of rods. We call it a developable domain.

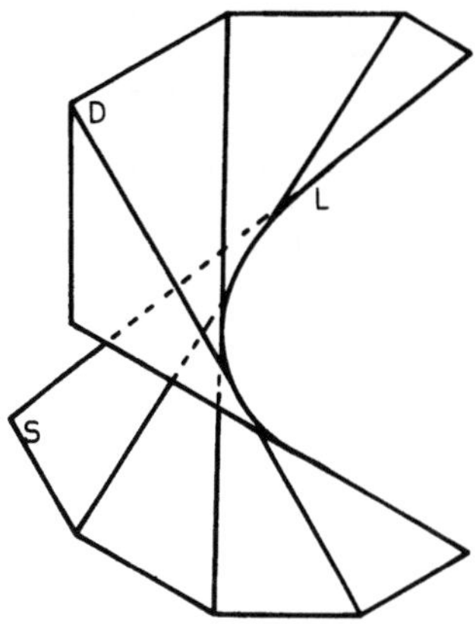

Fig.5 Scheme of a developable surface generated by the tangents D to a space curve L, the so-called cuspidal edge of the developable

3.2. Simple types of developable domains. First, let us consider the case when the developable is a circular cylinder of radius a; the curve L is reduced to a point at infinity in the direction of the generatrices of the cylinder. Consider a circular section of this cylinder; the planes Π cut this section along the tangents to the circle of radius a, and the rods are evolutes of this circle (Fig.6). Clearly, such a configuration is that one of a disclination line of strength S = + 1, of core radius r_c larger than a necessarily, so that we expect an empty core, or at least a core in which the rods take an orientation entirely different from that outside, for example parallel to the generatrix of the cylinder. Such objects are observed in very thin samples of discotic phases where rods orient parallel to the surface. (Fig.7 ; the sample is a C_5 hexaalcoxy derivative of the triphenylene and was kindly provided to me by Dr. J.C. DUBOIS and Dr. A. ZANN [1]). However S = 1/2 lines are more numerous than S = 1 lines, but it can be shown experimentally that their geometry is that of a half developable domain. Therefore our analysis is applicable (far from the core). There is in Fig.7 an example of a S = + 1 line, but the core looks extended and is not a perfect circular cylinder) (for a more detailed analysis, see [9]).

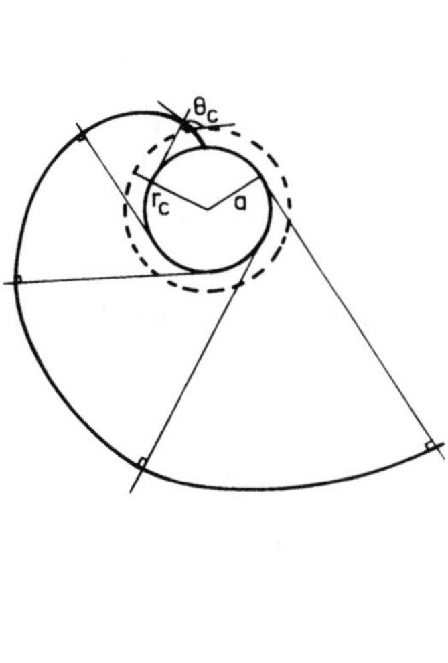

Fig.6 Cross section of a developable domain equivalent to a disclination of strength unity. a is the radius of the developable, r_c is the radius of the core. The rods are evolutes of the circle

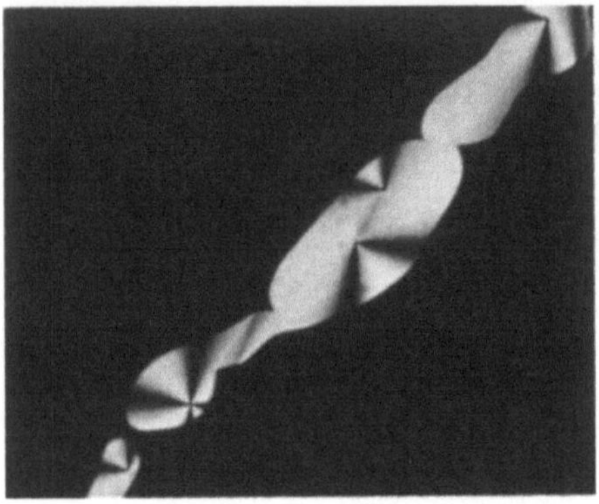

Fig.7 A thin droplet of a discotic hexagonal phase spread on a glass-plate shows the appearance of numerous disclination lines of strength half unity which have the geometry of half developable domains (the rods are partly along evolutes, partly straight)

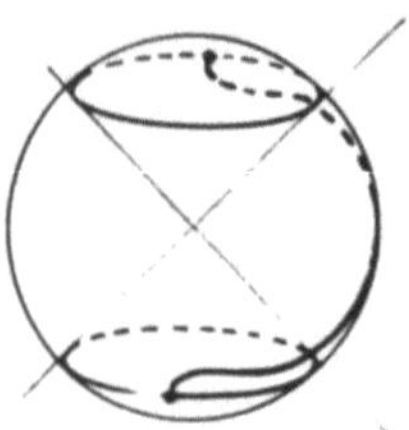

Fig.8 A singular point as a developable domain. Rods are spherical curves joining one point on a circle of intersection of the sphere with a fixed cone to another point on the other circle of intersection. Other rods on the same sphere obtained by axisymmetry. Then concentric spheres are to be considered

Among the other simple developable domains, let us quote that one in which the cuspidal edge L is a point at finite distance and the developable a cone of revolution. It is possible to show that the rods are spherical curves (the spheres centered on L are Monge's surfaces of the problem) which are orthogonal trajectories of a set of great circles (which are here the geodesic lines of the problem) drawn on the sphere and bitangent to the intersection of the cone with the sphere (Fig.8). The rods are easily obtained by equispacing, and symmetry about the axis of the core. Notice that the solution can be either right-handed (as in Fig.8) or left-handed, which means that it is possible to define a chirality for a set of rods without twist !

3.3 Classification of developable domains. A developable surface can of course be developed on the plane by pure bending without stretching. In such a mapping the cuspidal edge becomes a line $\mathcal{L}$ in the plane and the straight lines D of S become the tangents to $\mathcal{L}$. The straight lines of the two sheets of S (see Fig.5) map on one or the other of the half infinite segments $\mathcal{D}_1$ and $\mathcal{D}_2$ of $\mathcal{D}$ bounded by the point of contact $\mathcal{M}$ on $\mathcal{L}$.

Lengths and angles are preserved, which implies that the curvatures of $\mathcal{L}$ and L are the same at corresponding points. Hence $\mathcal{L}$ and L differ only by a torsion fonction $\tau(\sigma)$, which can be chosen at will.

Since any torsion $\tau(\sigma)$ can be chosen in order to build L, it is clear that another way of mapping the developable domain on the plane P is to "untwist" it, (rather than unbend it), i.e. decrease $\tau(\sigma)$ in a continuous manner without changing the curvature $\rho(\sigma)$, until $\tau(\sigma)$ vanishes. Any intermediary situation in the course of such a process is a developable domain which possesses the same representation in the plane P. In particular the final state in P is a developable domain in which all planes tangent to S are in P. This also means that the hexagonal pattern of rods of any plane Π maps on the same pattern in P (Fig.9), while at the same time each physical rod becomes a straight line perpendicular to P.

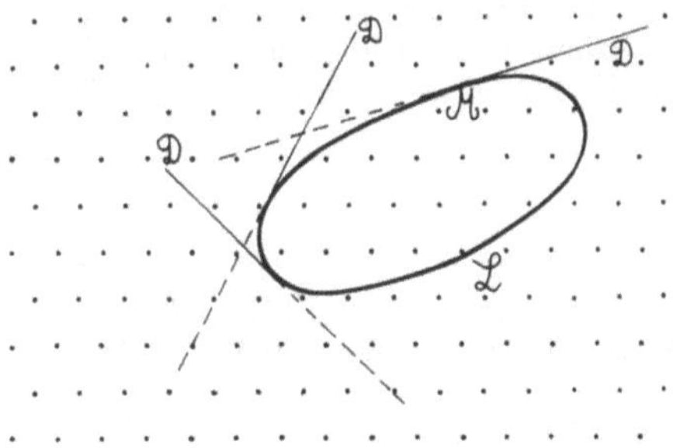

Fig.9 Representation of a developable domain on P. $\mathcal{L}$ is the image of L ; each plane Π maps on P, while its hexagonal pattern maps on a unique hexagonal pattern on P independent of the choice of Π

Notice that the reality of the analytical process of untwisting does not mean that it is possible to decrease continuously to zero the energy of a developable domain by a physical process of that sort, because untwisting is not a conservative process.

Fig.9 is drawn for a convex closed curve $\mathcal{L}$. The lifting to the two sheets of the corresponding developable in space is easy to imagine. The situation is a bit more involved if $\mathcal{L}$ contains asymptotes, points of inflexion, termini. Then it is necessary to divide $\mathcal{L}$ in simple convex arcs $\mathcal{L}_i$ and lift each of them. This is represented Fig.10

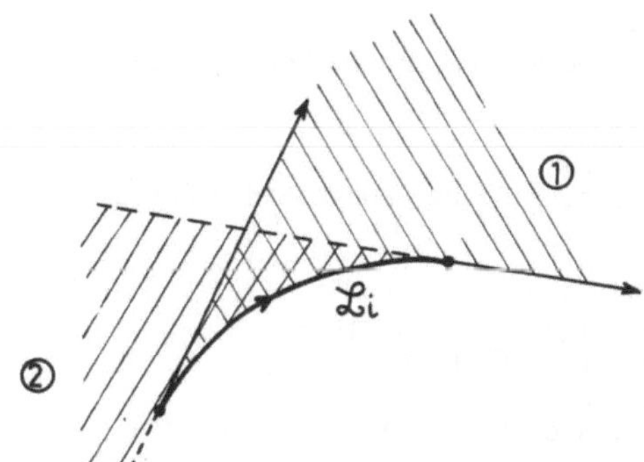

Fig.10 The convex arc $\mathcal{L}_i$ and its forward and reverse end tangents divide the exterior part in various sectors. Sector 1 is spanned by the forward half tangents and lifts to one sheet of S ; similarly for sector 2 (reverse half tangents)

4. Conclusion

We have in this paper briefly summarized the situation concerning defects which do not involve elastic distortions, but only curvature deformation. They are well known in crystalline solids, where they can be described in terms of densities of dislocations of the same sign ; the singularities of such sets are disclinations, which are then "hierarchically" related to dislocations. However, because of the necessary finite Burgers' vectors of the 3 d crystalline state, the vanishing of elastic distortions is not complete on a local scale. But this can be achieved in liquid crystals (smectics and canonics) for some sorts of deformations involving only curvature (like cofocal domains in smectics and developable domains in canonics), while the other types of defects of curvature described in ordinary crystals still exist (grain boundaries,..). An open question is the relation between the various types of objets of curvature in a liquid crystal. Developable domain is a new concept which will probably play a role in the future studies of canonics.

References

1. S.Chandrasekhar,B.K.Sadashivn and K.A.Suresh: Pramana 9, 471 (1977)
   J.Billard, J.C.Dubois, Nguyen Hun Tinh and A.Zann: Nouv. Journ. Chim. 2, 535 (1978)
2. M.Kléman: Points, lignes, parois dans les solides cristallins et les liquides anisotropes, ed. de Phys. Tome I, Orsay (1977)
   C.E.Williams and M.Kléman: J. Phys. Colloq. 36, C1-321 (1975)
3. G.Toulouse and M.Kléman: J. Phys. Lett. 37, L.149 (1976)
   D.Mermin: Rev. Mod. Phys. 51, 591 (1979)
4. Y.Bouligand: in preparation
5. M.Kléman: to be published in Journ. de Phys.
6. J.F.Nye: Acta Met. 1, 153 (1953)
7. E.Kröner: Kontinuumstheorie der Versetzungen und Eigenspannungen, Springer Verlag, Berlin (1958)
8. J.Friedel: Dislocations, Pergamon Press, London (1964)
9. P. Oswald: in preparation
10. G.Friedel and F.Grandjean: Bull. Soc. Franç. Minera. 33, 192 (1910) ; 33, 409 (1910)
11. G.Darboux: Théorie Générale des Surfaces, Chelsea Pub. Cy, Bronx, New York (1954)
12. Y.Bouligand: J. Physique 33, 525 (1972)
13. C.S.Rosenblatt, R. Pindak, N.A.Clark and R.B.Meyer: J. Physique 38, 1105 (1977)
14. M.Kléman: J. Physique 38, 1511 (1977)
15. Y.Bouligand: J. Physique 34, 1011 (1973)
16. R.Bidaux, N.Boccara, G.Sarma, L.de Sèze, P.G. de Gennes and O.Parodi: J. Physique 34, 661 (1973)
17. M.Kléman: Phil. Mag. 34, 79 (1976)
18. B.A. Bilby: Prog. Sold. Mech. 1, 329 (1960)
19. I.Dzyloshinskii and G.Volovik: J. Physique 39, 693 (1978)
20. B.Julia and G.Toulouse: J. Physique Lett. 40, L-395 (1979)
21. F.C.Frank and M.Kléman : in preparation
22. D.Hilbert and S.Cohn-Vossen: Geometry and the imagination, Chelsea Pub. Cy, Bronx, New York (1957)
23. G.Monge: Applications de l'Analyse à la Géométrie, Paris (1807)

AC and DC Mechanical Response of Smectic Liquid Crystals *

G. Durand, R. Bartolino[1], and M. Cagnon

Laboratoire de Physique des Solides associé au CNRS (LA n°2)
Université de Paris-Sud, F-91405 Orsay, France

[1] and Universita degli studi di Calabria, Dipartim. di Fisica, Arcavacata di rende cosenza, Italy

Two recent experiments made in Orsay are described, which show mechanical properties of smectic liquid crystal. One experiment measures the pressure transmitted through a smectic A sample squeezed normal to the layers. An AC extension of a previous model [1] is presented, to estimate the frequency dependent pressure head associated with permeation, which is superimposed to the one dimensional smectic elasticity. The observed relaxation of stresses is discussed to distinguish that part due to defect motion. A second experiment concerns the mechanical response of a smectic B to a low frequency AC shear parallel to the layers. In the present range of capability of the experimental set up (1 to 300 Hz), the sample used (40.8 at 40°C) shows an elastic reaction, characteristic of a real crystal (and not a liquid crystal). The shear modulus C_{44} is estimated to $\sim 5.10^7$ cgs, in absence of accurate calibration.

1) Orsay Group on Liquid Crystals. J. Phys. C 1, 305 (1975).

* Work supported by D.G.R.S.T. under contract n° 650/827

Molecular Statistical Model for Twist Viscosity in Smectic C Liquid Crystals

A.C. Diogo and A.F. Martins

Centro de Fisica da Matéria Condensada (INIC)
2 - Av.Gama Pinto, 1699 Lisboa Codex, Portugal

Abstract: We propose a molecular interpretation of the twist viscosity γ_1 in smectic-C liquid crystals. One main result is that the thermal dependence of γ_1 should be described by the following expression:

$$\gamma_1 \ (T) = a.\sin^2\theta.\exp \frac{u.\sin^2\theta}{kT} ,$$

where θ is the tilt angle and $\underline{a}$ and $\underline{u}$ are coefficients that may be derived from a molecular theory of the Sm-C phase. We also show that the elastic constants related to the curvature of the $\underline{\tau}$-director are proportional to $\sin^2\theta$.

1. Introduction

The order parameter for Sm-C liquid crystals is a complex quantity $\psi=\sin\theta.e^{i\phi}$, where θ is the tilt angle of the molecules relative to the normal $\underline{n}$ to the smectic layers, and ϕ is their azimuth in the plane of the layers [1]. The phase ϕ is an hydrodynamic variable, and γ_1 is the dissipative coefficient associated with the gradients of $\partial\phi/\partial t$ [2]. If we use a local set of unit vectors $\underline{n}$, $\underline{\tau}$, $\underline{p}$, where $\underline{n}$ is the normal to the layers, $\underline{\tau}$ (the τ-director) lies both on the plane of the layers and on the plane of tilt, and $\underline{p} = \underline{n} \times \underline{\tau}$, then $\partial\phi/\partial t=\partial\underline{\tau}/\partial t$.

The aim of this paper is to give a molecular interpretation of γ_1 in Sm-C liquid crystals. By way of this, we get the thermal dependence of γ_1, a result we present in the next section. In section 3 the tilt angle dependence of both γ_1 and the elastic constants B_1, B_2, B_3 and B_{13} is derived from a general thermodynamic argument. Section 4 contains a brief discussion about the present experimental situation concerning $\gamma_1(T)$ together with some suggestions for further work. In section 5 we give the conclusions of this work.

2. Molecular- statistical interpretation of γ_1

Viscosity, in general, is a rate process, and the discussion below is substantially developed in the framework of EYRING's theory [3]. We follow the simple lines of [4] where a molecular theory was proposed for the twist viscosity in nematics.

Let us consider the reorientation of the molecules in the Sm-C layers about the equilibrium position specified by the τ-director. This reorientation takes place against an intermolecular periodic potential $E(\phi)$ which has a minimum for $\phi=\phi_o$ (τ-director) and a maximum for $\phi_M=\phi_o+\pi$. In the absence of an external torque, the characteristic frequency ν_o of the molecular jumps that overcome the potential barrier is given by:

$$\nu_o = \frac{kT}{h} \cdot \Pi \cdot \exp\left(\frac{-U}{kT}\right) \tag{1}$$

where h is the Planck constant, Π is a probability factor depending on the partition functions of the molecule in the $\phi_M = \phi_o + \pi$ and ϕ_o orientations, and U is the height of the potential barrier (activation energy): $U = E(\phi_M) - E(\phi_o)$.

Under a constant external torque density $\underline{\Gamma}=\Gamma\underline{n}$, the molecular reorientations are favoured in one sense (+), so that the frequencies of the jumps that overcome the potential barrier in the forward (+) and backward (-) directions are no longer equal to ν_o, but given by:

$$\nu_\pm = \nu_o \exp\left(\pm\frac{\pi\Gamma V^*}{kT}\right) \tag{2}$$

where $W = \pi\Gamma V^*$ is the work given by Γ, and V^* is the volume needed by the molecule to jump. This volume is available after a local "lattice" expansion around the jumping molecule [4].

Let us now relate the <u>microscopic</u> frequencies $\nu_\pm$ to the (<u>macroscopic</u>) motion of the τ-director which is observed when $\underline{\Gamma}$ is applied. Consider the function $F(\phi)$ that counts the fraction of molecules with an azimuth between ϕ_o and ϕ; $f(\phi) = \partial F/\partial\phi$ is the distribution function. The total time-derivative of $F(\phi_M)$ is zero:

$$\frac{d}{dt}F(\phi_M) = \left(\frac{\partial F}{\partial t}\right)_{\phi_M} + \frac{\partial F}{\partial\phi_M}\cdot\frac{d}{dt}\phi_M = 0 \tag{3}$$

Here, $\partial F/\partial t$ is the number of molecules that cross ϕ_M per unit time as seen by a space-fixed observer. i.e., the difference between the fluxes of molecules in the positive and negative senses:

$$\frac{\partial F}{\partial t} = \nu_- f(\phi_M) - \nu_+ f(\phi_M). \tag{4}$$

Substituting (4) in (3) we find:

$$\frac{d}{dt}(\phi_M) = (\nu_+ - \nu_-) \tag{5}$$

which means that the distribution function f, and so the τ-director, rotate in the laboratory frame with an angular frequency

$$\Omega = 2\pi(\nu_+ - \nu_-) \tag{6}$$

From (2) and (5), and assuming W<<kT (newtonian flow), we have

$$\Omega \simeq 4\pi^2\nu_o\frac{\Gamma V^*}{kT} \tag{7}$$

and since

$$\gamma_1 = \Gamma/\Omega$$

We obtain:

$$\gamma_1 = \frac{kT}{4\pi^2 V^*} \cdot \frac{1}{\nu_o} \tag{8}$$

Calling Δp the difference in the internal pressure that changes V into $V^* = V + \Delta V$, and χ the isothermal compressibility of the layers, we have:

$$\Delta p = \left(\frac{\partial \langle E\rangle}{\partial V}\right)_\psi = \frac{\Delta V}{\chi . V^*} \tag{9}$$

where $\langle E\rangle$ is the mean energy per molecule. Now, from (1), (8) and (9) we find:

$$\gamma_1 = \frac{h\chi}{4\pi^2 . \Delta V . \Pi} \cdot \left(\frac{\partial \langle E\rangle}{\partial V}\right)_\psi \cdot \exp\left(\frac{U}{kT}\right) \tag{10}$$

Up to this point the arguments used were largely independent from the precise form of the potential E. Expressions for U and $(\partial\langle E\rangle/\partial V)$ may be found using PRIEST's model of the Sm-C phase [5]; we obtain:

$$U = u . \sin^2\theta \tag{11}$$

and

$$\frac{\partial \langle E\rangle}{\partial V} = b . \sin^2\theta \tag{12}$$

where <u>u</u> and <u>b</u> are phenomenological coefficients (see Appendix). From (10), (11) and (12) we get:

$$\gamma_1 = a . \sin^2\theta . \exp\left(\frac{u . \sin^2\theta}{kT}\right), \tag{13}$$

where

$$a = \frac{h . \chi . b}{4\pi^2 . \Delta V . \Pi} \tag{14}$$

A few remarks on this expression are in order: a) excluding the exp(...) term, γ_1 is proportional to the square of the order parameter, $|\psi|^2$; b) the activation energy is proportional to $|\psi|^2$ too; c) for a second order Sm-C-A phase transition, γ_1 goes to zero like $(T-T_{CA})^{2\beta}$ where β is the critical exponent of the tilt angle; d) the activation energy goes also to zero as T approaches T_{CA}.

The thermal dependence of γ_1 is implicit in the tilt angle and explicit in the term U/kT. Moreover, the factor <u>a</u> in (13) may be temperature dependent. To see this, we recall that the probability of a molecular jump over the energy barrier U is given by the probability that the molecule gets enough energy to overcome the potential barrier <u>times</u> the probability that there is a local lattice expansion such that it makes the jump possible. This last factor is included in the Π factor in (1). In some cases this factor may show a temperature dependence comparable to or even stronger than exp(U/kT). This point will be discussed in [6] in a more detailed form.

3. Tilt angle dependence of γ_1 and of some elastic constants

The proportionality of γ_1 to $\sin^2\theta$ can be proven from another more general argument, i.e., considering the entropy production Σ in a Sm-C in the absence of velocity and temperature gradients and neglecting permeation. We have [2,7]:

$$T\Sigma = \underline{K} \cdot \frac{\partial \tau}{\partial t}, \tag{15}$$

where $\partial\tau/\partial t$ is the velocity of the τ-director with respect to the background fluid, and $\underline{K}$ is the thermodynamic force conjugated to it, which may be called the τ-molecular field. Assuming linear response we have $\underline{K} = \gamma_1 . \partial\tau/\partial t$ and therefore:

$$T\Sigma = \gamma_1 . \left(\frac{\partial \tau}{\partial t}\right)^2 \tag{16}$$

Let us consider now the entropy production in terms of the order parameter ψ. In the situation described above, the only thermodynamic fluxes of interest are $\dot{\psi} = \partial\psi/\partial t$ and its complex conjugate. Calling F the free energy, the thermodynamic forces conjugated with these fluxes are $-\partial F/\partial\psi^*$ and its C.C. Linear response then implies:

$$-\frac{\partial F}{\partial \psi^*} = \eta \frac{\partial \psi}{\partial t} = \eta . i\psi . \frac{\partial \phi}{\partial t}, \tag{17}$$

where η has the dimensions of a viscosity. The entropy production now reads:

$$T\Sigma = -\frac{\partial F}{\partial \psi^*} \cdot \frac{\partial \psi^*}{\partial t} - \frac{\partial F}{\partial \psi}\frac{\partial \psi}{\partial t} = 2\eta . \sin^2\theta . \left(\frac{\partial \phi}{\partial t}\right)^2 \tag{18}$$

But $\partial\phi/\partial t = \partial\tau/\partial t$ so that comparison of (16) and (18) gives:

$$\gamma_1 = 2\eta . \sin^2\theta \tag{19}$$

The proportionality of γ_1 to $\sin^2\theta$ has another interesting consequence, namely, the proportionality to $\sin^2\theta$ of some of the elastic constants of a Sm-C liquid crystal. In Fourier components, the hydrodynamic equation for the τ-director reads [2]

$$-i\omega\, \gamma_1 . \delta\tau_p(\underline{q},\omega) = K_p(\underline{q},\omega) \tag{20}$$

where the subscript p refers to "projection onto p". Using the explicit form of the free energy of deformation for a smectic-C [7] we find

$$K_p(\underline{q},\omega) = -B(\underline{q}) . \delta\tau_p(\underline{q},\omega) \tag{21}$$

where

$$B(\underline{q}) = B_1 q_x^2 + B_2 q_y^2 + B_3 q_z^2 + 2B_{13} q_x q_z \tag{22}$$

and B_1, B_2, B_3 and B_{13} are the elastic constants related to τ-director deformations [7,8]. It follows from (20), (21) and (22) that if $\gamma_1 \propto \sin^2\theta$ so are B_1, B_2, B_3 and B_{13}.

4. Discussion and experimental results

We first notice the similarity between the order parameter dependence of the elastic constants related to deformations of the τ-director field in smectics-C, and the elastic constants related to deformations of the director field in nematics. Eq.(13) also brings some analogy with an expression for γ_1 in nematics proposed by us [4,9] which has been successfully contrasted with experimental data [4,6,9].

The present situation concerning experimental data on γ_1 in Sm-C liquid crystals is not clear. The only published data to our knowledge are those of MEIBOOM and HEWITT on di-heptyloxy-azoxybenzene (7OAB) [10] and TBBA [11]. These data are somewhat questionable. In fact, the same technique was used by them to measure γ_1 in the nematic phase of 7OAB, and gave values about 50% less than those obtained by the rotating magnetic field method [6] which has been widely tested.

An alternative method of measuring γ_1 consists in the observation of the motion of m-disclination lines in smectics-C: the motion of these disclinations is analogous to that discussed in [12] for nematics. We leave this as a suggestion for further work.

5. Conclusion

We propose here a theoretical expression for the thermal behaviour of γ_1 in Sm-C liquid crystals. It is shown that γ_1 depends on the square of the order parameter in two distinct ways: directly and through the activation energy. This behaviour implies that the elastic constants related to the curvature of the τ-director field also depend on $|\psi|^2$. We are not able to compare this theory with experiment because the lack of reliable data on $\gamma_1(T)$ for Sm-C materials. More experimental work is thus urgently required.

Appendix

In the framework of PRIEST's theory [5] the mean field energy of a molecule $E(\alpha,\beta,\gamma)$ is expressed as a function of the Euler angles (α,β,γ) that transform a molecular-fixed frame whose z-axis coincides with the long molecular axis to a space-fixed frame in which the z-axis gives the direction of the mean alignment of the molecules and the y-axis lays in the plane of the layers. $E(\alpha,\beta,\gamma)$ can be approximately written as:

$$E(\alpha,\beta,\gamma) = -\varepsilon_o\cdot\left(1-\frac{3}{2}\sin^2\beta\right) - \sin^2\theta\left[\varepsilon_1\cdot\left(1-\frac{3}{2}\sin^2\beta\right) + \varepsilon_2\cdot\sin^2\beta\cdot\cos 2\gamma\right]$$

where θ is the tilt angle and ε_o, ε_1 and ε_2 are phenomenological coefficients characteristic of the material. The activation energy is thus given by:

$$U = E(o,-2\theta,o) - E(o,o,o) = 6\varepsilon_o\cdot\sin^2\theta + \mathcal{O}(\sin^4\theta) \qquad \text{(A1)}$$

In order to compute $\partial\langle E\rangle/\partial V$ we notice that to a good approximation $\langle P_2(\cos\beta\rangle \simeq 1$ in the temperature range of the Sm-C phase |5|. Therefore:

$$- \frac{\partial\langle E\rangle}{\partial V} \simeq \sin^2\theta . \frac{\partial\varepsilon_1}{\partial V} \tag{A2}$$

Expressions (A1) and (A2) reduce to (11) and (12) if we define $u = 6\varepsilon_o$ and $b = - \partial\varepsilon_1/\partial V$.

References

1. De Gennes, P.G. - C.R.Acad.Sci.Paris,274-B(1972)758
2. Martin, P.C.; Parodi,O.;Pershan,P.S. - Phys.Rev.A-6 (1972)2401
3. Glasstone, S.N.; Laydler, K.; Eyring, H. - The Theory of Rate Processes, Mc Graw Hill, N.Y.(1941)
4. Martins, A.F.; Diogo, A.C. - Portgal.Phys. 9(1975)129
5. Priest, R.G. - J. Chem. Phys. 65(1976)408
6. Diogo, A.C.; Martins, A.F. - to be published
7. De Gennes, P.G. - The Physics of Liquid Crystals, Oxford University Press (1975)
8. Saupe, A. - Mol. Cryst. Liq. Cryst. 7(1969)59
9. Martins, A.F.; Diogo, A.C.; Vaz, N.P. - Ann. Physique 3(1978)361
10. Meiboom, S.; Hewitt, R.C. - Phys. Rev. Lett. 34(1975)1146
11. Meiboom, S.; Hewitt, R.C. - Phys. Rev. A-15(1976)2444
12. Imura,H.; Okano, K. - Phys. Lett. 42-A(1973)403

Dynamic Shear Properties near a Smectic A to Smectic B Phase Transition

P. Martinoty and Y. Thiriet

Laboratoire d'Acoustique Moléculaire,
Université Louis Pasteur (E.R.A. au C.N.R.S.),
F-67070 Strasbourg Cédex, France

Abstract : The real and the imaginary parts of the shear mechanical impedance were measured at 15 MHz near the smectic-A to smectic-B phase transition in N-(p-butoxybenzylidene)-p-n-octylaniline, for shear waves propagating along the normal to the layers. In both phases the shear reponse presents large relaxation effects. A pretransitional behavior in the real part of the shear impedance is found in a temperature interval of ∿ 2°C below the transition.

Introduction

Considerable work has been done in the smectic-B liquid crystals specially in order to establish the nature of the interlayer correlations [1]. The possibility of long range interlayer correlations (with $c_{44} \neq 0$) and of no interlayer correlation (with $c_{44} = 0$) has been suggested [2,3]. Recent X-ray data on thick films seem to show that the structure of the B phase is in fact that of a three-dimensional solid with a rather weak shear elastic modulus c_{44} [4].

Preliminary measurements of the real part of the shear impedance made by the authors in the smectic-B phase of N-(p-butoxybenzylidene)-p-n-octylaniline (40.8) yielded finite value for c_{44} at 5 and 14 MHz [5]. However these experiments showed the existence of a large dispersion in this frequency range and thus the observed shear modulus does not correspond to the hydrodynamic limit.

Eventhough the ultrasonic data do not represent the hydrodynamic regime, the effect of the interlayer correlations should appear at the smectic-A to smectic-B phase transition (A-B). We present here measurements of the real and imaginary parts of the shear impedance in the smectic-A and smectic-B phases of 40.8 for shear waves propagating along the normal to the layers. We confirm our previous observations in the B phase and we report an interesting pretransitional behavior of the real part of the shear impedance which appears in a temperature interval of ∿ 2°C below the transition.

Experimental details

Our sample of 40.8 was prepared at the Ecole de Chimie of Strasbourg and has an A-B transition on cooling at 49.2°C. The real part R and the imaginary part X of the shear impedance were measured by a pulse reflection technique where shear waves propagate in a fused-quartz bar and are reflected at normal or oblique incidence at the quartz-liquid crystal interface. Oblique incidence was used because the sensitivity of the method is increased by a factor of ∿ 8 and it was verified that normal incidence gives results which are basically identical to those obtained at oblique incidence. The main features of the set-up have been described elsewhere [6].

The samples were prepared in the homeotropic configuration between the reflecting surface of the bar and a cover glass, both coated with a thin layer of lecithin. The sample director was thus aligned in a direction perpendicular to the reflecting surface of the bar or, equivalently, in a direction perpendicular to the shear. By decreasing the temperature slowly from the nematic phase, sample alignment was maintained in the smectic phases. The quality of alignment was evaluated by examination between crossed polarizers. For the samples used it was possible to extinguish completely and uniformly the light transmitted between crossed polarizers.

Teflon spacers were used to obtain thick samples. For such samples the alignment in the bulk was obtained by applying a magnetic field of 6kG. We observed no change in R when the thickness of the sample was increased from a few μ to ∿ 150 μ. The sample temperature was measured by a quartz thermometer and temperature stability was better than 3 mK.

Results and Discussion

Our measurements of R and X at 15 MHz are reported in Fig.1. R and X were determined from the measured reflection coefficient $re^{-i\phi}$ by the relations

$$R = Z_q \cos\Theta \frac{1-r}{1+r}$$

$$X = Z_q \cos\Theta \frac{2r\phi}{(1+r)^2}$$

where Z_q is the shear mechanical impedance of the fused-silica bar and Θ is the angle of incidence of the shear wave.

In the hydrodynamic regime the shear elastic modulus c_{44} is related to R by $c_{44} = R^2/\rho$. In this regime R is frequency independent and X is equal to zero.

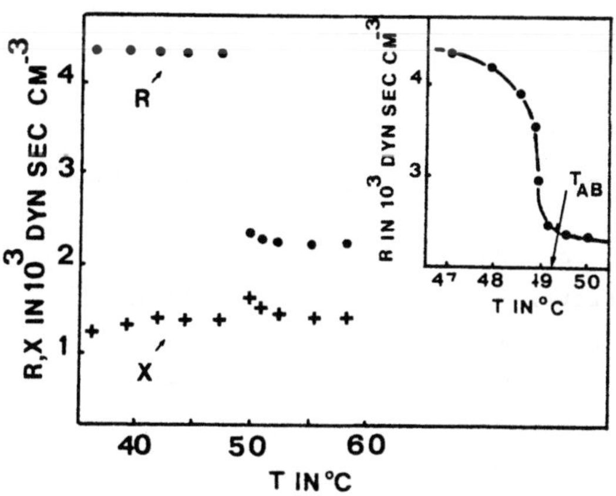

Fig. 1. The real part R and the imaginary part X of the shear impedance in the smectic-A and smectic-B phases of 40.8. The insert shows the detailed experimental data near the AB transition. The frequency was 15 MHz.

In the presence of a relaxation process, R increases with increasing frequency and X becomes non zero. Furthermore, R and X are different. This is exactly the behavior observed in Fig.1 and thus the measured shear modulus G' does not correspond to the hydrodynamic value (i.e. c_{44}). The dispersion is also at the origin of the dynamic shear modulus which is observed in the smectic-A phase where c_{44} = 0. Further evidence of the dispersive behavior in the A and B phases was provided by measurements of R between 3 and 100 MHz showing that R is strongly frequency-dependent ([7]).

Our results are in contradiction with the behavior reported by Bhattacharya and Letcher at this conference. They found that R is frequency independent in the B phase and they concluded that they have observed a propagating shear mode with elastic constant $c_{44} = R^2/\rho \sim 10^8$ dyn/cm^2. The reason for the discrepancy between our results and those of Bhattacharya and Letcher is not clear but it might be related to some difficulties in the calibration of their set-up ([8]).

From our measurements we deduced the dynamic shear modulus G' which is given by $G' = (R^2 - X^2)/\rho$. We find G' = 3 10^6 dyn/cm^2 in the A phase and G' = 1,7 10^7 dyn/cm^2 in the B phase with no significant temperature dependence. Near the transition we observe a softening of R (or G') which appears below T_{AB}. The occurence of a pretransitional behavior is not surprising in view of the small latent heat of the A-B transition ([9]) and is consistent with the behavior of the birefringence observed by Lim and Ho ([10]). Eventhough our data do not represent the hydrodynamic limit the pretransitional behavior of R (or G') reflects the softening of c_{44}.

A more complete and detailed account of this work including results between 3 and 100 MHz will be reported in a forthcoming paper. Here we simply wish to point out that the value of G' in the B phase - $G' \sim 10^7$ dyn/cm^2 - is three order of magnitude smaller than the shear modulus of a conventional crystal. This indicates , since c_{44} < G', that the B phase of 4O.8 has a rather weak interlayer elasticity.

Acknowledgments

We are very grateful to M. Franck Neumann for the synthesis of the compound and to S.V.Letcher for stimulating discussions.

REFERENCES

1. See for example P.G.de Gennes, The Physics of Liquid Crystals (Oxford U.P., London 1974).
2. P.G.de Gennes and G.Sarma, Phys. Lett. 38 A (1972) 219.
3. A.M.Levelut, J.Doucet, and M.Lambert, J. Phys. (Paris) 35, (1974) 773.
4. D.E.Moncton and R.Pindak, Phys. Rev. Lett. 43 (1979) 701.
5. Y.Thiriet and P.Martinoty, J. Phys. Lett. (Paris) 36 (1975) L-125.
6. F.Kiry and P.Martinoty, J. Phys. (Paris) 38 (1977) 153.
7. Y.Thiriet and P.Martinoty, to be published.
8. S.V.Letcher, private communication.
9. G.W.Smith and Z.G.Gardlund, J.Chem. Phys. 59 (1973) 3214.
10. K.C.Lim and J.T.Ho, Phys. Rev. Lett. 43 (1979) 1167.

Convective Instabilities in Cholesteric and Smectic A Liquid Crystals

H. Pleiner and H. Brand

Universität Essen, D-4300 Essen, Federal Republic of Germany

We predict the occurrence of an oscillatory convective instability when a planar texture of Cholesteric or Smectic A liquid crystals is heated from below.

1. Introduction

We consider horizontal layers of Smectic A or Cholesteric liquid crystal between an upper and a lower plate at z=d and z=O, respectively. A temperature difference ΔT is applied across the layers so that $T(z=d)-T(z=O)=\Delta T$.

In the xy-plane the layers of Smectics A or Cholesterics are taken to be infinite, for simplicity. The preferred direction $\vec{p}$ of the liquid crystals is assumed to be perpendicular to the layers i.e. in our configuration parallel to the z-axis.

For Smectics A $\vec{p}$ is identical with the averaged axis of the molecules, usually called $\vec{n}$, whereas for Cholesterics $\vec{p}$ denotes the pitch axis which is perpendicular to the spiraling molecular axis $\vec{n}$. Furthermore we assume the presence of an external magnetic field $\vec{H}$ parallel to $\vec{p}$.

For the conserved quantities ρ (mass density), $\vec{g}$ (momentum density) and σ (entropy density) and for $\vec{R}$, the displacement vector of the layers, which characterizes the broken translational symmetry along $\vec{p}$, we use hydrodynamic equations whose linearized version to lowest order in k was derived by MARTIN, PARODI and PERSHAN [1] and which were recently generalized by the authors to include nonlinear terms [2]. In the following we neglect a static and a dynamical coupling between $\vec{R}$ and T. Only at the end of the paper we will study the influence of these couplings.

When the applied temperature gradient is small there exists a steady, pure heat conduction regime (velocity $\vec{v}=0$) with the same distribution of pressure $p=p(z)$ and mass density $\rho=\rho(z)$ as in simple fluids; in addition there is $R_z=0$, which means constant distance of the layers (or constant pitch). The stability of this heat conduction regime is analysed by standard linear stability analysis within the Boussinesq approximation [3].

2. Linearized Equations

For linear deviations from the heat conduction state we get the following system of equations ($\vec{v}=(u, v, w)$):

$$
\begin{aligned}
&\frac{\partial u}{\partial x} + \frac{\partial w}{\partial z} = 0 \\
&\dot{T} + d^{-1}\,\Delta T\, w = (\varkappa_{\parallel}\frac{\partial^2}{\partial z^2} + \varkappa_{\perp}\frac{\partial^2}{\partial x^2})\, T - \varkappa_a d^{-1}\,\Delta T\,\frac{\partial^2}{\partial x^2} R \\
&\rho_o\dot{u} + \frac{\partial p}{\partial x} - q_o g_1 \frac{\partial}{\partial x}\chi_{op} R = \nu_{xjkl}\nabla_j\nabla_l v_k \\
&\rho_o\dot{v} + (2q_o)^{-1}\frac{\partial}{\partial x}\chi_{op} R = \nu_{yjkl}\nabla_j\nabla_l v_k \qquad (1)\\
&\rho_o\dot{w} + \frac{\partial p}{\partial z} - \alpha g_E T - \chi_{op} R - q_o g_2 \frac{\partial}{\partial z}\chi_{op} R = \nu_{zjkl}\nabla_j\nabla_l v_k \\
&\dot{R} - w - (2q_o)^{-1}\frac{\partial v}{\partial x} + q_o g_a \frac{\partial w}{\partial z} = \zeta\chi_{op} R
\end{aligned}
$$

with

$$\chi_{op} = \frac{\partial^2}{\partial z^2} \pm \chi_a H^2 \frac{\partial^2}{\partial x^2} - K_1 \frac{\partial^4}{\partial x^4}$$

where the upper (lower) sign refers to Smectics A (Cholesterics). Dissipative processes are heat conduction ($\varkappa_{\parallel}$, $\varkappa_{\perp}$, $\varkappa_a=\varkappa_{\parallel}-\varkappa_{\perp}$), diffusion ($\nu_{ijkl}$ cf. [4]) and permeation (ζ). The buoyancy force is described by the term $\alpha g_E T$. We have put $\frac{\partial}{\partial y} = 0$ without loss of generality.

As boundary conditions for z=0, d we choose

$$T = w = \frac{\partial^2}{\partial z^2}w = R = 0 \text{ and } \frac{\partial u}{\partial z} = \frac{\partial v}{\partial z} = 0$$

in order to get analytic results.

In equation (1) all terms containing q_o (the wave vector of the pitch) are absent in Smectics A thus characterizing the more complex hydrodynamic behaviour of Cholesterics. We will discuss the influence of these q_o-terms in section 4. Section 3 is devoted to Smectics A and can be regarded as a first approximation for Cholesterics.

3. Instabilities for Smectics A

The pure heat conduction state gets unstable if ΔT is increased beyond a certain threshold. The onset of instability is characterized by solutions $\sim\exp(i\omega t - \sigma t)$ with $\sigma=0$ and ω real. There are two such solutions of (1) at two different values of ΔT, thus indicating two different kinds of instabilities. One is stationary (hereafter called TAI) $\omega=0$

$$\text{at}\quad \Delta T(k) = \frac{\rho_o d}{\alpha g_E}\,\frac{\hat{\chi}^2\hat{\varkappa}^2}{k_\perp^2}\,\frac{\zeta\hat{\mu}^4+k_\perp^2\rho_o^{-1}}{\varkappa_a k_\perp^2-\zeta\hat{\chi}^2} \qquad (2)$$

the other is oscillatory (hereafter called OSI)

$$\omega^2(k) = (\varkappa_{\parallel}k^4 + \hat{\mu}^4)^{-1} [\zeta\hat{\chi}^2(\hat{\mu}^4 + \hat{\varkappa}^2k^2)(\varkappa_a k_{\perp}^2 - \zeta\hat{\chi}^2) + \varkappa_a k_{\perp}^2 \hat{\mu}^4 \hat{\varkappa}^2 + (\varkappa_{\parallel}k^2 - \zeta\hat{\chi}^2) \times k_{\perp}^2 \hat{\chi}^2 \rho_o^{-1}] \quad (3)$$

at

$$\Delta T(k) = -\frac{\rho_o d}{\alpha g_E} \frac{\hat{\mu}^4 + \zeta\hat{\chi}^2 k^2}{\hat{\mu}^4 + \varkappa_{\parallel}k^4} [k_{\perp}^{-2}(\hat{\mu}^4 + \hat{\varkappa}^2k^2)(\zeta\hat{\chi}^2 + \hat{\varkappa}^2) + \hat{\chi}^2 \rho_o^{-1}]$$

We used the abbreviations

$$\hat{\mu}^4 = \rho_o^{-1}(\nu_3[k_{\parallel}^2 - k_{\perp}^2]^2 + 2[\nu_1 + \nu_2]k_{\perp}^2 k_{\parallel}^2)$$

$$\hat{\varkappa}^2 = \varkappa_{\parallel}k_{\parallel}^2 + \varkappa_{\perp}k_{\perp}^2$$

$$\hat{\chi}^2 = \chi_{\parallel}k_{\parallel}^2 \pm \chi_a H^2 k_{\perp}^2 + K_1 k_{\perp}^4$$

where $k_{\parallel}$ and $k_{\perp}$ are the components of the wave vector $\vec{k}$, parallel and perpendicular to $\vec{p}$, respectively.

The stationary instability (TAI) was already discussed by DUBOIS-VIOLETTE [5] and PARSONS [6] for Cholesterics. It is dominated by the anisotropy of the heat-conduction ($\sim\varkappa_a$) and by the permeation effect ($\sim\zeta$). It is obtained by heating from above ($\Delta T>0$) if $\varkappa_a k_{\perp}^2 - \zeta\hat{\chi}^2 > 0$ or from below ($\Delta T<0$) if $\varkappa_a k_{\perp}^2 - \zeta\hat{\chi}^2 < 0$. It should be noticed that the sign of $\varkappa_a k_{\perp}^2 - \zeta\hat{\chi}^2$ can be changed by the magnetic field strength.

The oscillatory instability (OSI) is a novel prediction. It is obtained only by heating from below ($\Delta T<0$). The mechanism involved consists mainly in an elastic oscillation of the layers driven by the temperature. For details cf. [7].

The region of stability of the heat-conduction state is sketched in the following diagram.

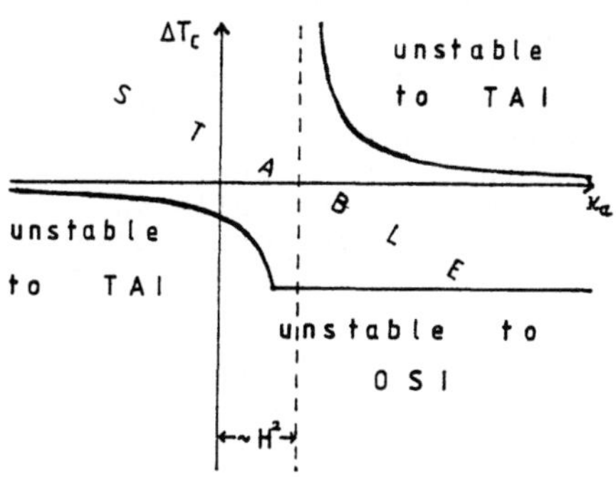

In order to get numerical results for ΔT and ω, we have to specify the actual values $k_{\parallel}^{c}$ and $k_{\perp}^{c}$ of the wave vectors $k_{\parallel}$ and $k_{\perp}$. For the free-free boundary conditions $k_{\parallel}^{c}=\pi/d$. $k_{\perp}^{c}$ is obtained by minimization $\Delta T(k)$ with respect to $k_{\perp}$. This leads to algebraic equations for $k_{\perp}^{2}$, which have to be solved for their lowest root $k_{\perp}^{c2}$ numerically. Generally one expects $k_{\perp}^{c} = \sqrt{g}\ \pi/d$ where g is a number of order unity. For $\hat{\mu}^4 k^{-2} \gg \hat{\varkappa}^2 \gg \bar{\zeta}\hat{\chi}^2, H^2 \neq 0$ (i.e. viscous damping is the dominating dissipative process) g is determined for the TAI by

$$\varkappa_{\perp}\chi_a H^2(\varkappa_a - \zeta\chi_a H^2)g^2 - 2\varkappa_{\perp}\chi_a H^2\zeta\chi_{\parallel}g - \zeta\chi_{\parallel}^2\varkappa_{\perp} - \chi_{\parallel}\varkappa_{\parallel}\varkappa_a = 0$$

and for the OSI by

$$2\nu_3\varkappa_{\perp}g^3 + g^2(\nu_3\varkappa_{\parallel} + 2[\nu_1 + \nu_2 - \nu_3]\varkappa_{\perp} + d^2/\pi^2\ \chi_a H^2) - \nu_3\varkappa_{\parallel} = 0$$

Thus the critical temperature difference ΔT and the critical frequency ω_c are $\Delta T_c = \Delta T(k_{\parallel}^{c}, k_{\perp}^{c})$ and $\omega_c = \omega(k_{\parallel}^{c}, k_{\perp}^{c})$.
For TAI numerical values were given by DUBOIS-VIOLETTE [5] for Cholesterics (MBBA): $\Delta T_c \sim 1^{o}K$ with d=1 cm, p = 50μ.
Under the same conditions we find for OSI $\Delta T_c = 1^{o}K$ and the critical frequency $\omega^c = 10^{-1}$Hz; both ω_c and ΔT_c will rise, if d and p_o are lowered.

For Smectics A the elastic energy of the layers ($\sim\hat{\chi}^2$) is much greater than that for Cholesterics. Thus, the critical temperatures are greater in Smectics A. In order to reach the instability experimentally, it is favourable to use a destabilizing magnetic field i.e. $\vec{H}\,||\,\vec{p}$ if $\chi_a<0$ or $\vec{H}\perp\vec{p}$ if $\chi_a>0$.

The eigenfunctions, both for TAI and OSI, allow for two-dimensional convection rolls (in the latter case oscillatory). It should be mentioned, that the linear stability theory presented here is not able to derive the space-time pattern which occurs beyond the threshold; to this aim a nonlinear theory is required. Therefore, the eigenfunctions due to (2) and (3) denote those fluctuations, which become undamped at ΔT_c, but not necessarily the space-time pattern above ΔT_c.

4. Instabilities for Cholesterics

The results of the preceding section are valid for Cholesterics only as a first approximation. As already discussed above, the hydrodynamic behaviour of Cholesterics is more complicated than that of Smectics A. The difference is two-fold:
First, there are the terms $\sim (2q_o)^{-1}$ in the Eqs. for $\dot{v}$ and $\dot{R}$ (cf. (1)). These terms describe the connection between a homogeneous rotation of the helices and a homogeneous translation of the layer planes defined by the helices. Since $k_{\perp}/q_o \ll 1$, the influence of these terms on the numerical values of ΔT_c and ω_c of TAI and OSI is small (especially ω_c of TAI remains zero). However, by the three-dimensional nature of a rotation of the helices, the eigenfunctions are no longer two-dimensional convection rolls.
Second, the terms containing the phenomenological parameters g_1, g_2 ($g_a \equiv g_2 - g_1$) change the nature of TAI from a stationary into an oscillatory instability with $\omega_c \sim g_a$. Although we do not

know numerical values of g_a one may expect ω_c (of TAI) to be small and the influence of the g_a-terms on ΔT_c (of TAI and OSI) and on ω_c (of OSI) to be negligible ($\sim g_a^2$).

5. Temperature Dependence of the Pitch (Layer Spacing)

Up to now we have neglected the coupling between R and T ("temperature independent pitch"). There is a static coupling by the susceptibility $\gamma_2 \equiv \partial_\varepsilon^2/\partial T \partial \nabla_z R$ and a dynamical coupling by the dissipative transport parameter ξ ($\dot{R} \sim \xi \nabla_z T$ and $\dot{T} \sim \xi \nabla_z R$). The pure heat conduction state now shows non-constant layer distances with $R \sim z^2 (\xi \zeta^{-1} + \gamma_2 T_o^{-1} C_p) \Delta T$.

Although these cross-coupling terms are small, for the instabilities they do no longer allow for stationary solutions. Thus, TAI, too, is oscillatory with $\omega_c \sim \xi$ and $\sim \gamma_2$; parts of the γ_2-dependence where already given by PARSONS [6]. For Cholesterics this ω_c adds to the $\omega_c \sim g_a$ discussed in the last section. For ΔT_c of TAI and for ΔT_c, ω_c of OSI the ξ- and γ_2-terms have only a weak influence.

It should be noticed that TAI and OSI, although now both oscillatory, are quite different in nature, since the former is mainly stationary with small moving contributions whereas in the latter case the whole structure oscillates. An extended discussion of these topics will be given elsewhere.

References

[1] P.C. Martin, O. Parodi, and P.S. Pershan, Phys. Rev. A6, 2401; 1972

[2] H. Brand and H. Pleiner submitted for publication

[3] S. Chandrasekhar, Hydrodynamic and Hydromagnetic Stability, Clarendon, Oxford; 1961

[4] D. Forster, T.C. Lubensky, P.C. Martin, J. Swift, and P.S. Pershan, Phys. Rev. Lett. 26, 1016; 1971

[5] E. Dubois-Violette, J. de Phys. 34, 107; 1973

[6] J.D. Parsons, J. de Phys. 36, 1363; 1975

[7] H. Pleiner and H. Brand, to appear in Phys. Lett. A

Part IV

Special Phase Transitions: Smectic A_1 - Smectic A_2, Reentrant Nematic, and Nematic-Isotropic

Reflections on the Reentrant Nematic and the S_A-S_A Phase Transitions

J. Prost

Centre de Recherche Paul Pascal, Domaine Universitaire,
F-33405 Talence, France

1. Introduction

In the usual classifications of liquid crystals, the nematic phase is said to occur at higher temperature than the smectic A one. Although it is well known that no definite relation between macroscopic symmetry and entropy exists, the general character of this sequence seemed to make sense since smectic A really looks intermediary between the nematics and the crystals. Thus when P.E. CLADIS reported the observation of a nematic phase occurring at a lower temperature than the smectic [1], the liquid crystal community was quite interested and puzzled. The experimental findings were as follows : consider a solution of 4-n-hexyloxybenzylidene-4'-cyanoaniline (HBAB) in 4-cyanobenzylidene-4'-n-octyloxyaniline (CBOOA), with a 15 (mole) % of HBAB in CBOOA, you will observe a rather usual Isotropic-Nematic-Crystal sequence ; however with mixtures containing less than 10 % HBAB, you will observe upon cooling the sequence Isotropic - Nematic - Smectic A - Nematic again and eventually crystal. The low temperature nematic phase is monotropic, but stable enough to allow the measure of very characteristic nematic properties such as the bent elastic constant K_{33}. This last quantity exhibits the same pretransitional phenomena when approaching the smectic A phase either from the high or from the low temperature side [1]. This set of observations seemed to be a real challenge to our understanding of the smectic layering. In particular no microscopic theory of the Kobayashi-Mac Millan type predicted this so-called reentrant behavior [2,3]. In fact, one could argue that a theory designed for a pure compound should not be compared to results obtained with a mixture. It was thus of prime importance to show that the same type of behavior could be obtained with a pure compound. This was done two years later by P.E. CLADIS et al. [4], with the study of the pressure- temperature phase diagram of octyloxycyanobiphenyle (8OCB) (fig. 1). Part of the same diagram was independently obtained by L. LIEBERT and W.B. DANIELS [5]. The curvature of its NS_A phase boundary is such that the nematic domain of stability may be found both at higher and lower temperatures than the S_A phase in a suitable pressure range.

One should point out that not only our understanding of the smectic layering is questioned, but also the validity of any liquid crystal classification which would rest upon the sequential appearance of such a phase. Indeed a very slight modification of the diagram (fig. 1) can result in a complete inversion of the sequential appearance of the nematic and smectic A phases at at-

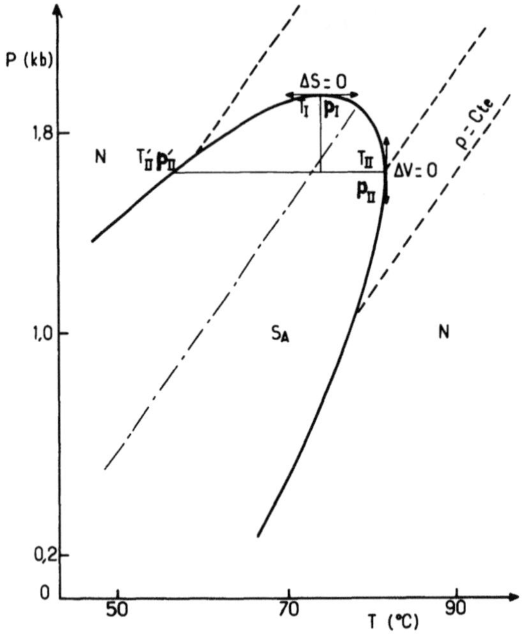

Fig. 1 : Schematic representation of the P.T. phase diagram of 8OCB.
The solid line represents the experimental boundary [4] as well as theoretical fits according to [15] or [17].
The dotted lines are isodensity lines, parallel to the dashed dotted line (parabola axis) according to [17]. The equalities ΔS=0 at $P_I T_I$, ΔV=0 at $P_{II} T_{II}$ are easily derived from the Clapeyron relation

mospheric pressure, as shown fig. 2. Note that part of this modification has already been achieved : thermotropic reentrant nematics have been found [6]. More recently, the full inversion of the nematic phase and a tubular one have been discovered [7].

On the other hand if this phenomenon raises basic questions for the understanding of liquid crystals, the "reentrance" of a high symmetry phase at a temperature lower than a less symmetric one is encountered in several other systems. The pressure dependence of the He^3 melting temperature was perhaps the first example of a similar behavior [8]. The reentrant normal phase in superconductors doped with Kondo impurities was predicted theoretically [9], and practically immediately found experimentally [10]. Other systems such as the metmyoglobin, ribonuclease or chymotrypsinogen water systems [11] [12] exhibit similar P.T. diagrams (at constant pH).

More recently, a reentrant Nematic-Isotropic phase boundary has been predicted in polyγ benzyl-L-glutamate dissolved in denatured solvents [13]. This list does not pretend to be exhaustive. There is no universal explanation suited to all these different systems ; for some of them the understanding stick to a basic thermodynamical level [8,11,12], in the two other cases specific models are developed [9,13].

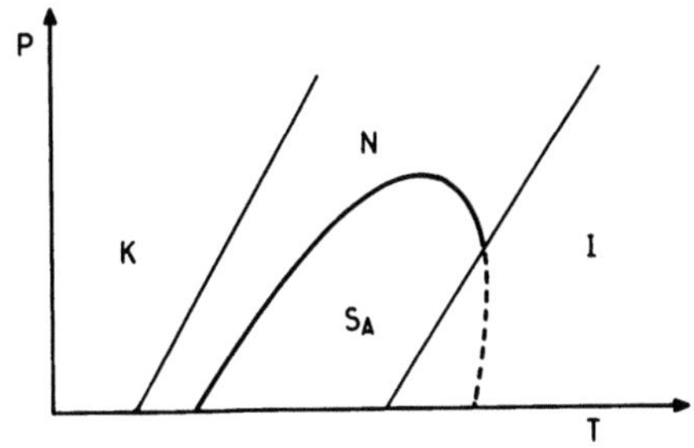

Fig. 2. Hypothetic phase diagram obtained from Fig. 1, by shifting the N-I and the crystallization phase boundaries toward lower temperatures. Note at atmospheric pressure the inverted sequence

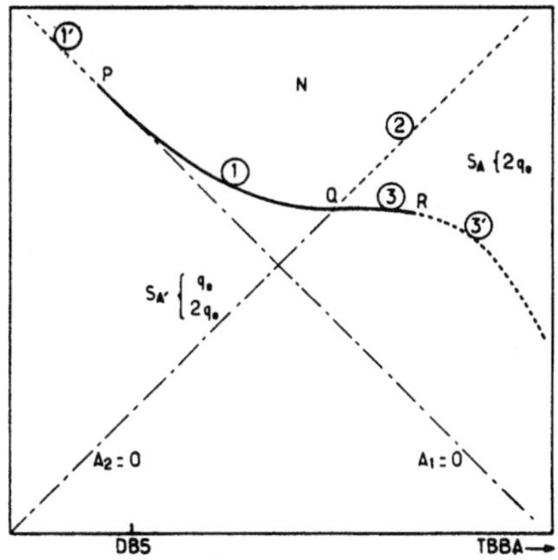

Fig. 3 : Topology of the N-S_A-$S_{A'}$ diagram according to [23]. This corresponds also to part of the DB5-TBBA isobaric diagram [14]. The full lines 1 and 3 represent first order transition boundaries, and the dotted lines 1' , 2 and 3' second order ones. Note the possibility for two tricritical points and one pseudotriple point (only this last one has been observed experimentally). The dashed dotted lines correspond to the $A_1=0$ and $A_2=0$ conditions

Another possibility which is usually discarded in classification schemes, is that of a transition between two phases of the same nature. Such an occurence would somehow imply that there is some hidden symmetry which is not accounted for in our current description of this phase. Such an interesting discovery has been reported at the Bordeaux meeting, by G. Sigaud et al. [14]. If you look under the microscope at a mixture containing a few mole percent of terephthalidene-bis-(4-n-butylaniline) (TBBA) in 4-cyanobenzoyloxy-(4'-n-pentylphenyl)benzoate (DB5 for short) and you find that the N-S_A boundary exhibits an angular point at (12 mole % TBBA, 120°C) (branches 1 and 2 of the diagram fig. 3). Branch 3 is not observed under the microscope, but this unexpected S_A to S_A phase transition is evidenced from calorimetric measurements. Thus although the two smectic A phases must be considered as isomorphous as far as the contact method is concerned, they are separated by a weak first order transition.

The reentrant nematic, and the S_A-S_A transitions have in common to challenge our understanding of the smectic phase, and classification schemes based only on microscopic observations and sequential appearance versus temperature. In the last section of this paper, we show that the introduction of two order parameters, one of the smectic the other of the antiferroelectric type (eg: linked to the bilayer tendency) provides an unified framework for discussing these two phenomena , and further allows the prediction of novel behaviors such as commensurate to incommensurate transitions.
In the third section we discuss the experimental evidence which precisely supports the introduction of these two order parameters. We focus our attention on X-rays studies which currently provide the most direct information

on the local and long range order prevailing in these systems. Eventually, in the second section, we present the theoretical approaches which were shown to describe the experiments and discuss the possibility of observation of a new type of critical behavior linked to the unusual approach to regular critical points allowed by the curvature of the phase boundaries. The first three sections have essentially the character of the (non exhaustive) review that was required for an invited lecture, the last section is original.

2. First Theoretical Approaches

The first questions which come to mind when dealing with the reentrant phenomenon are : is it really consistent with regular thermodynamics and what is the requirement for such an observation ? The answer given by CLARK [15] is in fact quite similar to the one imagined in the case of biochemical system [11,12]. At any transition point, the Gibbs free energy of the nematic phase equals that of the smectic phase : $\Delta G = G_N - G_S = 0$. Expand, G_N and G_S up to second order around a reference point T_o, P_o and you get the phase boundary equation in the case of a first order transition (with obvious notations)

$$0 = \Delta G_o + \Delta V_o(P-P_o) - \Delta S_o(T-T_o) + \Delta\beta(P-P_o)^2/2 + \Delta\alpha\,(P-P_o)(T-T_o) - \Delta C_p\,(T-T_o)^2/2T_o \qquad (1)$$

This is the general expression of a conic section, the best fit is obtained with an ellipse. This requires that $|\Delta\alpha^2 T_o/\Delta\beta\Delta C_p| < 1$ which is not particularly unusual. With a clever choice of P_oT_o (at the center of the ellipse) the number of adjustable parameters drops down to four, their ratio ($\Delta\beta/\Delta G_o$, $\Delta\alpha/\Delta G_o$, $\Delta C_p/2T_o\Delta G_o$) only being relevant. In principle all these quantities could be determined experimentally independently and the validity of these considerations checked without adjustable parameters. This task has not been achieved yet to our knowledge.

The next question is : is there a simple modification to the DE GENNES-MAC MILLAN picture [3,16] of the NS_A which allows the description of the reentrant behavior. The answer can be in fact obtained simply by looking at the phase diagram fig. 1 [17]. If one approaches from the nematic phase the NS_A boundary, along isodensity lines (dotted lines fig. 1), one clearly sees that there is a density for which the transition temperature is the highest (T_{II}). In other words there is an optimum density favoring the layering process. Translating these words into the DE GENNES-LANDAU picture of this transition gives :

$$F_S - F_N = A/2\,\rho_{q_o}^{\;2} + B/4\,\rho_{q_o}^{\;4} + C/6\,\rho_{q_o}^{\;6} \qquad (2)$$

ρ_{q_o} = that Fourier component of the mass density which undergoes a condensation at the NS_A transition.

$\rho(\underset{\sim}{r}) = \rho + \sum_{q>0} \rho_q \cos(\underset{\sim}{q}\underset{\sim}{r} + \psi_q)$

B,C as usual are supposed weakly temperature dependent.

$A = a(T-T^*)$ optimum density : $T^* = T_o - t_2(\rho-\rho_o)^2$ (3)

To be rigorous $F_S - F_N$ should express a Helmoltz free energy, and working at constant pressure requires the use of the Gibbs free energy. Switching from one to the other does not give dramatic effects and the phase boundary is given by :

$$T-T^*(\rho_N) \left\{ \begin{array}{ll} 0 & \text{if } B>0 \text{ (continuous transition)} \\ 3B^2/16aC & \text{if } B<0 \text{ (discontinuous transition)} \end{array} \right. \quad (4)$$

with $(\rho_N-\rho_o)/\rho_o = -\alpha_N(T-T_o) + \beta_N(P-P_o)$ (5)

Combining (4) and (5) one obtains for the phase boundary a parabola which can be expressed in terms of very simple experimental values :

$$(T_{II}-T'_{II})(T_{II}-T) = \left[(T_{II}-T) + \beta_N(P-P_{II})/\alpha_N\right]^2 \quad (6)$$

The slope of the parabola axis (α_N/β_N) is the ratio of the thermal expansion coefficient over the compressibility in the nematic phase. The significance of the other parameters is obvious on fig. 1. α_N/β_N can easily be estimated to lie between 20 and 25°K/Kbar which compares well with the experimental value 23°K/Kbar. The good quality of the fit suggests that at least for pure compounds the ellipticity of the contour is quite weak. This approach allows for a qualitative description of pretransitionnal phenomena as well [17]. Some of the points of the phase boundary have remarkable properties : for instance along the P_I isobar (in case of a second order transition), the free energy expansion reads :

$$F_S - F_N = a(T-T_I)^2|\rho_{qo}|^2/(T_{II} - T'_{II}) + \ldots \quad (7)$$

This is a quite unusual Landau expansion in which the coefficient of the quadratic term goes to zero without changing sign. Fluctuations go to infinity but no condensation occurs. In fact, the (P_I,T_I) point has nothing particular, but only its approach along a tangent to the phase boundary leads to this unusual behavior (equation (7) predicts $\tilde{\gamma}=2$, $\tilde{\nu}=1$). One can raise the question more generally:what kind of critical behavior does one observe, when the curved second order phase boundary is approached along a tangential line? The answer follows simply from the existence of one correlation length in the critical domain for one state of the physical system :

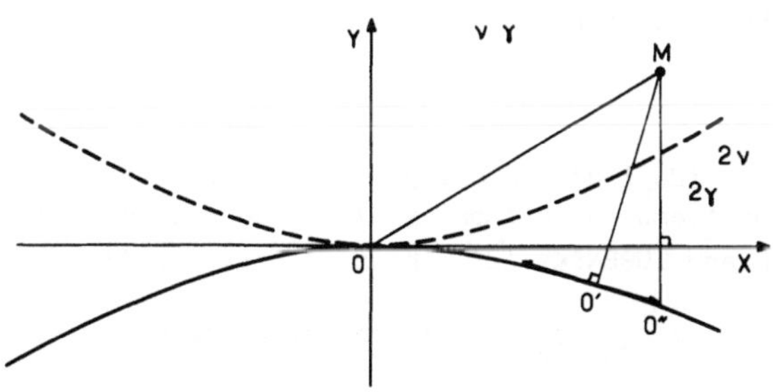

Fig. 4 : Unusual approach to a regular critical point. Full heavy line : locally parabolic phase boundary. Dotted line cross over from ν,γ to 2ν, 2γ behavior. The X parameter may be the temperature close to the T_I, P_I point, the pressure close to the T_{II},P_{II}, or a suitable combination at any other point

The state of the point M(X,Y) is essentially characterized by its correlation length ξ. The second order phase boundary, being a usual critical line one can write :

$$\xi \propto (MO')^{-\nu} \propto (MO'')^{-\nu} \propto (Y + a\,X^2)^{-\nu} \qquad (8)$$

if one can approximate the phase boundary with $Y = aX^2$. Thus, if one approaches O, along the Y=0 line, one observes $\xi \propto X^{-2\nu}$ leading to an apparent doubling of the exponent ν. Since the critical line in itself, has nothing special, the exponent η defined at T_c is unchanged, and in turn $\gamma = \nu(2-\eta)$ exhibits exactly the same apparent doubling. Note that equation (8) defines the domain in which this behavior is observable, and that only the exponents corresponding to the "outside" of the phase boundary can be affected. Further note that this result does not depend on the particular critical behavior that the system might exhibit (one would observe $\tilde{\gamma} = 2.66$ $\tilde{\nu} = 1.33$ in the helium analogy ! One can further check directly on the Wilson's [18] recursion relation that such a doubling would occur). It is our hope that some experiments will be designed to check these features [19].

The same kind of question may be asked for the S_A-S_A transition : can one find a Landau theory suited to this case ? The problem is to somehow connect a bilayer smectic (the DB5 has been shown to have a layer periodicity d of about twice the molecular length l [20]) to a monolayer one (in the case of TBBA : $d \simeq l$). We show on fig. 5, that the monolayer order parameter appears

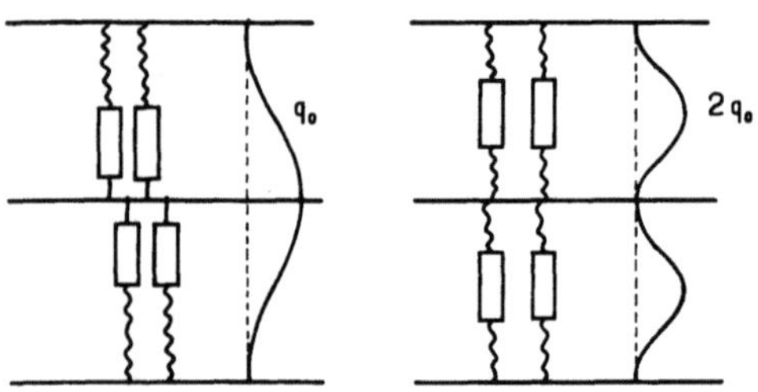

Fig. 5 : Bilayer smectic of the DB5 type (left) compared to a monolayer smectic of the TBBA type (right)

to correspond to the first harmonic ($2q_o$) of the bilayer order parameter (q_o). The coupling between the fundamental and the harmonics of the density modulation have been discussed by MEYER and LUBENSKY [21]. Keeping only the first one gives :

$$2(F-F_o) \simeq A_1\, \rho_{q_o}^{\;2} + A_2\, \rho_{2q_o}^{\;2} - D\, \rho_{q_o}^{\;2}\, \rho_{2q_o} + \text{fourth order terms} \qquad (9)$$

$$\rho(r) = \rho + \rho_{q_o} \cos(q_o z + \psi_{q_o}) + \rho_{2q_o} \cos(2q_o z + \psi_{2q_o}) + \ldots$$

Again $A_1 = a_1(T-T_1)$ and $A_2 = a_2(T-T_2)$. On the basis of microscopic considerations MEYER and LUBENSKY suggested that $T_2 \simeq T_1/2$. In such a case, a Rodbell Bean type of analysis of (9) is possible [22], and one gets a redefinition of the fourth order terms with the possibility of a tricritical point. This explains why the NS_A transition of the pure DB5 is first order although the Mac Millan criterion would predict a second order one. On the other hand in the TBBA-DB5 diagram the two temperatures T_1 and T_2 have to cross each other since the DB5 is a bilayer and the TBBA a monolayer smectic. Working out the minimization procedure in this case [23] leads to the phase diagram fig.3 the topology of which agrees fairly well with the experiment. Its most novel feature is the existence of a S_A-S_A transition. Whenever the fundamental ρ_{q_o} condenses it drives a non zero ρ_{2q_o} and one obtains a bilayer smectic in which the first harmonic of the layer modulation is expected to be strong. This N-S_A' phase transition may be first or second order depending on the value of D^2/A_2 relative to the fourth order coefficient, and the helium analogy ought to be correct. On the other hand, when ρ_{2q_o} condenses first, no bilayer order appears (ρ_{q_o}=0) and this NS_A transition is always continuous (except if one includes the coupling with the nematic-isotropic order parameter in which case the Mac Millan criterion is obeyed). It corresponds exactly to DE GENNES original description of the N-S_A problem if one renames $2q_o$ as q'_o. The S_A-S_A' transition, which can be either first or second order corresponds to the onset of the fundamental q_o in a matrix already condensed at the harmonic periodicity $2q_o$. If one keeps the phase variable, the coupling term between ρ_{q_o} and ρ_{2q_o} may be written :

$$D\ \rho^2_{q_o}\ \rho_{2q_o}\ \cos(2\psi_{q_o} - \psi_{2q_o}) \qquad (10)$$

This expression is minimized for $\psi_{q_o}=\psi_{2q_o}/2$; (π). The binary choice (0,π) is equivalent to the possibilities spin up or down in the Ising case, and thus the S_A-S_A' transition belongs to the n=1, d=3 universality class. In particular topologically stable defects are surfaces, in agreement with the general arguments of TOULOUSE G. and KLEMAN M. [24](see fig. 6).

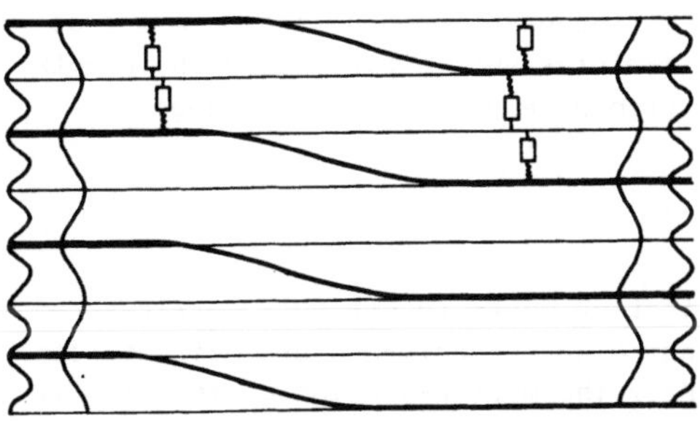

Fig. 6 : Example of a topologically stable defect : the ρ_{2q_o} order keeps unchanged from the left to the right of the drawing, whereas ρ_{q_o} has shifted by an amount of π. Note that in the core of this defect the (q_o) wavevector is not perpendicular to the average plane of the layers and will contribute to diffuse XR scattering off the axis of the structure, as pointed out to us by A. LEADBETTER

This interpretation of the S_A-S_A' transition leads to clear predictions on the experimental level such as a Bragg condensation of ρ_{q_o}, whereas ρ_{2q_o} is already condensed, and we will see in the next section how the experiment confirms them. On the other hand with just the binary diagram as experimental clue, one certainly has to wonder if this interpretation is unique [23]: the topology of the diagram and the existence of the S_A-S_A' transition is entirily due to the coupling term (10). In a Landau picture one could replace ρ_{q_o} by any order parameter P_{q_o} (vector, second rank tensor and so on) and get the same basic result. In particular this led us to discuss the possibility of a vector order parameter which would give a better image of the experimental situation : indeed each DB5 molecule bares a strong dipole and the bilayer has an antiferroelectric symmetry which better corresponds to a $P_z(q_o)$ than a ρ_{q_o} order (z being defined as the normal to the layers). In [23], we pointed out that if that was to be the case, there would be no reason why the bilayer and monolayer orders should be commensurate. We will develop this idea further in the fourth section and show a number of new features that such a point allows to predict. If including the coupling of the smectic order parameter to the mean value of the density on one hand, and to its first harmonic Fourier component on the other helps giving a macroscopic framework in which the reentrant S_A-S_A' transitions can be discussed, it does not help much in understanding the microscopic origin of these phenomena. At least one can say that the idea of an optimum density for stabilizing the smectic layers agrees fairly well with CLADIS et al. microscopic picture of the reentrant phase [25]. Their main point is to stress that whereas the nematic order may be obtained with repulsive forces only (in general many kinds of forces are involved in sustaining the nematic order, but all of them are not necessary [26]), the smectic layering involves necessarily attractive ones [3]. At low density a regular nematic may be formed in suitable temperature conditions, increasing the density will first increase the attractive forces and thus favor the smectic layering, but past a given density, repulsive forces will take over and thus favor the nematic phase again. A microscopic theory taking into account these features and in particular the dipolar character of attractive forces has not been done yet to our knowledge.

3. Recent Experimental Advances

In our introduction, we have stressed that we would report only on X-ray results. This does not imply of course that other experiments are not interesting, but the X ray technique is currently the most revealing of the smectic structure. For instance, F. HARDOUIN et al. work on the DB5-TBBA mixture provides a neat confirmation of the nature of the S_A-S_A' transition [27] (fig.7). They show that the pure DB5 is the first compound for which the layer spacing is essentially twice that of the molecular length in its most extended configuration ($d \simeq 1.95\ l$), and that a strong first harmonic is to be seen on its X ray pattern. Furthermore, they show quite clearly that the S_A-S_A' transition corresponds to the Bragg condensation of the fundamental ρ_{q_o} in a matrix S_A in which the harmonic ρ_{2q_o} was already condensed, in agreement with our theory [23]. The ρ_{q_o} condensation reinforces the ρ_{2q_o} intensity, and the ρ_{q_o} and ρ_{2q_o} intensities are of the same order of magnitude, again in agreement with [23] (in the domain in which $S_AS_{A'}$ is first order). The diffuse scattering spots around q_o, in the S_A phase, may be split off axis in certain cases in [28]. As stressed to us by A. LEADBETTER, this is a signature of the existence of a large number of defects of the type described fig. 6. They destroy the long range order of the bilayer type, but do not perturb the harmonic

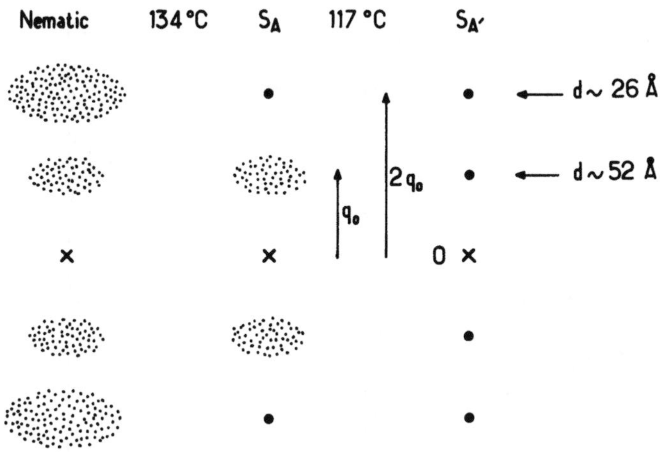

Fig. 7 : Schematic representation of the X ray pattern given by the 16.6% mole TBBA in DB5 according to F. HARDOUIN et al. [27]:

left : the nematic phase (note a diffuse scattering at both q_0 and $2q_0$)

center : the S_A phase, the Bragg condensation on $2q_0$ is clearly seen whereas the q_0 order only contributes to diffuse scattering (in the case of this mixture the diffuse area is centered on the z axis, but with other compounds it may be off axis [28], indicating that this S_A phase is probably obtained via a large number of π discommensurations of the type described fig. 6.

right : the S_A' phase. Note the Bragg reflections at q_0 and $2q_0$, which have typical intensities of the same order of magnitude

periodicity (eg monolayer). Other recent results by F. HARDOUIN and A.M. LEVELUT which raise questions of incommensurability will be discussed in the fourth section.

X-rays are equally instructive of the basic questions relevant to the reentrant case. D. GUILLON et al. have studied the CBOOA, HBAB binary system, and also the behavior of pure CBOOA under pressure [29,30].In all cases, they find that the layer spacing varies very little in the smectic A phase (except when the S_A phase evolves toward a crystal) ($d \simeq 1.3\ l$), in the reentrant nematic smectic like fluctuations are clearly seen, the corresponding periodicity decreases indicating that our theoretical approach [17] is somehow not sufficient (fig. 8). The authors also mention the importance of another length connected in their discussion to a better packing in the nematic phase and in the underlying solid phase. R. SHASHIDAR and K.V. RAO, also performed high pressure measurements with a somewhat different experimental set up: they find that at constant $T_{NA}-T$ = 4°K, the 8OCB layer spacing goes through a minimum at P=1.4 Kbar. Similar observations have been made, at atmospheric pressure versus temperature by CHANDRASEKHAR et al. [32], on three benzene rings compounds (4-n-undecyl- and 4-n-dodecylbenzoyloxy(4'-cyanophenyl)-3-methylbezoates). Another interesting set of experiments have been reported by

Fig. 8

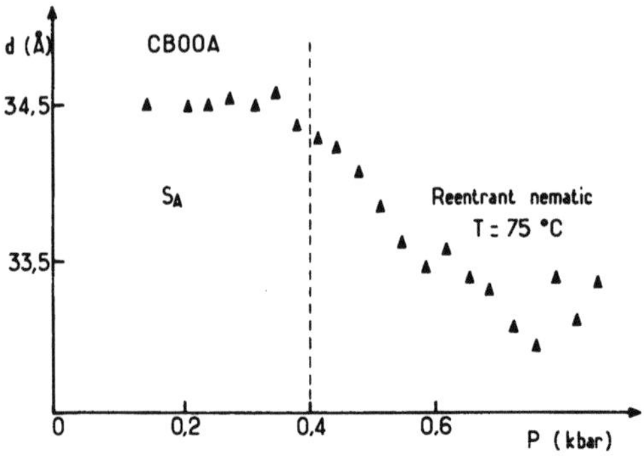

Fig. 9

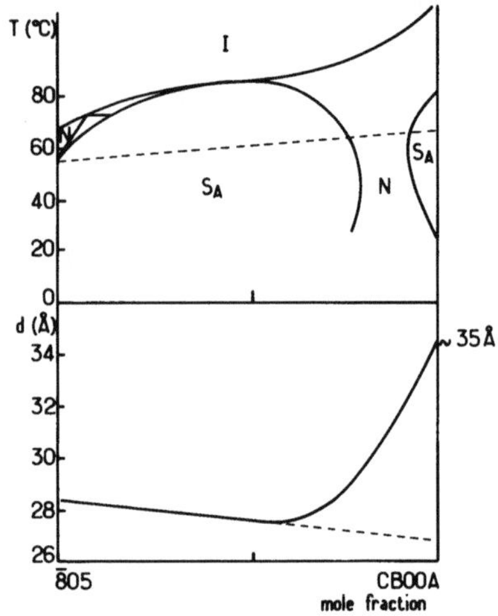

Fig. 8 : Pressure dependence of the CBOOA layer spacing, at 75°C according to D. GUILLON et al. [29]. Note its constancy in the S_A phase and the sudden decrease in the nematic

Fig. 9 : Isobaric binary diagram of $\bar{8}05$ and CBOOA (upper part) and layer spacing of the mixture taken along the dotted line of the diagram (lower part) according to B. ENGELEN et al. [33]. Note the sudden change of behavior around 70 mole % of CBOOA corresponding to a stabilization of the nematic phase

ENGELEN et al . For instance they measure the smectic layer spacing along the dotted line of fig. 9, of the $\bar{8}05$-CBOOA isobaric phase diagram ($\bar{8}05$: 4 n octyloxy-|4-n-pentyl| benzoate.

On the $\bar{8}05$ side of the phase diagram, the layering is typically one molecular length and obeys a simple linear mole fraction dependence extrapolating to a one molecular CBOOA length on the right side ; somewhere around 70% mole CBOOA, very suddenly the layer spacing changes abruptly, whereas the nematic phase is strongly stabilized, and eventually another S_A phase is reached, for which the periodicity is about 1.35 time the CBOOA molecular length. This suggests that there may be more than one smectic phase likely to exist, and that they are somehow not compatible.

In their theoretical work on Kondo impurities, in superconductors, MULLER HARTMANN & ZITTARTZ [9] had not only predicted a reentrant normal behavior, but at even lower temperature a reentrant superconducting phase (eg the succession : Normal-Superconducting-Normal-Superconducting as T decreases). This complete succession has not been evidenced yet in superconducting systems to our knowledge but its equivalent in the N-S_A system has been discovered by HARDOUIN et all [34,35]. The (octyloxy-4'-benzoyloxy)4-cyanostilbene)(T8) exhibits the surprising thermotropism : Isotropic $\xleftarrow{283°C}$ Nematic $\xleftarrow{247°C}$ Smectic A $\xleftarrow{139°C}$ Nematic $\xleftarrow{96.5°C}$ Smectic A $\xleftarrow{94°C}$ Crystal

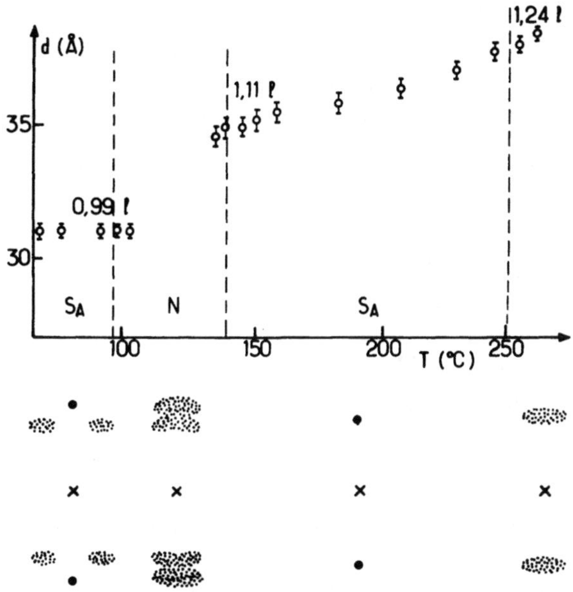

Fig.10 : Temperature dependence of the T8 layer spacing (upper part). Schematic representation of the X ray pattern successively from right to left in the high temperature nematic, the high temperature smectic A, the reentrant nematic and the reentrant smectic A phases, according to F. HARDOUIN and A.M. LEVELUT [36].Note in the reentrant nematic phase, the existence of diffuse scattering spots corresponding to the two incommensurate periodicities ($d \simeq 1.2l$ and $d \simeq l$).The diffuse scattering in the reentrant smectic, appears to be a phase modulation of the layers [36], and probably corresponds to soliton like fluctuations as discussed in the fourth section

The low temperature nematic phase was shown to exhibit pretransitional behavior on γ_1 both by increasing and by decreasing the temperature |35|. Although thermotropic in a very narrow range the low temperature smectic phase is easily supercooled. The corresponding X ray work (F. HARDOUIN, A.M. LEVELUT) is perhaps the most revealing clue to date (fig. 10) [36]. In the high temperature S_A phase the layer spacing is of the order 1.2 times a molecular length, whereas in the low temperature S_A phase it is almost exactly one molecular length (.99l) : in the reentrant nematic phase two types of cybotactic groups are seen, some centered around the high temperature S_A periodicity others around the low temperature one. The idea of two smectic phases competing for condensation and yielding to the reentrant phenomenon is thus strongly suggested by this experiment, as stressed by the authors [36]. A further, striking point, is the observation in the low temperature phase, of a diffuse scattering which corresponds to a phase modulation of the layers. We will see in the next section how this different aspects can be included in a single free energy functional.

4. Bilayer Smectics: A Frustated System

Most of the experimental evidence that we have reported in section 3, point toward the recognition that in the smectics made of cyanoderivatives, two microscopic lengths are important (fig. 11) namely that of the single molecule and that of the pair. Depending on the dipole delocalization on the molecule, the pair length may correspond to A. LEADBETTER's model [37] or be roughly twice that of the isolated molecule [27].

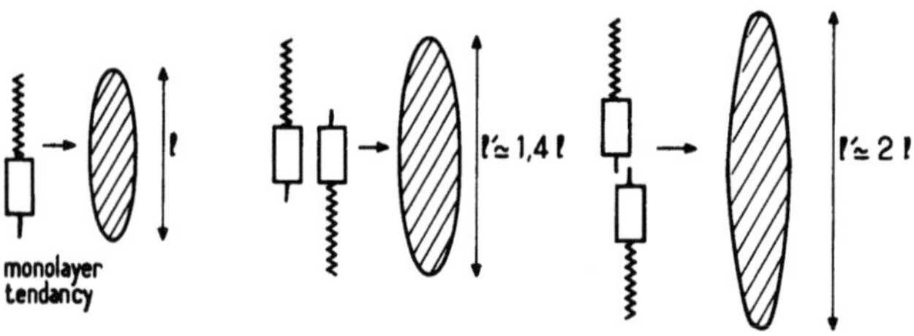

Fig. 11 : The two important lengths in the polar smectic problem : l the molecular length, l' the pair length which can range from 1.2 to 2 times l

. If, in a gedanken experiment, each of these entities, could be separately considered, they would lead to the condensation of layered phases with respectively $d \simeq l$, $d \simeq 1.3l$, $d \simeq 2l$. In the first case, the relevant order parameter is the usual Fourier component of the density modulation $\rho(k)$ [16], in the last two, one may choose the antiferroelectric order parameter $P_z(k)$ introduced in [23] and already alluded to in section 2.

The corresponding free energy in k space reads :

$$\begin{aligned} 2F = V \sum_{k,k',k''} & A_2(k)\,|\rho(k)|^2 + A_1(k')\,|P_z(k')|^2 \\ & + \{C\, k'_z\, P_z(k')\rho(-k) + C^*\, k'_z P_z(-k')\rho(+k)\}\,\delta_{k-k'} \\ & - \{D\, P_z(k')\, P_z(k'')\rho(-k) + D^*\, P_z(-k')\, P_z(-k'')\rho(k)\}\,\delta_{k'+k''-k} \\ & + \text{4th order terms} \end{aligned} \qquad (11)$$

C and D are weakly temperature dependent, complex coefficients.

$$A_2(k) = a_2(T-T_2) + (k_z-k_2)^2/M_2 \qquad k_2 = 2\Pi/l$$
$$A_1(k) = a_1(T-T_1) + (k_z-k_1)^2/M_1 \qquad k_1 = 2\Pi/l' \qquad (12)$$

l molecular length ; l' pair length

T_2 condensation temperature of the monolayer order alone ($a_2>0$)

T_1 condensation temperature of the antiferroelectric order alone ($a_1>0$)

The divergence of P_z, has a scalar symmetry which imposes the existence of a harmonic coupling term between $\nabla_z P_z$ and ρ. The cubic coupling term $P_z^2\rho$, is analogous to the one used in (10). The k dependence of A_1 and A_2, expresses that the optimum periodicity of ρ and P_z is respectively that of the molecule, and that of the pair.

. Although first conjectured on purely speculative ground [23], this k dependence of A_1 and A_2 is strongly suggested by the recent results of HARDOUIN and LEVELUT in the reentrant nematic phase of the T8 compound [36] and also by similar observation of A. LEADBETTER in a "non-reentrant nematic" [37] .

4.1 : $l' \simeq 2l$. The S_A-S_A' and a commensurate to incommensurate phase transition.

Only those wavevectors $k \simeq k_2$ and $k' \simeq k_1$, will lead to values of the free energy which will not be prohibitively large. Thus, the second order coupling term which imposes $k=k'$ and thus either $A_1(k)$ or $A_2(k')$ large, may be dropped out of (11). With :

$$\rho(r) = \psi_2(r)\exp(i2k_o z) + \psi_2^*(r)\exp(-i2k_o z) \qquad \psi_2(r)=|\psi_2(r)|\exp(i\beta(r))$$
$$P_z(r) = \psi_1(r)\exp(ik_o z) + \psi_1^*(r)\exp(-ik_o z) \qquad \psi_1(r)=|\psi_1(r)|\exp(i\alpha(r))$$
$$2k_o = k_1 + k_2/2 \quad q_o = k_2 - 2k_1 \qquad D=|D|\exp(i\delta) \tag{13}$$

One gets in real space (forgetting about the x,y dependence of the order parameters) and with $A_i = a_i (T-T_i)$ (i : 1,2) :

$$2F = \int_V \{A_1|\psi_1|^2 + A_2|\psi_2|^2 - 2|D|\,|\psi_1|^2\,|\psi_2|\cos(2\alpha-\beta+\delta) \tag{14}$$
$$+ M_1^{-1}|(i\nabla_z - q/4)\psi_1(r)|^2 + M_2^{-1}|(i\nabla_z + q_o/2)\psi_2(r)|^2\}\,dv$$
$$+ \text{4th order term}$$

If one forgets the elastic term, this is exactly the free energy used to describe the S_A-S_A' transition. In other words, if elasticity plays a minor role we will get phenomena similar to those described [23] and in the second section. Note however that the S_A phase has a pure smectic character whereas the S_A' has both a smectic and an antiferroelectric character. The qualitatively new feature, brought in by the elastic term, may be evidenced by looking at solutions keeping the amplitude of ψ_1 and ψ_2 homogeneous in space, but allowing α and β to be a function of z.

With : $\psi_1(z) = \psi\cos\theta\exp(i\alpha(z)) \qquad \gamma = 2\alpha - \beta + \delta$

$\psi_2(z) = \psi\sin\theta\exp(i\beta(z))$

The free energy after minimization with respect to a certain mean phase variable ($\mu = (\alpha\cos^2\theta + 2\beta\sin^2\theta)/(1+3\sin^2\theta)$) assumes the simpler form :

$$F = \int_V \{2^{-1}A(\theta)\psi^2 + 4^{-1}B(\theta) + \psi^2(2M(\theta))^{-1}(\nabla z\gamma + q_o)^2 - D(\theta)\psi^3\cos\gamma\}\,dv \tag{15}$$

$$A(\theta) = A_1\cos^2\theta + A_2\sin^2\theta \; ; M^{-1}(\theta) = M^{-1}\sin^2\theta\cos^2\theta/(1+3\sin^2\theta) ; D(\theta) = D\cos^2\theta\sin\theta \tag{16}$$

The last two terms, have the typical structure often used in the discussion of commensurability problem [38,39,40,41,42]. The simplest presentation may be given in terms of an analogy with the magnetic field induced cholesteric to nematic transition. The twist elastic constant K_{22} corresponds to $\psi^2/M(\theta)$ and the magnetic energy $\chi_a H^2$ to the lock in term $4D(\theta)\psi^3$; one knows [39,40] that beyond a critical field ($H_c = \pi K_{22}^{1/2}/2q_o\,\chi_a^{1/2}$) the cholesteric structure is unwound, and that one obtains a nematic ($\gamma=0$). For field values, slightly smaller than H_c, the structure is a periodic stack of Bloch walls, the separation of which tends to infinity when $H \to H_c$ (fig.12). In the present case, if the lock in term overcomes the elastic one ($|D|\psi > (\Pi^2 q_o^2\sin\theta)/16M(1+3\sin^2\theta)$), $\gamma=0$ everywhere and one gets a unique periodicity defined by the minimization of μ. Apart from minor modifications, this is exactly the S_A-S_A' problem as already pointed out. For $|D|\psi < \Pi^2 q_o^2\sin\theta/16M(1+3\sin^2\theta)$ a new situation is reached : the ψ_1 and ψ_2 order parameters are locked in large regions of space, but at some localized places equivalent to the Block walls of the cholesteric case they uncouple : the $\gamma = 2\alpha - \beta + \theta$ phase difference suddenly undergoes a 2Π jump (fig. 12 and 13). A new period appears in the system : the separation between two such singularities : $P = 8K(k)E(k)/\Pi q_o$ (K(k), E(k) elliptic integrals of first and

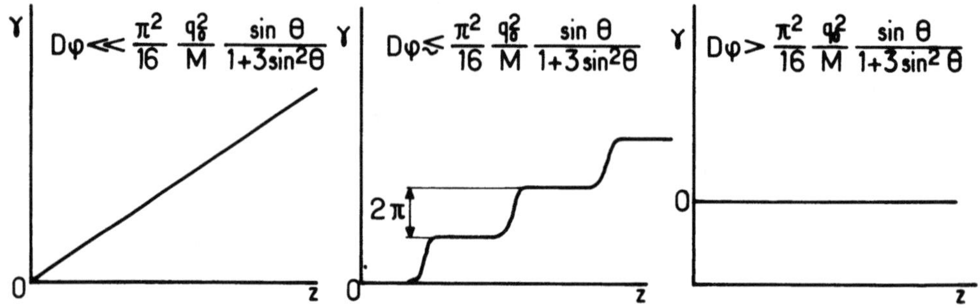

Fig.12 : Typical aspect of the commensurate-incommensurate transition : right : when the coupling term exceeds a critical value analogous to the critical field of the cholesteric to nematic transition , the phases 2α and β are locked everywhere ($\gamma=0$)
middle: For values slightly smaller than the the threshold value, the phases 2α and β are locked in large areas, but undergo 2Π changes in restricted regions which are called, solitons or discommensurations in other branches of physics. The regular stack of such solitons tends to restore the two independent periodicities.
left : in the (inprobable) case of a negligible lock in term, the two phases 2α and β would be completely independent, which would be the equivalent of the cholesteric phase

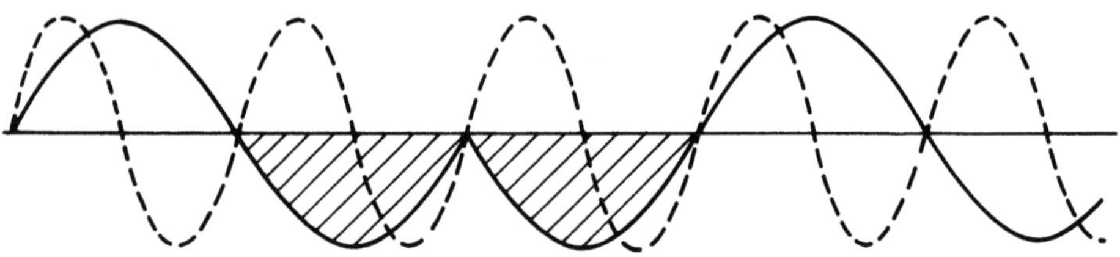

Fig.13 : Example of discommensuration in a case when the ρ elastic constant is much larger than the Pz one (___ Pz order parameter ; --- ρ order parameter). Note that it clearly bares a net dipole moment. However the resulting structure is not, in general, ferroelectric. Electric properties of this system will be discussed in a forthcoming paper

second kind. k is determined by the relation $|D|\psi = k^2\Pi^2 q_0^2 \sin\theta/16E^2$ (k) $M(1+3\sin^2\theta)$). This type of transition is well known in solid state physics and commonly referred to as the commensurate to incommensurate phase transition [43]. The assumption $|\psi_1|$ and $|\psi_2|$ independent of r is known to be only qualitatively correct [44] : at each location of the singularity $|\psi_1|$ and $|\psi_2|$ should vary, but this does not perturb the essential result that there should exist a second order commensurate to incommensurate transition. The best chances for its observation are when $\theta \simeq 35°$, that is in the S_A, phase.

On the other hand, in the S_A phase $\theta = \Pi/2$ ($\Psi_1=0$) and $\psi = (-A_2/B)^{1/2}$. The mean field lock in parameter reads then, $(|D|(-A_2/B)^{1/2} - \Pi^2 q_o^2/64M)$. Very close to the NS_A transition, bilayer fluctuations should be incommensurate with the monolayer periodicity, but as the smectic order builds up, they should progressively lock. In fact, this mean field parameter is certainly not sufficient [42], and the main assertion we can put forward is the existence of soliton like phase fluctuations, which can be commensurate or not with the monolayer periodicity. Such observations have been recently made by F. HARDOUIN and A.M. LEVELUT [45] , but the corresponding commensurate to incommensurate transition remains still to be discovered.

4.2 : $l' \simeq l$. Conjectures on the reentrant nematic, an other S_A-S_A transition and also a commensurate to incommensurate transition.

In this case, the third order coupling term leads to prohibitively large $A_1(k)$ or $A_2(k')$ terms, and hence cannot contribute significantly to the ρ, Pz fluctuation spectrum. The free energy can be simplified :

$$2F \simeq v \sum_k A_2(k) \; |\rho(k)|^2 + A_1(k) \; |P_z(k)|^2 - 2|C|k|P_z(k)|\,|\rho(k)|\cos\gamma k + \text{4th order terms} . \qquad (17)$$

The existence of the harmonic coupling term implies that the condensed phase will always have both a smectic and an antiferroelectric character. When the fourth order terms are positive, the transition is second order, and in a mean field approximation given by :

$$A_1(T_{NA}^{MF}, k_{NA}) \; A_2(T_{NA}^{MF}, k_{NA}) = k^2_{NA} |C|^2 \qquad \begin{cases} A_1(k)A_2(k) - k^2|C|^2 > 0 & \text{Nematic phase} \\ A_1(k)A_2(k) - k^2|C|^2 < 0 & \text{Smectic phase} \end{cases} \qquad (18)$$

It is clear from (18) that this transition occurs at temperatures for which both A_1 and A_2 are still largely positive, and thus at a temperature much larger than both T_1 and T_2. The more polar smectics should have a greater tendency for Pz condensation, and thus a higher T_2. This in turn implies, a higher T_{NA}^{MF} , in good qualitative agreement with P. CLADIS' comparison between alkyl and alkoxy transition temperatures |25|.

We have plotted on fig. 13, the temperature and wavevector dependence of the smallest eigen value $s_1(k)$, of the quadratic form contained in (17) under the assumption that the $A_2(T)$ and $A_1(T)$ coefficients cross each other at a temperature T_{AA} larger than T_1 and T_2. As already pointed out the smectic condensation is obtained at a temperature significantly higher than T_1 and T_2, and on a wavevector close to k_1, that is close to the natural periodicity of the smallest A coefficient. At a temperature T_{AA}, there are two minima of equal depth in the $s_1(k)$ curve, and at lower temperature the smallest minimum shifts toward the k_2 values (which again corresponds to the smallest of the A values). This implies the existence at T_{AA} of a discontinuous transition (without symmetry change) from one layer spacing to another layer spacing ($d \simeq 1.3l$ to $d \sim l$ in the example). With different orders of magnitude (smaller elasticity) the change from k_1 to k_2 could be smooth. One could conjecture that close to T_{AA} , the elastic energy could be so large that the system prefers to revert back in the nematic phase, however mean field does not predict such a behavior. One can indeed show that as long as A_1 and A_2 are linear in T, no choice of the parameters involved in the free energy (17) will ever give the reentrant behavior.

On the other hand, the layer spacing predicted by this mean field analysis, agrees fairly well with the experimental observations of HARDOUIN and LEVELUT on the T8 compound, and this comparison suggests that the reentrant phase occurs in the T_{AA} region in which two periodicities are competing for condensation. It is possible to give a heuristic argument suggesting that the fluctuating contribution of the fourth order terms, omitted up to now, is able to lead to this reentrant behavior. Indeed, if we call Φ_1 and Φ_2 the two linear combinations of ρ and P_z competing for condensation (centered around the wavevectors k_1' and k_2') the part of the free energy which will most contribute to the fluctuation spectrum may be approximated by (although omitted for clarity sake, the gradient terms are implicit in the following equation) :

$$2F' = \int_V \{s_1(T)|\Phi_1|^2 + s_1'(T)|\Phi_2|^2 + \frac{B_1}{2}|\Phi_1(r)|^4 + B|\Phi_1(r)|^2|\Phi_2(r)|^2 + \frac{B_2}{2}|\Phi_2(r)|^4 dv \quad (19)$$

s_1 becomes negative at T_{NA}^{MF}, and s_1' becomes negative at a temperature a little higher than T_{AA}. A Ginzburg type of analysis, would give for the susceptibilities :

$$\chi_{\Phi_1}^{-1}(k_1') \simeq s_1(T) + 3B_1 \langle|\Phi_1^2(r)|\rangle + B\langle|\Phi_2(r)|^2\rangle$$
$$\chi_{\Phi_2}^{-1}(k_2') \simeq s_2(T) + 3B_2 \langle|\Phi_2^2(r)|\rangle + B\langle|\Phi_1(r)|^2\rangle \quad (20)$$

In these two expressions, the second terms define the usual Landau-Ginzburg criterion for non trivial critical region ; on the other hand the third terms may explain the reentrant behaviour :

- close to T_{NA}^{MF} there is just one minimum in the $s_1(k)$ curve, which implies that $\langle|\Phi_2(r)|^2\rangle$ is very small in this region and that the only correction to mean field is the usual one, given by the renormalization group theory (in this region there is even no reason to single out Φ_2 from the overall fluctuation spectrum)
- as T is lowered toward T_{AA}, however $B\langle|\Phi_2(r)|^2\rangle$ will increase, and drive $\chi_{\Phi_1}^{-1}$ to positive values although $s_1(T)$ is negative. The situation being symmetric in Φ_1 and Φ_2, around T_{AA}, one can understand that with both $s_1(T)$ and $s_1'(T)$ negative, one can have however $\chi_{\Phi_1}^{-1}$ and $\chi_{\Phi_2}^{-1}$ positive, and thus a stable reentrant nematic phase. Another way to look at this problem, is to remark that the temperature drop $(T_{NA}^{MF}-T_{NA})$ is proportional to (n+2), where n is the number of components of the order parameter [46] ; close to T_{NA}^{MF}, n=2 , but close to T_{AA} the situation is very close to a problem in which n=4, which implies a larger drop and hence a stabilization of the nematic phase in appropriate conditions.
- eventually, as T is further lowered, the Φ_1 well disappears, $\langle|\Phi_1|^2\rangle$ has now no reason to be singled out, and a condensation toward the $|\Phi_2|$ smectic phase occurs with the usual critical behavior.

Up to now, we have explained (although with hand waving arguments) the stability of the reentrant nematic phase, but not the instability of the high temperature smectic. In fact, it does follow from the same type of argument if one integrates out the Φ_2 fluctuations one finds for the mean value of Φ_1 (leaving out the part leading to the definition of the Ginzburg criterion)

$$B_1\Phi_1^{\circ 2} = +|s_1| - B\langle|\Phi_2^2|\rangle$$

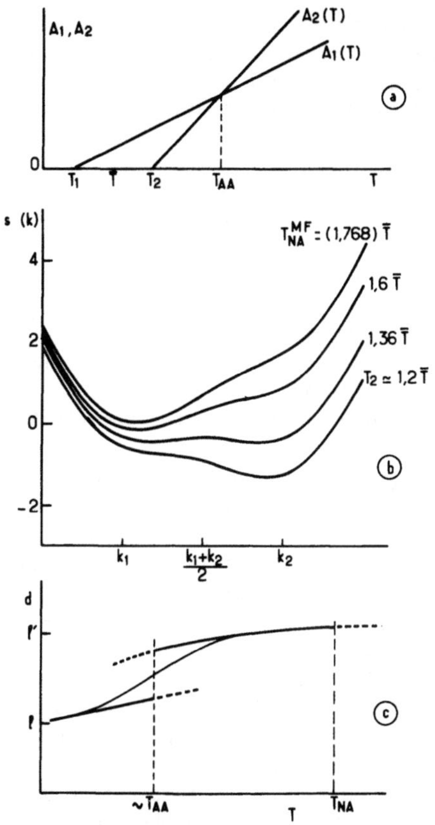

Fig.14:a) Assumed temperature dependence of the coefficients A_1 and A_2. The choice here is such that the resulting layer spacing agrees with the T8 reults of F. HARDOUIN and AM. LEVELUT, but one can imagine cases in which 1 and 2 could be inverted (the essential point being the existence of T_{AA} for which $A_1=A_2$) b) Plot of the smallest eigen value of the quadratic form versus k for different temperatures. Note that the transition occurs at a value much larger than T, and the prediction of a discontinuous jump from $k_1' \simeq k_1$ to $k_2' \simeq k_2$ close to T_{AA}. c) Resulting temperature dependence of the layer spacing (full line); with a different choice of the parameters (smaller elasticity) one could have a smooth evolution from l' to l without the discontinuous jump (light line)

which shows precisely that Φ_1° goes to zero, at the very limit at which the reentrant phase becomes stable (again omitting the critical region). Note that taking account of the critical fluctuations is known to stabilize the most symmetric phase, and thus will decrease the stability range of the high temperature smectic. Note also that the possibility for reentrant behavior depends on the value of the coefficients involved in the free energy : one may imagine cases in which the reentrant phenomenon is replaced by the S_A-S_A transition predicted by the Landau theory. An other important point is that within this interpretation in case of a reentrant behavior, one should always have the sequence (upon lowering temperature) Nematic-Smectic A - Nematic - Smectic A. The observation of the whole sequence may of course be hidden by the occurrence of other transitions such as crystallization. At last, the incommensurability of k_1 and k_2 did not enter explicitly in this picture ; in principle any couple of order parameters leading to a free energy expansion isomorphous to (18) has the potentiality of giving the same kind of behavior. The recent observation by A. LEADBETTER of a smectic F phasse somehow quenched between the two more ordered phases S_B and S_G (tilted B) may perhaps be relevant of the same kind of explanation [47].

If the incommensurability of k_1 and k_2 did not play an important rôle in the previous picture of the reentrant phenomenon, it does lead to the possibility of a commensurate to incommensurate transition. The case is very similar to the one exposed in 4.1. Again looking for phase fluctuations, restricting the problem to the z dependence, with the further approximations $kC \simeq k_o C$ and $M=M_1=M_2$ and the following variable change :

$\rho(r) = \psi_2(r)\, e^{ik_o z} + \psi_2^*(r)\, e^{-ik_o z}$ $\quad \psi_2(r)=e^{i\beta(r)}\, \psi \sin\Theta$

$P_z(r) = \psi_1(r)\, e^{ik_o z} + \psi_1^*(r)\, e^{-ik_o z}$ $\quad \psi_1(r)\, e^{i\alpha(r)}\, \psi \cos\Theta$

$C = |C|\, e^{i\delta'}$ $\quad 2k_o = k_1 + K_2$ $\quad q_o = k_1 - k_2$

One can isolate in the free energy, a part which involves the phase difference $\gamma = \alpha - \beta + \delta'$ only

$$2F = \psi^2 \int \left[\sin^2\Theta\cos^2\Theta M^{-1}\left(\frac{\partial\gamma}{\partial z} - q_o\right)^2 - 2k_o|C|\sin\Theta\cos\theta\cos\gamma\right] dz \qquad (21)$$

This problem is again isomorphous to the cholesteric to nematic transition with $K_{22} = 4\sin^2\Theta\cos^2\Theta/M$, $\chi_a H^2 = \Pi^2 q_o^2 \sin\theta\cos\theta/16M$.

When $|C|k > \Pi^2 q_o^2 \sin\theta\cos\theta/16M$, the stable solution is $\gamma=0$, the smectic A phase already described is the most stable.

When $|C|k < \Pi^2 q_o^2 \sin\Theta\cos\Theta/16M$ an instability is reached which leads to a new phase in which 2Π discommensurations are periodically piled up. The new period is again $P=8K(k)E(k)/\Pi q_o$ and appears experimentally as a phase modulation of the smectic layers. The threshold condition is favored close to $\Theta \simeq \Pi/4$: for a large enough elastic term, the mean field $S_A(d\simeq 1.3l)-S_A(d\simeq l)$ transition should be replaced by the sequence $S_A(d\simeq 1.3l)$ - incommensurate - $S_A(d\simeq l)$. In fact, a commensurate to incommensurate phase transition of the type described here has been observed in the S_mE phase of 4 n Octyl 4' n cyanoterphenyl by A. LEADBETTER |48|; the stiffening of the elastic constant (1/M) is probably responsible for this transition, which can explain that it occurs in the S_mE phase rather than in the simpler Smectic A.

We believe that the diffuse scattering observed in the low temperature S_A phase of the T8, by HARDOUIN and LEVELUT [36] and described as a phase modulation of the layers corresponds to thermally excited solitons or discommensurations of the type described above. They are the counterpart of the existence of two diffuse scattering regions in the reentrant nematic phase.

The picture could however be improved by taking account of the in plane variables x,y allowing thus for fluctuations at an angle with the z axis, as they show up experimentally.

Conclusion

The review of the current knowledge on the reentrant and the smectic A - smectic A phenomena , have led us to conjecture that two order parameters with characteristic incommensurate periodicities, were the key feature of these problems. We believe that we have been able to switch the question of understanding the physics of these two systems to the more formal one of understanding the statistical properties of a rather simple free energy functional. Although we have not been able to solve this problem rigorously, we think that we have provided arguments for the following points :

- the S_A-S_A' and reentrant phenomena are relevant to the same problem : when the pair length is 1.4, 1.3 times the molecular length, one has chances to observe the reentrant phenomenon. When the pair length is close to 2 times the molecular length, the S_A-S_A' possibility takes over. One case or the other is obtained depending on how much the dipoles are delocalized along the molecule (strongly localized $d \simeq 2l$; delocalized : core overlap $d \simeq 1.3l$)

- the reentrant behavior is linked to the mutual exclusion of two phases competing for condensation, via the fluctuating part of the fourth order coupling term. There is no direct relation between the layer spacing and the reentrant behavior, except for the fact that it can provide, with appropriate order of magnitudes, a signature for the existence of the two modes.

- there should always be a doubly reentrant phenomenon as reported in [34,35,36] (sequence N-S_A-N-S_A) only the prior occurrence of other transitions such as crystallization or S_A-S_B, may prevent the observation of the low temperature S_A phase.

One can understand when the second order Bragg scattering will be strong or not in the smectic A phases : smectic A' should intrinsically exhibit a strong harmonic (corresponding to the monolayer order in the fundamental bilayer matrix) whereas the regular Smectic A ($d \simeq l$ or even $d \simeq 1.3L$) have no reason for showing up a similar behavior.

We furthermore predict the existence of two commensurate to incommensurate transitions in smectic A phases, one of which have already been observed although in a smectic E [48]. Soliton like fluctuations in frustrated S_A phases are also predicted in agreement with recent observations [36,45,46].

Eventually, it is clear that a more refined analysis of the proposed free energy is needed. The main merit of our presentation is to provide a common framework for apparently quite different observations, its main drawback is its very crude account of the molecular level.

ACKNOWLEDGEMENTS

Many people have influenced my understanding of the above evoked problems in many ways, among them all the members of the Bordeaux Group. I am also quite gratefull to A.M. LEVELUT and A. LEADBETTER for explaining many of their results prior to publication, and to P. DELHAES for introducing me to the litterature on commensurability. Last but not least, it is a pleasure to acknowledge the many enriching discussions I had with F. HARDOUIN and C. COULON.

1- P.E. Cladis, Phys. Rev. Lett. 35, 48, 1975
2- K.K. Kobayashi, Phys. Letters 31A, 125, 1970;J.Phys. Japan 29,101,1970
3- W.L. Mc Millan, Phys. Rev. A4, 1238, 1971
4- P.E. Cladis, R.K. Bogardus, W.B. Daniels, G.N. Taylor, Phys. Rev. Lett. 39, 720, 1977
5- L. Liebert, W.B. Daniels, J. de Phys. 38, L-333 1977
6- in the case of a mixture see : E.P. Raynes, R.D. Holden, Bordeaux Int. Conf. (1978) ; for the first observation of thermotropic reentrant behavior at atmospheric pressure see ref. |34|
7- C. Destrade, J. Malthete, Nguyen Huu Tinh, H. Gasparoux, to be published
8- A.C. Anderson, W. Reese, J.C. Wheatley, Phys. Rev. 130, 1644, 1963 and references therein
9- E. Müller-Hartmann, J. Zittartz, Phys. Rev. Lett. 26, 428, 1971

10- G. Riblet, K. Winzer, Solid State Comm. 9, 1663, 1971
For a review see:P.Schlottmann, J. of low temps physics 20, 123, 1975
11- A. Zipp; W. Kauzmann, Biochem. 12, 4217 1973
12- S.A. Hawley, Biochem. 10, 2436, 1971
13- V.T. Rajan, C.W. Woo, Physics Letters 73A, 224, 1979
14- G. Sigaud, F. Hardouin, M.F. Achard, H. Gasparoux, J. de Physique, Colloq 40, C3-356 , 1979
15- N.A. Clark, J. Physique, Colloq. 40, C3-345, 1979
16- P.G. De Gennes, Solid State Commun. 10, 753, 1972
17- P.S. Pershan, J. Prost, J. de Physique 40, L 27, 1979
18- For instance : K.G. Wilson,Kogut J., Physics Reports 12C, 75,199,1975
See in particular ; pages 107-111
19- The Bell Laboratories group seems to have the equipment required for such an experiment as pointed out to us by P.E. Cladis
20- F. Hardouin, A.M. Levelut, J. Benattar J.J., G. Sigaud, Sol. State Commun. 33, 337, 1980
21- R.B. Meyer, T.C. Lubensky, Phys. Rev. A14, 2307, 1976
22- C.P. Bean, D.S. Rodbell, Phys. Rev. 126, 104, 1962
23- J. Prost, J. Physique 40, 581, 1979. The theoretical existence of a S_A-S_A transition has been confirmed by a statistical molecular treament : G. Vergoten, B.W. Van der Meer, Physica A 99, 237, 1979
24- G. Toulouse ., M. Kleman, J. de Physique 37, L-149, 1976
25- P.E. Cladis, R.K. Bogardus, D. Aadsen, Phys. Rev. A 18, 2292, 1978
26- P.G. De Gennes, "The physics of liquid crystals" Clarendon
27- F. Hardouin, A.M. Levelut, J.J. Benattar, G. Sigaud, these proceedings, p. 147 see also ref. [20]
28- A. Leadbetter, Heptyl cyanocyclohexylcyclohexane X-ray pattern, private communication
29- D. Guillon, P.E. Cladis, J.Stamatoff, Phys. Rev. Lett. 41, 1598, 1978
30- D. Guillon, P.E. Cladis, D. Aadsen, W.B. Daniels, Phys. Rev. appearing 1980
31- R. Shashidar, K.V. Rao, Bangalore Int. Liquid. Cryst. Conf., 1979
32- S. Chandrasekhar, K.A. Suresh, K.V. Rao, Bangalore Int. Liq. Cryst. Conf. , 1979
33 - B. Engelen, G. Heppke, R. Hopf and F. Schneider, Mol. Cryst. Liq. Cryst. 49 193, 1979
34 - F. Hardouin, G. Sigaud, M.F. Achard, H. Gasparoux, Phys. Lett. 71A, 347,1979
35 - F. Hardouin, G. Sigaud, M.F. Achard, H. Gasparoux, Solid State Commun. 30, 26 1979
36 - F. Hardouin, A.M. Levelut, J. de Physique 41, 1980
37 - A.D. Leadbetter, J.L.A. Durrant, M. Rugnan, Mol. Cryst. Liq. Cryst. 34, 231, 1977. One can observe two distinct diffuse scattering spots in the nematic phase of the octylcyanoterphenyl (A. Leadbetter, private communication)
38 - F.C. Frank, J.H. Van der Merwe,Proc. Roy. Soc. London A. 198, 216, 1949 ; A 200, 125, 1949
39 - P.G. De Gennes, Solid State Commun. 6, 163, 1968
40 - R.D. Meyer, Appl. Phys. Lett. 14, 208, 1968
41 - W.L. Mc Millan, Phys. Rev. B 12, 1187, 1975 ; Phys. Rev. B 14, 1496, 1976
42 - Y. Okwanoto, H. Takayama, H. Shiba, J. of Phys. Soc. Japan 46, 1420, 1979

43 - For a general review see for example : J.P. Pouget, Phase transformation in Solids , Edit. de physique, (Aussois) 1978
44 - For instance see : H. Shiba, Y. Ishibashi, J. of Phys. Soc. Japan 44, 1592, 1978
45 - F. Hardouin, A.M. Levelut, these proceedings p. 154
46 - G. Toulouse, P. Pfeuty, "Introduction au Groupe de renormalisation" p. 129, P.U.G., France, 1975
47 - A. Leadbetter, M.A. Mazid, R.M. Richardson, Bangalore Int. Conf., 1979
48 - A. Leadbetter, Private communication

Experimental Evidence of Monolayers and Bilayers in Smectics

P. Seurin, D. Guillon, and A. Skoulios

Centre de Recherches sur les Macromolécules,
6, rue Boussingault, F-67083 Strasbourg-Cédex, France

All even terms (6<n<18) of the homologous series of

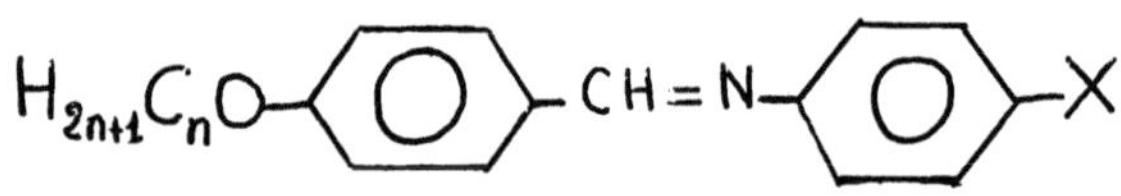

have been synthesized for X=F, Cl, Br, I and CN. Optical microscopy and isomorphy have been used to identify the mesophases occurring as a function of temperature. While the cyano series exhibits only smectic A the halogenated series exhibit both B and A smectic phases.

Dilatometric experiments have then been performed. They (i) confirmed the melting of the paraffin tails when the liquid-crystalline phases appear on heating and (ii) they allowed for the measurement of the molar volumes of the aliphatic and aromatic parts of the molecules.

Finally, X-ray diffraction studies of the structure of the smectic phases observed have been carried out. In the case of the halogenated derivatives, the aromatic sublayer corresponds to a single-layer of interdigitated aromatic stems oriented presumably perpendicular to the planes. On the contrary, in the case of the cyano-compounds, the aromatic sublayer corresponds to a double layer of aromatic stems highly tilted on the planes, the uniaxial character being due to rotational disorder.

To be submitted to Molecular Crystals and Liquid Crystals

X-Ray Investigations of the Smectic A_1 - Smectic A_2 Transition *

F. Hardouin[1], A.M. Levelut[2], J.J. Benattar[2], and G. Sigaud[1]

[1] Centre de Recherche Paul Pacal, Université de Bordeaux I, F-33405 Talence, France

[2] Laboratoire de Physique des Solides, Université Paris-Sud F-91405 Orsay, France

The existence in a binary system constituted of a three phenyl ring cyano compound mixed with TBBA (p=1atm) of a smectic A-smectic A transition line has been reported [1]. By microscopic observation, it is impossible to detect this S_A-S_A transition because the two smectic A, called S_{A_1} and S_{A_2} are perfectly isomorphous. In fact,this line was first revealed by means of Differential Scanning Calorimetry and then by diamagnetic susceptibility measurements [2].

Using X-ray diffraction technique,we find that the 001 Bragg diffraction spots observed for the smectic A_1 correspond to a layer spacing d close to the molecular length l(d/l $\sim$ 0.97).In the case of the smectic A_2,the X-ray pattern exhibits undoubtedly the characteristics of a smectic A. Nevertheless, in contrast with the S_{A_1}, the layer spacing d is doubled and equals practically to two molecular lengths (d/l $\sim$ 1.95) and we find that the second order reflection intensity I(002) is of same order of magnitude as I(001). Thus, a phase transition between two smectic A phases is confirmed by X-ray study and from the higher temperature phase S_{A_1} to the lower temperature phase S_{A_2} the long range layer periodicity is doubled. This result supports the theoretical arguments proposed by J. Prost to describe this transition [3].

In addition, in smectic A_1, the analysis of diffuse scattering intensities at small diffraction angles indicates the existence of pretransitional effects when we approach the smectic A_2 and also suggests an analogy between this transition and a para-antiferroelectric one.

At last,we are performing additional X-ray measurements on a new "S_A-S_A" binary system constituted with cyano derivatives. In the S_{A_1} phase the S_{A_2} fluctuations are centred around values incommensurate with the long range S_{A_1} periodicity (it is not the case of the system including TBBA). At the S_{A_1}-S_{A_2} transition and in the S_{A_2} phase the two periodicities lock themselves leading to commensurate Bragg reflections. To conclude as discussed elsewhere by J. Prost [4],these unusual anomalies of layer periodicities seem to be the fundamental common factor connecting the S_A-S_A transition to the reentrant nematic and reentrant S_A phenomenon [5].

1- G. Sigaud, F. Hardouin, M.F. Achard, H. Gasparoux,J. de Phys. 40,C3-356,1979
2- G. Sigaud, F. Hardouin, M.F. Achard, Phys. Lett. 72 A, 24, 1979
3- J. Prost, J. de Phys. 40, 581, 1979
4- J. Prost, these proceedings p. 125
5- F. Hardouin, A.M. Levelut, J. de Phys. 41, 41, 1980

---

x published in Solid State Communications 33, 337, 1980

Electron Spin Resonance: Structure of C.B.O.O.A. and B.O.B.O.A.[1]

F. Barbarin, E. Boulet, J.P. Chausse, C. Fabre[1], and J.P. Germain

Laboratoire d'Electronique et Résonance Magnétique,
F-63170 Aubière, France, E.R.A. 90 du C.N.R.S.

[1] Permanent address: G.R. n°12 du C.N.R.S.,
Laboratoire de Chimie Organique,
2 rue Henri Dunant, F-92320 Thiais, France

Introduction

In order to study the local order parameter in C.B.O.O.A. and in B.O.B.O.A. [408] E.S.R. measurements are performed using three nitroxide probes chosen, as follows :

- probes I :(3 spiro- [2' N-oxyl-3',3' dimethyloxazolidine] 5-α androstane-17- β-ol) and II : perdeuterated (2,2',6,6' tetramethyl 4- [p-ethyloxy] benzolamino piperidine 1-oxyl) both labeled on their rigid "core", give information about the structures and the motions of the "core" of the molecules of the mesophase .
- probe III : (2- [4-(8 oxyazolidinyl-N-oxy) hexadecyl oxybiphenyl] - quinoxaline) labeled on the lateral chain gives a knowledge about the aliphatic chain motions.

Structures and motions of the aliphatic chains

The results are obtained from probe III . The order parameter associated with the nitroxide group decreases at the $N \rightarrow S_A$ transition, in C.B.O.O.A. as well as in B.O.B.O.A. . In the S_A phase the nitroxide environment is really a chain one while in the N phase it is both a "core" and a "chain" environment because of the translational diffusion . Moreover, in C.B.O.O.A., as well as in B.O.B.O.A., the mean direction of the aliphatic chain becomes more colinear with the molecular "core" in the S_A phase than in the N one . In the S_A phase the lateral free area for chains decreases . In B.O.B.O.A. no change in the mean conformation of aliphatic chains is observed when one goes through the $S_A \rightarrow S_B$ transition . But an increase of the chain order parameter is measured at this transition .

Structure and motions of the molecular "core"

Experimental results show that the diffusion axis of the probe I lies approximatively along the director in each of C.B.O.O.A. and B.O.B.O.A. . No decrease of the order parameter of the nitroxide group is observed at the $N \rightarrow S_A$ transitions . The motion of this group is governed by the lateral orientational interactions of the "cores" of the solvant that increase at the $N \rightarrow S_A$ transition . A large discontinuity of the order parameter is observed at the $S_A \rightarrow S_B$ transition and in this phase a very good orientation of the "cores" is observed ($S \sim 0.9$) .

[1] Final publication submitted to Molecular Crystal and Liquid Crystals .

Influence of a Smectic Phase on Thermodynamical Behaviour of the Nematic to Isotropic Phase Transition

M.F. Achard, G. Sigaud, F. Hardouin

Centre de Recherche Paul Pascal, Université de Bordeaux I,
F-33405 Talence, France

1. Introduction

From an experimental point of view it does not seem to exist a straightforward relationship between the transition entropy ΔS_{NI} and the orientational order parameter discontinuity $\Delta\eta_{NI}$ at the nematic-isotropic transition even if, for example, the even-odd effect simultaneously arises on the transition temperature and on these both properties. Although a simple Landau theory (1) predicts a $\Delta S_{NI} \sim (\Delta\eta)^2_{NI}$ variation, this proportionality failed in several cases and these experimental discrepancies lead to put forward different explanations :

- recently, within nematic homologous series F. Leenhouts et al. (2) have shown that with long aliphatic chains the difference in population of chain conformations between the nematic and isotropic phase enhances the nematic-isotropic transition entropy without influence on the order parameter η.

- besides this fact, within series exhibiting a nematic-smectic polymorphism ΔS_{NI} increases anomalously when appears at lower temperature a smectic phase (3,4). The existence of smectic short-range ordering contributes to the N-I transition entropy and does not disturb in the same extent the orientational order (5,6).

Referring to earlier works (7,8,9), the cybotatic smectic groups (i.e. smectic fluctuations in the nematic or even in the isotropic phases) also seem to play a role in the pretransitional behaviour in the isotropic phase. In addition, in Schiff's bases series (9,10) we note that generally the more first order the nematic-smectic transition is, the stronger the anisotropic intermolecular correlations near nematic-isotropic transition are. This last point leads us to study the influence of the nematic-smectic transition on the NI transition entropy in binary mixtures, in particular if the NS excess entropy decreases due to a more second order character of the NS transition but also to diminutions of the critical heat capacity.

2. Experimental Results

We report here results concerning some binary systems with nematic-smectic A or nematic-smectic C transition lines giving a triple point :

- In all these diagrams the shape of the nematic-isotropic phase boundary indicates a monotonous evolution of the nematic-isotropic transition temperature T_{NI} (fig. 1a, 2a, 3a, 4a).

For short the different compounds which appear in the various diagrams are labelled as follows :

- 7 one = 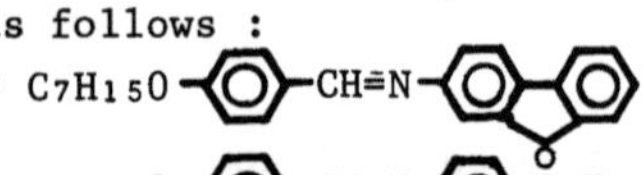
- 804 = 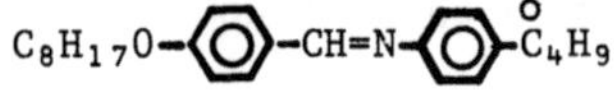
- 8OCB = 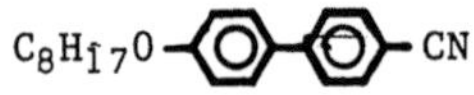
- Cno = cholesteryl nonanoate
- DB5 = 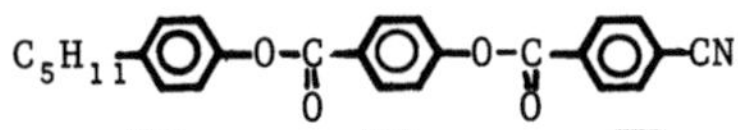
- TBBA = 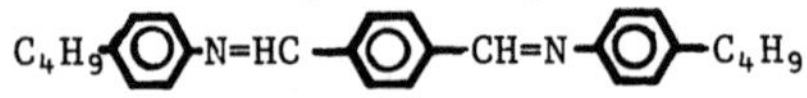

- Transition enthalpies and entropies have been systematically determined using a DUPONT 990 DSC and under the same experimental conditions (temperature program rate : 5°C min^{-1}, sample weight : 7 mg) for all binary mixtures; our method of determination of enthalpies takes in account the whole energy of transformation.

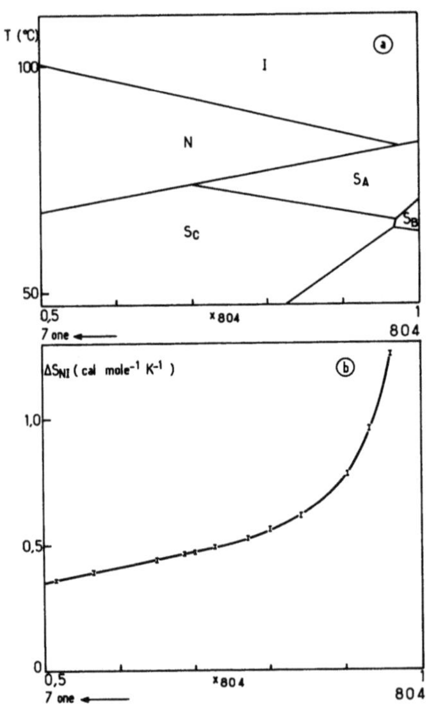

Fig. 1: 1a) Binary isobaric (1 atm) diagram between 804 and 7 one 1b) Variation of the nematic-isotropic transition entropy versus the molar fraction of 804 in mixture with 7 one (A similar figure caption applies to figures 1a,b-3a,b and 4a,b).

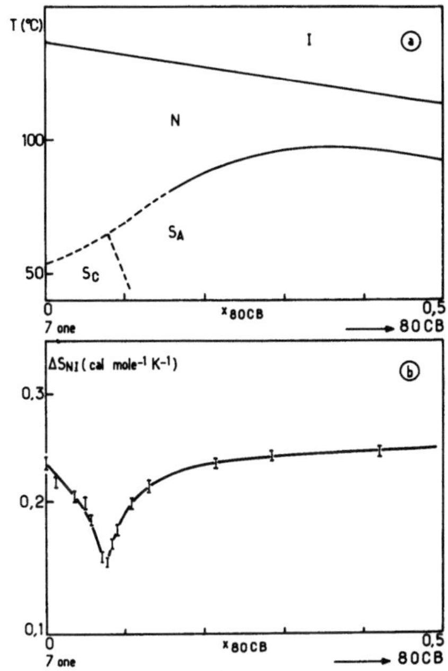

Fig 2

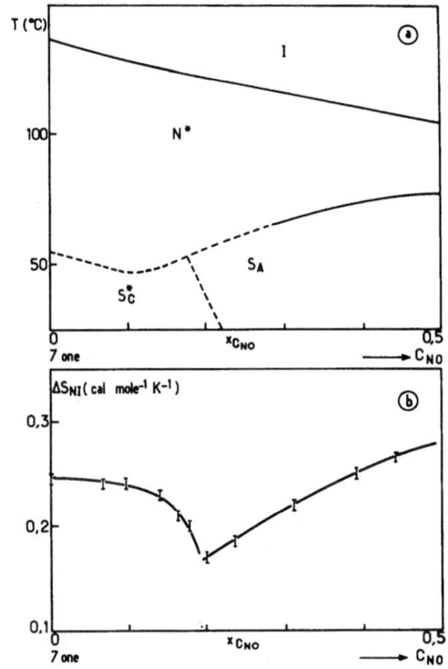

Fig. 3 : 3a) * indicates twisted nematic and smectic C phases

- As depicted in figure 1b, in most binary systems a regular evolution of ΔS_{NI} with respect to T_{NI} is observed. We note that, in the "7one-804" binary system the increase of ΔS_{NI} (fig. 1b) in the vicinity of pure 804 coincides with the strong increase of ΔS_{NA} entropy and with the simultaneous diminution of the nematic domain up to give a smectic A-isotropic transition. This confirms the influence of the smectic ordering stability on ΔS_{NI} as we have above noticed for homologous series.
For the other three systems, that we have yet studied by another way (11,12, 13,14), the NI transition entropy exhibits an obvious anomaly (fig. 2b,3b, 4b) which in any case cannot be included in the experimental uncertainties: we observe an abrupt decrease of ΔS_{NI} for mixtures the molar fractions of which are in the vicinity of the concentration of the triple point NS_AS_C (fig. 2a), $N{*}S_AS_C{*}$ (fig. 3a), N $S_{A1}S_{A2}$ (fig. 4a). As a matter of fact, contrary to the binary system "7one-804" for which we observe a regular evolution of the nematic-smectic transition entropy on both sides of the NS_AS_C triple point, these last triple points are singular. Indeed, they correspond to the intersection of two nematic-smectic lines along which we have found that the transition entropy decreases regularly when we approach the point (11,12,13). For the "7one-80CB" and "7one-Cno" systems these transition entropies tend to zero giving a polycritical point. Thus, for binary mixtures we can observe a ΔS_{NI} minimum depending intrinsically on nematic-

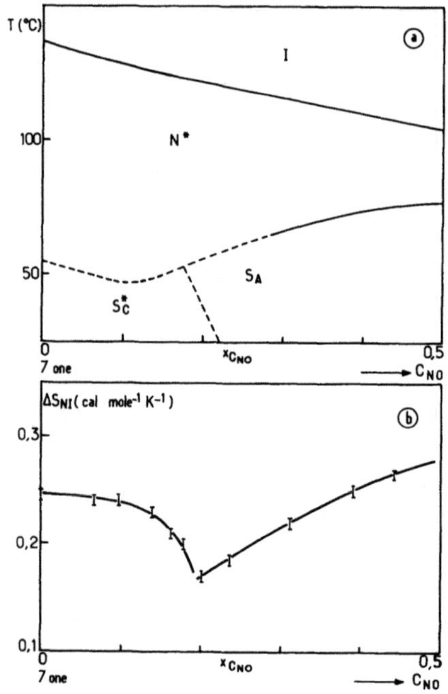

Fig. 4: 4a) A_1 and A_2 distinguish two types of smectic A of different layer spacings (respectively one molecular length and two molecular lengths)

smectic-smectic singular triple point, in spite of a regular evolution of T_{NI}.

3. Conclusion

To conclude, we give evidence by our investigations on binary systems of the influence of the evolution of the thermodynamical behaviour of the lower temperature nematic-smectic transition on the NI transition entropy : more the nematic-smectic transition weakly first-order is, weaker the additional contribution to ΔS_{NI} of the smectic short-range ordering is. This suggests an important coupling between these two parameters.

At last the values of the nematic-smectic entropy are very small close to polycritical points (11,12,13) and sometimes unobservable for N-S_A transition in our experimental limit (11,12), nevertheless the ΔS_{NI} entropy is markedly influenced. This can mean that the low temperature smectic phase effect is not reduced to a linear coupling between ΔS_{NI} and ΔS_{NS}, and that the shrinking of specific heat anomalies which can be associate to the decrease of the N-S excess entropy influences also the magnitude of the NI entropy.

References

1 - L.D. Landau and E.M. Lifshitz, Statistical Physics, Pergamon Press Ltd. London (1958)

2 - F. Leenhouts, W.H. De Jeu, A.J. Dekker, J. de Phys. 40,989 (1979)

3 - H. Arnold, Z. Chem. 4, 211 (1964)

4 - J. Van Der Veen, W.H. De Jeu, M.W.M. Wanninkhof, C.A.M. Tienhoven, J. Phys. Chem. 77, 2153 (1973)

5 - W.H. De Jeu, W.A.P. Claassen, J. Chem. Phys. 68, 102 (1978)

6 - F. Hardouin, M.F. Achard, G. Sigaud, H. Gasparoux, Mol. Cryst. Liq. Cryst. 39, 241 (1977)

7 - H.J. Coles, Mol. Cryst. Liq. Cryst. Lett. 49, 67 (1978)

8 - H.J. Coles, C. Strazielle, Mol. Cryst. Liq. Cryst. Lett. 49, 259 (1979)

9 - A. Gohin, Thesis, Bordeaux (1979)

10- G.W. Smith, Z.G. Gardlund, J. Chem. Phys. 59, 3214 (1973)

11- G. Sigaud, F. Hardouin, M.F. Achard, Solid State Commun. 23, 35 (1977)

12- F. Hardouin, G. Sigaud, M.F. Achard, H. Gasparoux, Ann. Phys. 3, 381 (1978)

13- G. Sigaud, F. Hardouin, M.F. Achard, H. Gasparoux, J. de Phys. 40, C3-356 (1979)

14- G. Sigaud, F. Hardouin, M.F. Achard, Phys. Lett. 72 A, 24 (1979)

X-Ray Studies of Reentrant Nematic and Smectic A Phases in Pure Compounds at Atmospheric Pressure

F. Hardouin[1] and A.M. Levelut

Laboratoire de Physique des Solides, Université Paris-Sud, Bât.510, F-91405 Orsay, France

[1] Permanent address: Centre de Recherche Paul Pascal, Université de Bordeaux I, F-33405 Talence, France

In the 4-N-alkyloxybenzoyloxy-4'-cyanostilbenes series for the n=8 compound has been reported the evidence of the original tetramorphism (1) :

$N \xleftrightarrow{247°C} S_A \xleftrightarrow{139°C}$ Reentrant Nematic $\xleftrightarrow{96°C}$ Reentrant Smectic A

It is the first example in condensed matter physics of a system with two reentrant phases. Recently we have published a detailed structural X-ray study on this long cyano derivative (2). Following these results other similar compounds exhibiting a reentrant polymorphism have been investigated (3).The 4-N-octanoate benzoyloxy - 4'-cyanostilbene is the last material known with two reentrant phases (4) : 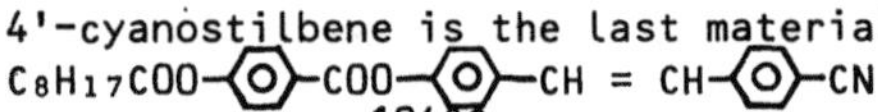

$N \xleftrightarrow{214°C} S_A \xleftrightarrow{177°C}$ Reentrant Nematic $\xleftrightarrow{104°C}$ Reentrant Smectic A

We intend here to sum up a comparison between the structural behaviours of the two compounds above mentioned. In both cases throughout the higher temperature S_A phase the layer spacing d is larger than the molecular length L. Nevertheless, for the octanoate compound no appreciable change in d with temperature is found in this phase ($d/L \simeq 1,27$) whereas for the octyloxy compound we had observed that the layer spacing decreased with decreasing temperature ($1,10 < d/L < 1,25$). On the other hand, with the octanoate compound we confirm the "monolayer" structure of the reentrant smectic A phase ($d \simeq L$ with no apparent thermal dependence). Concerning the short range order,in the whole reentrant nematic range of the octanoate compound, additional singular diffuse lines appear distinctly (in contrast with the octyloxy compound) and at low temperature the inner diffuse line split into two spots. In the reentrant S_A phase these spots are indicative of a two dimensional superlattice periodicity connected to the fundamental "monolayer" S_A periodicity (as for the octyloxy compound near the recrystallization). Finally, we confirm that the doubly reentrant nematic and smectic A phenomenon is a consequence of the competition between two different kinds of S_A arrangement (5).

1 - F. Hardouin, G. Sigaud, M.F. Achard, H. Gasparoux, Phys. Lett. 71A, 347, 1979 ; Solid State Commun. 30, 265, 1979
2 - F. Hardouin, A.M. Levelut, J. de Phys. 41, 41, 1980
3 - G. Sigaud, Nguyen Huu Tinh, F. Hardouin, H. Gasparoux, these proceedings, p. 155 and submitted to J. Chem. Phys.
4 - Nguyen Huu TInh, M. Joussot-Dubien, C. Destrade, Submitted to Mol. Cryst. Liq. Cryst. Lett.
5 - J. Prost, these proceedings, p. 125

Occurrence of Reentrant Nematic and Reentrant Smectic A Phases in Some Liquid Crystal Series

G. Sigaud, Nguyen Huu Tinh, F. Hardouin, and H. Gasparoux

Centre de Recherche Paul Pascal, Université de Bordeaux I,
F-33405 Talence, France

The first pure compound exhibiting at atmospheric pressure a stable reentrant nematic but also a reentrant smectic A phase was discovered a few months ago [1].The nature of these phases was confirmed in particular by miscibility diagrams [2],magnetic measurements [2] and X-ray experiments [3].From the squeletton of this previous molecule, the octyloxybenzoyloxy-cyanostilbene,we have synthetized numerous other compounds with three phenyl rings.A comparative study of these liquid crystals series (including also some materials synthesized in other laboratories) in which the molecular chemical structure is not randomly modified but keep some common features with the original compound has been undertaken. The properties of these rather similar substances can be very different since in some cases the reentrant phenomenon is ruled out.In this manner, we can define some essential characteristics for families apt to give reentrant phases at 1 atm :

- the cyano end group is essential.This seems to be the consequence of a strong dipole connected with relatively small dimensions.
- the electronic distribution along the molecule plays also an important role as schematized : 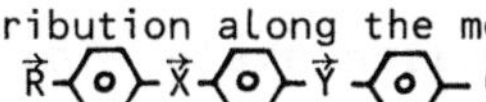 (→sense of dipoles),the absence of donor group, a linkage which stops the electronic transfert throughout the core, a strong transverse dipole withdraws the reentrant phases.
- an alkoxy chain is more favourable to the reentrance than an alkyl chain for which no reentrant smectic A phase is observed.
- the comparison of the various series allows to classify the different Y linkages (with a constant X=COO linkage) with regard to their ability to induce reentrant phases: CH=CH > N=N, C≡C > CH=N,nothing > COO. The role of X=COO linkage remains to be specified.

For short, we give some arguments in favour of the reentrant phases with regard to the molecular structure. But other phenomena, such as the existence of unusual S_{A_2} phase or ordered smectic phases are to be included in a more general comparative study of properties and molecular structure on these fruitful long cyano compounds.

1 - F. Hardouin, G. Sigaud, M.F. Achard, H. Gasparoux, Phys. Lett. 71A, 347, 1979

2 - F. Hardouin, G. Sigaud, M.F. Achard, H. Gasparoux, Solid State Commun. 30, 265, 1979

3 - F. Hardouin, A.M. Levelut, J. Phys. 41, 41, 1980

*Submitted for publication to J. Chem. Phys.

NMR Proton Relaxation Investigation of the Nematic-Isotropic Phase Transition on Homologues of the PAA Series

W. Wölfel, V. Graf, and F. Noack

Physikalisches Institut der Universität,
Pfaffenwaldring 57, 7 Stuttgart 80, Federal Republic of Germany

It is by now well-established that proton spin T_1 relaxation in both the nematic and isotropic phase of liquid crystals is mainly determined by director order fluctuations (OFD) and by self-diffusion (SD) [1]. However, in the vicinity of the nematic to isotropic phase transition temperature the relaxation process shows a more complex behaviour and is less well understood. Evidently there exists another T_1 mechanism! We have performed extensive measurements of the Larmor frequency and temperature dependence of the proton T_1 for several homologues in the PAA series (PAA, PAP, PAB, HAB) below the transition points, to find out the properties of the addidional 'third' relaxation rate in the nematic state. The third contribution was found to depend critically upon the temperature, whereas its connection with Larmor frequency and molecular chain length proved rather small. Deuteration of PAA considerably enhanced its magnitude. The experimental data cannot be described satisfactorily in terms of molecular rotations in the nematic potential as suggested by MARTINS [2], but are in agreement with the idea of critical fluctuations of the scalar order parameter S (OFS) developed by FREED [3] and recently analyzed quantitatively by GRAF [4]. A theoretically predicted relationship between the OFD and OFS contributions is approximately verified experimentally. Details of this study will be presented in a forthcoming paper [5].

1. e.g. CH.G.WADE, Ann.Rev.Phys.Chem. 28,47 (1977)
2. A.F.MARTINS, Portugaliae Physica 8,1 (1972)
3. J.H.FREED, J.Chem.Phys. 66,4183 (1977)
4. V.GRAF, Dissertation Universität Stuttgart (1979)
5. G.NAGEL, W.WÖLFEL, V.GRAF, and F.NOACK, in preparation

Pulse Acousto-Optic Modulator Using a Nematic Liquid Crystal in Its Isotropic Phase*

P. Martinoty and M. Bader

Laboratoire d'Acoustique Moléculaire, E.R.A. au C.N.R.S.,
Université Louis Pasteur,
4, rue Blaise Pascal, F-67070 Strasbourg Cédex, France

In simple liquids the birefringence induced by ultrasonic waves is quite small. However, the birefringence can be considerably higher for liquids exhibiting a strong local order such as liquid crystals in their isotropic phase.

Observations in the isotropic phase of pentylcyanobiphenyl (PCB) shows that a large birefringence can be obtained with a very low acoustic field. The birefringence is proportional to the square root of the acoustic intensity and increases as one approaches T_c. These results show that, besides their interest as electro-optic materials, liquid crystals in their isotropic phase are also very interesting as acousto-optic materials.

* to be published in Applied Physics Letters.

Part V

Cholesterics and Electrooptical Applications of Nonnematics

Experimental Results and Problems Concerning „Blue Phases"

H. Stegemeyer and K. Bergmann

Department of Physical Chemistry, University of Paderborn
4790 Paderborn, Federal Republic of Germany

1. Introduction

Most of the cholesteric mesophases formed by esters of cholesterol exhibit selective reflection of circularly polarized light in the u. v. region at temperatures close below the clearing point /1/. The well known reflection colours in the visible either occur near phase transitions to the smectic phase or in so-called compensated mixtures /1/. However, cooling down cholesteric esters only a few tenths of a Kelvin below the clearing point bright colours appear (mostly blue colours) which disappear on further cooling /2/. As early as in 1906 LEHMANN /3/ noted that these colours belong to an optically isotropic modification. In his opinion this modification represents a stable phase distinct from that of the ordinary cholesteric one.

Since the reinvestigation of this modification beginning in 1956 with a paper of GRAY /4/ the name "Blue Phase" has been used because of the striking light reflection effect. Sometimes, however, the term "blue apparition" has been preferred.

For many years there has been a confusing situation regarding the question of the nature of the blue phase (BP): Are BPs thermodynamically stable phases or nothing but special textures of cholesterics?

Different opinions of this problem which are mentioned in the literature originate from the fact that BPs may be supercooled easily for several degrees gradually transforming into the classical cholesteric texture. This effect of supercooling has not been observed in the case of all other liquid crystal phase transitions.

The first striking experiment to show the validity of the assumption of a distinct stable phase was performed by ARMITAGE and PRICE in 1975. By differential scanning calorimetry (DSC) they determined a finite value of transition enthalpy at the focal conic to BP transition /5/. Eight years before, however, as early as in 1967 BARRALL, PORTER, and JOHNSON reported results of differential thermal analysis of cholesteryl myristate (CM) indicating two forms of the cholesteric phase /6/. One year later, ARNOLD and ROEDIGER performed precise calorimetric measurements on cholesteryl esters and observed a distinct maximum in the $C_p(T)$ curve below the clearing point /7/. But they did not assign this effect with a transition to the BP.

Additionally, ARMITAGE and PRICE measured a very small change in density of about 0.004 per cent at the focal conic to BP transition /8/.

These results on one hand confirm the assumption of a stable phase existing between the isotropic liquid and the cholesteric state and show that the BP is only stable in a very narrow temperature range of less than 1 Kelvin. On the other hand these results indicate a discontinuous phase transition (1st order) between cholesterics and the coexisting BP. Because of the smallness of the effects, however, the structures of the two phases should not be too different. But the question of the molecular arrangement within the BP remained unsolved hitherto.

2. Optical Isotropy of BP

The optical isotropy of BPs has been demonstrated by the striking photographs in the paper of SAUPE /9/ showing bulk BP samples of cholesteryl p-nonylphenyl carbonate of high transparency. Quantitative measurements have been carried out by PELZL and SACKMANN /10/ and by DEMUS and co-workers /11/ conforming a zero birefringence in the BP state. Results obtained for cholesteryl nonanoate (CN) are shown in Fig. 1: Like in the isotropic liquid only one boundary of total reflection occurs in the BP state in the stable region as well as in the supercooled one. The very small discontinuity of refractive index at the clearing point is due to the density change /8/. This can be calculated by means of the LORENZ-LORENTZ equation /12/.

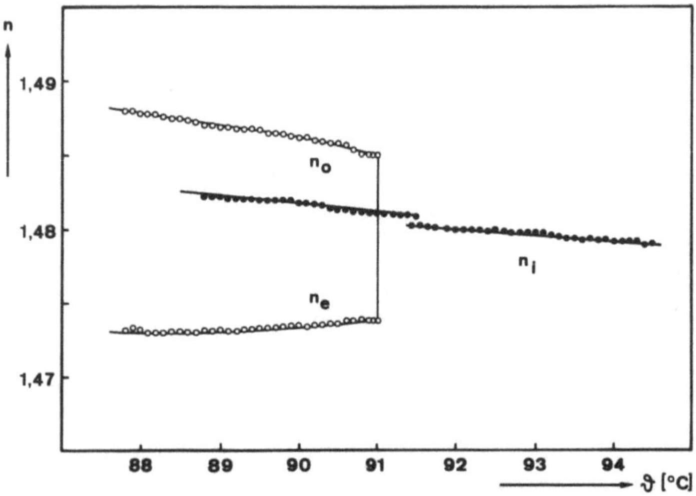

Fig. 1 Refractive indices of CN vs. temperature at λ = 589 nm
n_i: isotropic liquid state and BP
n_o, n_e: birefringent cholesteric state, ordinary and extraordinary ray
(o o) increasing temperature; (● ●) decreasing temperature
Solid line below 91.5°C: theoretical curve according to (4)

The curve of the refractive index of the BP in the supercooled state has been calculated from the data of the ordinary and extraordinary ray of the cholesteric state using the VUKS /13/ formula:

$$n_i^2 = 1/3\,(n_1^2 + n_2^2 + n_3^2) \tag{1}$$

Using the approximation $n_2 \approx n_3$ it follows for cholesterics

$$n_1^2 = 2n_o^2 - n_e^2 \tag{2}$$

$$n_2 = n_e \tag{3}$$

and

$$n_i = \left\{1/3\,(2n_o^2 + n_e^2)\right\}^{1/2} \tag{4}$$

This is shown by the solid line in Fig. 1. Consequently, n_i for the BP implies the values of n_j (j = 1, 2, 3) by equal statistical weight according to (1) /12/.

3. Light Reflection of BP

Our own work in the field of BPs started with the determination of the nature of light reflected from BPs. Registration of the spectral distribution has been performed by transmission scanning in a Cary 17.

In Fig. 2 the reflection curve of CN is shown in the supercooled BP state 0.4 K below the phase transition point exhibiting two completely separated reflection bands. The band at 350 nm can be assigned with the reflection of the classical cholesteric phase. The long wavelength band, however, occurs in a spectral region corresponding to the colours visually observed in the BP state of CN. The polarization of the reflected light has been determined

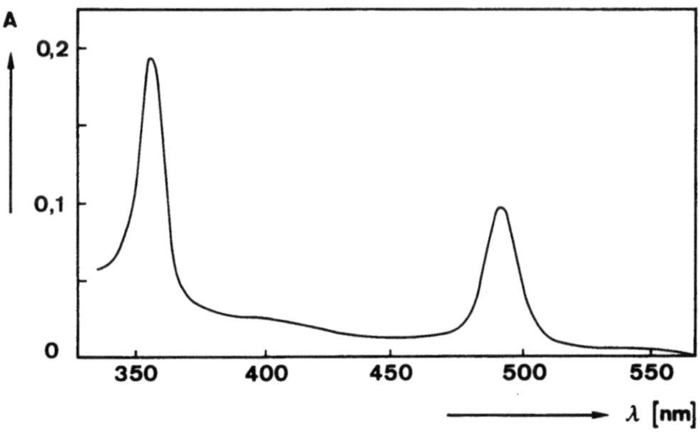

Fig. 2 Selective reflection of CN in the supercooled BP (90.6°C, 12 μm)

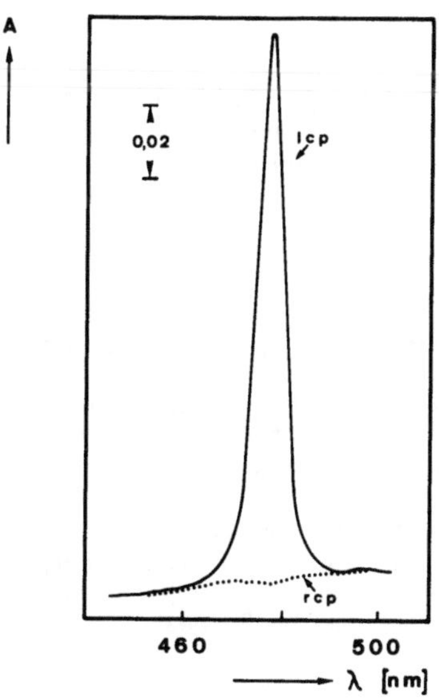

Fig. 3 Selective reflection of left (lcp) and right (rcp) circularly polarized light of CN in the BP (91.2°C, 12 μm)

by means of Polaroid circularly polarizing filters. As it can be derived from Fig. 3 the reflected light of the BP state of CN is left circularly polarized the same property as found in the corresponding cholesteric state of CN.

In cholesterics it is well known that the selective reflection of circularly polarized light is accompanied by an anomalous dispersion of optical rotatory power (ORD) within the same spectral region. Can we expect the same effect for the BP? Fig. 4 shows the ORD of the BP of CM in the supercooled state. As a parameter the observation time is given. One can see that the ORD effect of the cholesteric phase gradually comes out in the region of about 300 nm. In the blue spectral region a second effect is to be seen due to the supercooled BP of CM gradually disappearing on time.

From these preliminary results we learn that selective reflection and ORD of the BP is quite comparable with that of the cholesterics but red-shifted and of somewhat smaller intensity. The small half-width of the reflection band (cf. Fig. 3) should be emphasized as well as the fact that the sense of the reflected circularly polarized light is the same in the BP as in the cholesteric state. We will come back to these optical properties later on.

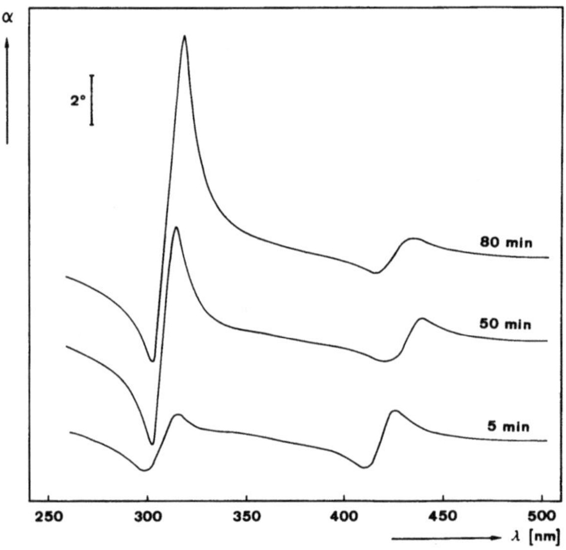

Fig. 4 Optical rotatory dispersion of CM in the supercooled BP (83.7°C, 12 μm)

4. Polymorphism of BP

In this section evidential results for the existence of *two* polymorphic forms of the BP are presented.

Reinvestigating the DSC measurements of ARMITAGE and PRICE /5/ we found a large peak at the clearing temperature and a small one for the chol/BP transition (Fig. 5) in the case of CM and CN. The values of transition enthalpies are given in Table 1.

The value for chol/BP transition is only 3 per cent of the clearing enthalpy /14/. - Surprisingly, in our highly resolved DSC thermograms using gold pans we observed a *third* peak partially hidden at the low temperature side of the clearing peak, about 0.3 to 0.4 K apart from the BP peak.

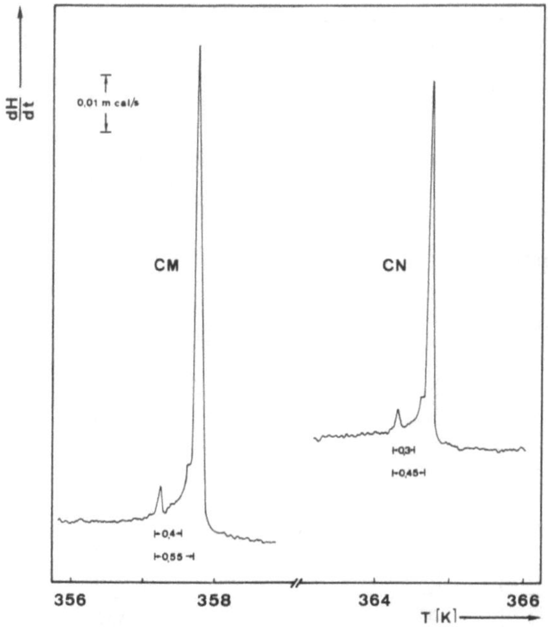

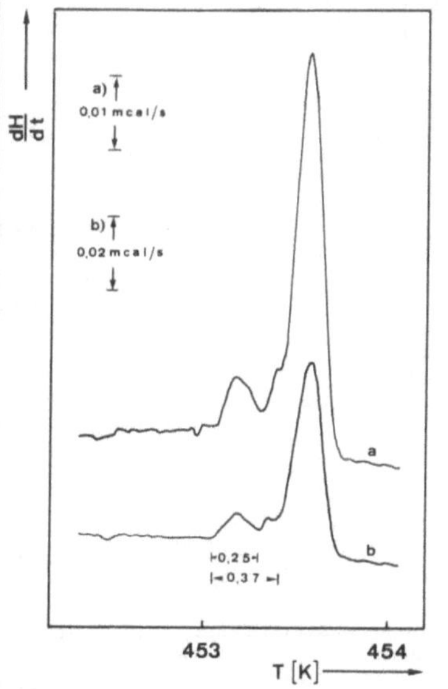

<u>Fig.5</u> DSC traces of CM (2.756 mg) and CN (2.919 mg); heating rate 0.625°/min

<u>Fig.6</u> DSC traces of CB. a) 2.54 mg, b) 1.86 mg; heating rate 0.625°/min

Similar results have been obtained for other cholesteric esters /15/ as shown in the Fig. 6 for benzoate (CB). The transition enthalpy resulting from the not completely resolved peaks is of the order of the first BP peaks (cf. Table 1).

<u>Table 1</u> Transition temperatures and enthalpies

		CH	BP I	BP II	I
CM	ϑ [°C]	84.0	84.4	84.55	
	ΔH [J/mole]	34	≈ 34	1100	
CN	ϑ [°C]	91.0	91.3	91.45	
	ΔH [J/mole]	17	≈ 17	530	

The previous DSC results of ARMITAGE and PRICE / 5 / show that an additional transition close below the clearing point has not been resolved by their experimental conditions. In their measurements of the expansion coefficient, however, one can see /8/ a *third* peak near the clearing temperature which may be due to an additional transition as found by our highly resolved DSC experiments /14/.

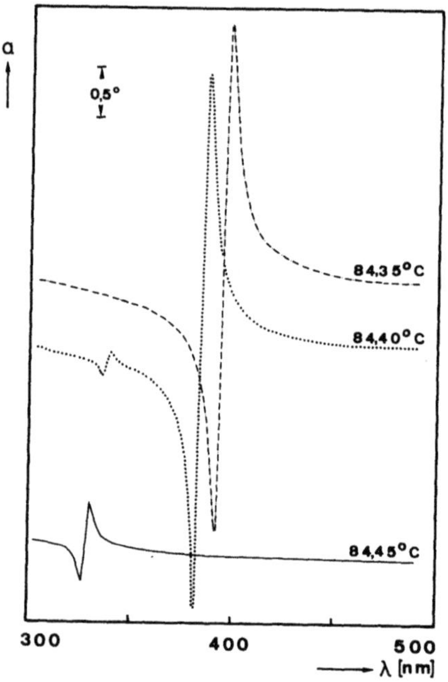

Fig. 7 ORD of CM

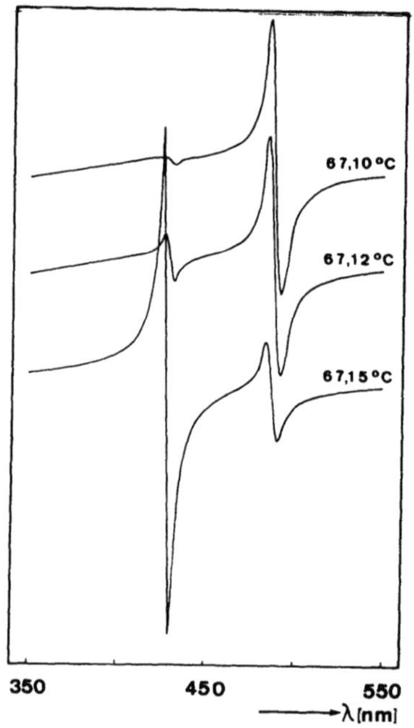

Fig. 8 ORD of CC

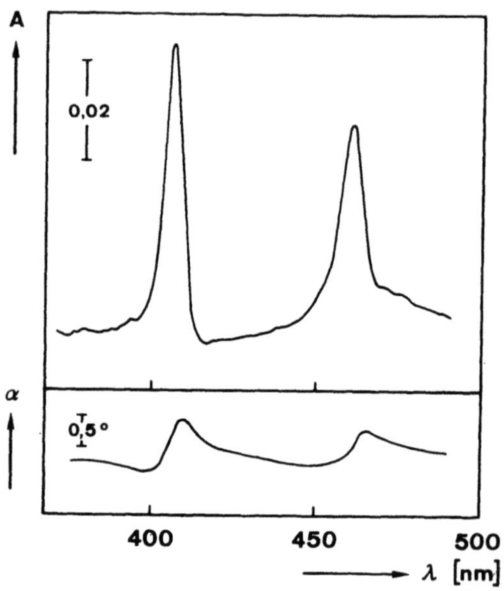

Fig.9 Figure caption see opposite page

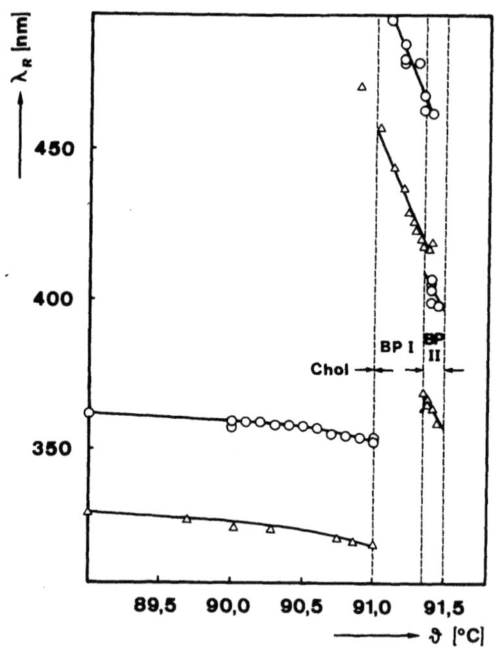

Fig.10 Figure caption see opposite page

To confirm the existence of another liquid crystalline phase close below the clearing point we reinvestigated the spectral properties of the BP as a function of temperature.

First the ORD of the BP of CM at different temperatures will be discussed in Fig. 7. At 84.35°C only one ORD effect is to be seen at about 400 nm. Slightly increasing the temperature additionally a second one appears at 330 nm. At still higher temperatures the amplitude of the second curve increases while that of the first one vanishes. It should be emphasized that all temperatures used in this experiment are above the phase transition to the cholesteric phase (no supercooling). A similar behaviour is shown in Fig. 8 for cholesteryl chloride (CC): On increasing temperature a second ORD curve appears at shorter wavelengths whereas the amplitude of the first one diminishes. Additionally, another result should be discussed by Fig. 8: The sign of the anomalous ORD curve of CC is reversed compared with that of the other cholesteryl esters (compare Fig. 7 for CM). This points out that right circularly polarized light is reflected by the BP of CC as it will be done from the corresponding cholesteric state which is known to possess a right-handed helical structure.

Returning to the problem of BP polymorphism we may see that our ORD results confirm the existence of nothing but *two* blue phases. The same result is derived from selective reflection measurements /16/: In Fig. 9 the occurrence of two separated reflection bands in the case of CN is shown at 91.35°C, a temperature where the two BPs coexist. In the lower part the corresponding ORD curves are given.

In Fig. 10 the temperature dependence of the wavelengths of reflection bands $\lambda_R(\vartheta)$ is shown for CN. At first, only the upper curves should be discussed obtained for normal incident light. Close below the clearing point reflection occurs at about 400 nm. At a lower temperature the wavelength is discontinuously shifted to 460 nm. On further cooling a continuous shift to longer wavelengths is observed which proceeds into the supercooled BP region. At 91°C the reflection band of the classical cholesteric phase occurs. This behaviour clearly confirms the assumption derived from the calorimetric measurements: actually *two* different polymorphic forms of BP exist. An analogous result has been observed for all other cholesteryl esters /15, 17/. Recently, MEIBOOM /18/ and COLLINGS /19/ reported optical measurements of cholesteryl esters confirming our results of two BPs.

5. Angular Dependence of the BP Selective Reflection

At this stage some comments should be given about the angular dependence of the BP selective reflection. This fact seems to be an important information regarding possible structures of BPs.

As well known the reflection bands of cholesterics are shifted to shorter wavelengths for oblique incidence /20, 21/:

$$\lambda_R^{\varphi} = \lambda_R^{0} \cos \varphi \qquad (5)$$

φ is the angle of incidence to the normal, λ_R^0 denotes the wavelength at normal incidence.

◄Fig.9 SR and ORD of CN at 91.35°C (coexistence temperature of BP I and BP II)

Fig.10 Temperature dependence of maximum wavelength of SR for CN.
Circles: normal incidence; Triangles: oblique incidence

In 1970 GOLDBERG AND SCHNUR /22/ mentioned that the colour of reflected light in BPs does not depend on the observation angle. Regarding the DE VRIES theory of cholesterics /20/ this statement seems to be in contradiction to the very small half-width of the BP reflection bands observed (cf. Fig. 3). Thus, we decided to reinvestigate the angular dependence.

By a simple experiment it can easily be demonstrated that the reflection colour of BPs *does* depend on the observation angle: Cooling down CN in a thermostatted capillary tube close below the clearing point a bright green colour appears if the sample is illuminated in the viewing direction (a so-called "green phase"). If the observation angle is increased to 90° against the direction of incident light, however, the colour changes to blue at constant temperature.

For quantitative determination of the angular dependence we measured the reflection by means of a Perkin-Elmer spectrofluorimeter MPF-4 /17/. The results for CN are presented in Fig. 10. For oblique incidence the reflection bands of all three phases cholesteric, BP I, and BP II are shifted to shorter wavelengths compared with that of normal incidence. The same result has been obtained for CM /17/.

Table 2 Angular dependence of BP selective reflection in CN

	Chol	BP I		BP II
ϑ [°C]	91	91	91.35	91.35
λ_R^0 [nm]	352.5	505.5	466	406.5
λ_R^φ [nm]	318	456.5	418	368.5
$\lambda_R^0/\lambda_R^\varphi$	1.108	1.107	1.115	1.103

From Table 2 it can be seen that the quotient of the wavelengths for normal and oblique incidence is always of the same value for the three phases under consideration (chol., BP I, BP II).

In contradiction to GOLDBERG and SCHNUR /22/ the angular dependence of BPs is analogous to that of cholesterics.

6. Pressure Dependence of BP Selective Reflection

It has been found by POLLMANN and SCHERER /23/ that the maximum wavelength of the BP reflection bands not only depends on temperature but additionally is very sensitive to pressure. This is demonstrated by Fig. 11 for CN.

At pressures below 81.5 bar no reflection occurs. Increasing the pressure sligthly at 82 bar a sharp reflection band was found at 400 nm which shifts to longer wavelengths on increasing pressure. At 84 bar a second band additionally appears at 465 nm. On further increased pressure the first band disappears whereas the new one remains. As increasing pressure is equivalent to decreasing temperature the pressure dependence shown here resembles the

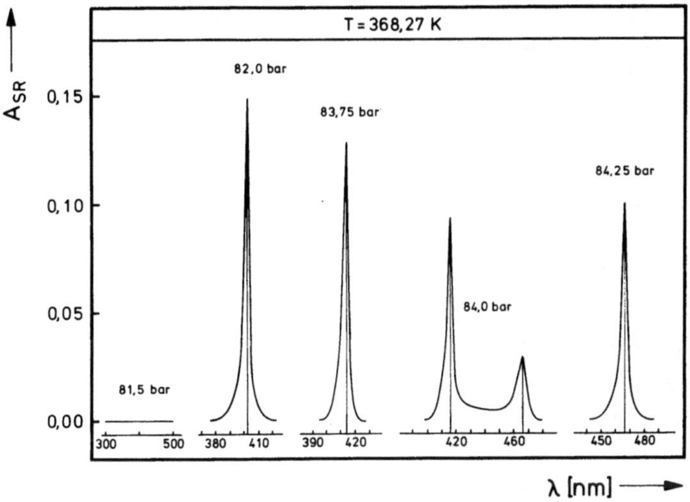

Fig. 11 Pressure dependence of selective reflection of CN at 95.12°C /23/

temperature effect discussed before: At low pressures BP II is stable. At a pressure of 84 bar both BPs coexist. At higher pressure only BP I is stable.

From these results POLLMANN and SCHERER /23/ derived a phase diagram of the BPs. The equilibrium curves between the cholesteric phase and BP I, between the two BPs, and between BP II and the isotropic liquid surprisingly are strictly parallel. The slope of the coexistence curve BP I/cholesteric has been found to be 22.1 bar/K. Considering the very small transition enthalpy /15/ (cf. Table 1) and the small volume change found by ARMITAGE /8/ one can verify the CLAUSIUS-CLAPEYRON equation to be satisfied by these experimental values.

7. Textures of BP

In this section some results regarding microscopic investigations of BPs are reported. In 1973 GRAY /24/ noted: "Whenever platelets are observed microscopically then the blue phase also exists and occurs over exactly the same temperature region". Such platelets are shown in Ref. /25/. For our experience, platelets are a special texture of the BP. CHISTIAKOV /26/ found that the platelet colour does not change on rotation of the stage indicating that the optical axis is parallel to the viewing direction. Rotation of the analyser alters the colour intensity indicating that the plane of polarization is rotated by the platelets. From these observations GRAY /24/ concluded:
i) platelets possess the structure of the classical cholesteric type,
ii) the orientation of the helical axes in different platelets differs with respect to the surface. Actually, we never could observe sharp reflection bands of the BP as shown before when the sample existed in form of a platelet texture. On mechanical disturbation the platelet texture transformed into a more unique texture. Only if samples of this texture have been used sharp reflection bands or ORD curves could be obtained (cf. Fig. 3 and 7). From this result we conclude that the helical axes now are approximately aligned parallel to each other in the whole sample.

This result must be seen in context with an observation made by KUCZYNSKI /36/. In a wedge-shaped sample of the BP he found GRANDJEAN-CANO lines. This

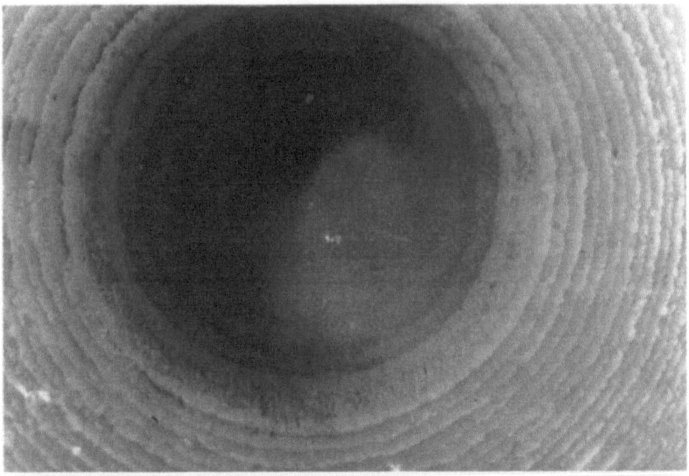

Fig. 12

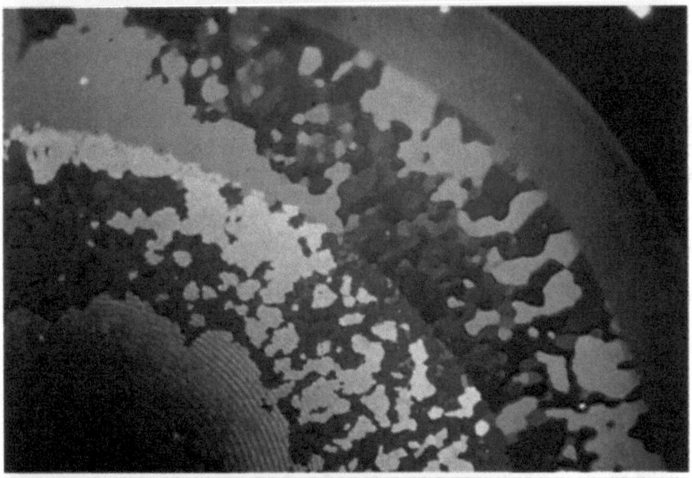

Fig. 13

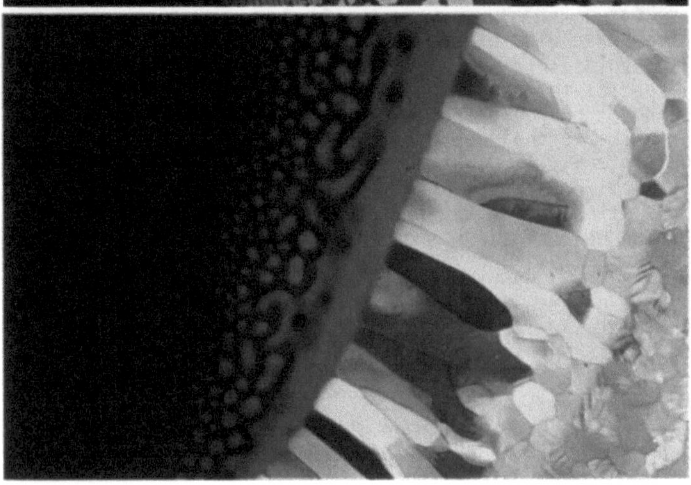

Fig. 14

Figs.12-14.
Captions see
opposite page

is demonstrated by Fig. 12 showing the behaviour of a mixture of CB/HOAB between a rubbed convex lense and a plane plate (HOAB: 4,4'-Diheptylazoxybenzene).

On increasing temperature the green colour changes more and more to the violet. Close below the clearing point the CANO texture is destroyed and partially substituted by a platelet texture (Fig. 13).

The existence of BPs is not restricted to derivatives of cholesterol. BPs also occur in chiral nematics /25, 27/. We detected BPs in the case of a chiral SCHIFF base MMBC /15/.

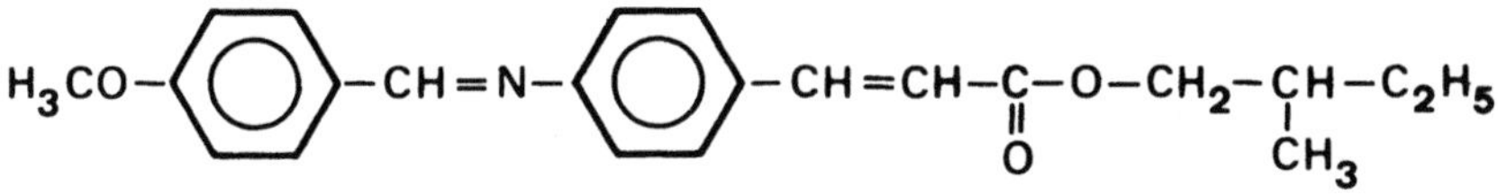

MMBC

The blue phases appear in form of platelet textures (Fig. 14). At temperatures between the isotropic liquid and the platelet BP there exists another "grey texture" in a very small temperature range whose properties are not known until now. It cannot be excluded that there exists additionally another type of liquid crystalline phase.

8. BP of Mixed Systems

In all systems in which BPs have been detected the corresponding cholesteric phase is characterized by very small helical pitches. In so-called induced cholesterics /28/ formed by optically active guest molecules soluted in nematic host phases which exhibit rather large pitches we never found any BP. Consequently, one should expect a critical pitch of the cholesteric helix to exist above which the corresponding BPs vanish.

This has been checked by investigation of BPs in a mixed system of CM and a nematogenic component PCPB.

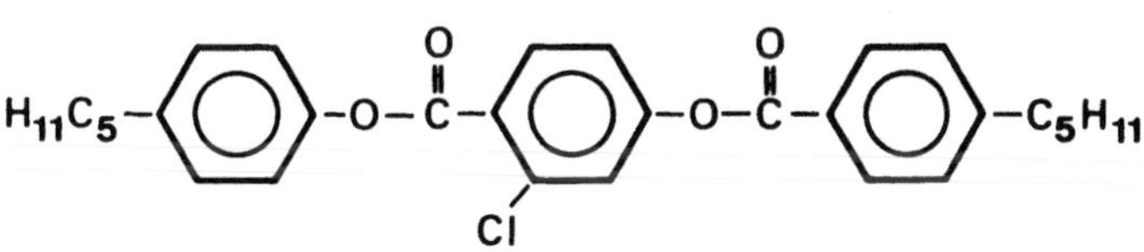

PCPB

The phase diagram of this mixed system has been determined by means of DSC and polarizing microscopy /15/. Generally, in mixed systems we have to take into account the occurrence of two-phase regions whose range may be comparable with the small region of BP existence. The phase diagram obtained is shown in Fig. 15: As the clearing points of CM and PCPB differ by 39 K we preferred to plot the temperature differences between the coexistence line of the cho-

◄ Fig.12 CANO lines in a wedge-shaped BP sample of CB/HOAB (42:58 mole%) at 135.7°C; magnification x 200

Fig.13 The same BP sample as in Fig.12 close below the clearing point. CANO lined structure is partially replaced by a platelet texture

Fig.14 Sample of chiral MMBC at $\vartheta \lesssim 100.4^{o}$C. Increasing temperature from right to left showing BP I, BP II (both as platelets), "grey phase", and isotropic liquid. Crossed polarizers; magnification x 200

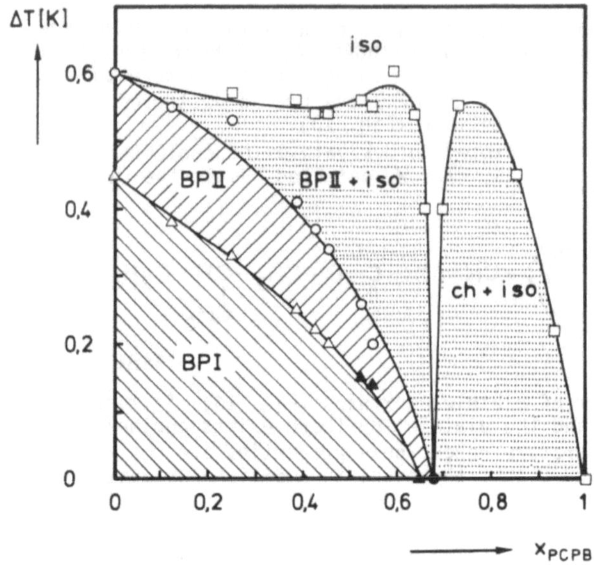

Fig. 15 Phase diagram of the mixed system CM/PCPB

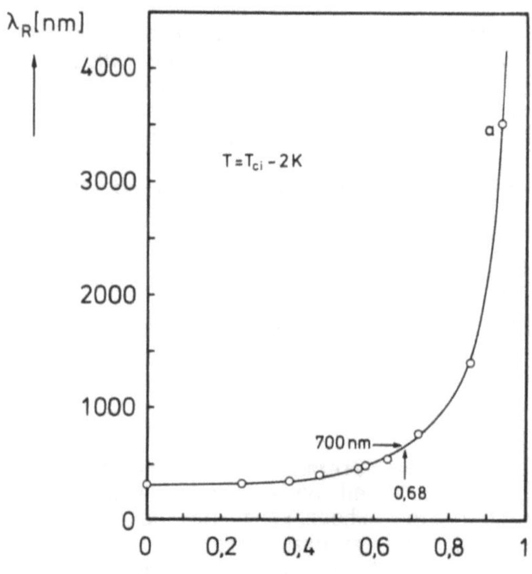

Fig. 16 Wavelength of SR bands of the cholesteric state in the mixed system CM/PCPB vs. mole fraction of PCPB at T = T_C - 2 K. a: value obtained by pitch measurement

lesteric phase and the other phase regions. The range of the two BPs decreases sharply with increasing mole fraction of the nematogenic component PCPB. At about 68 mole % both BPs vanish. Below the isotropic liquid we observed two-phase regions in which BP II and the isotropic liquid or cholesteric and the isotropic liquid occur, respectively.

In Fig. 16 the wavelength of selective reflection of the cholesteric phase of the system CM/PCPB is given as a function of mole fraction of PCPB. As it can be seen from Fig. 15 the BPs vanish at 68 mole % PCPB. At this point the cholesteric reflection wavelength amounts about 700 nm. At higher mole fractions the wavelength diverges strongly. From this result we conclude that a critical twist of the cholesteric structure exists above which the formation of BPs is no longer possible.

By means of the DE VRIES relation between the reflection wavelength λ_R and the pitch p ($\lambda_R = \bar{n} \cdot p$) taking a mean refractive index of $\bar{n} \approx 1.5$ a critical pitch of approximately 450 nm results.

9. Model of BPs

In the discussion of possible BP structures the experimental results reported in the preceding sections have to be taken into account: i) existence of *two* polymorphic BPs, ii) discontinuous phase transitions between the cholesteric, BP I, BP II, and isotropic liquid state, iii) selective reflection of circularly polarized light depending on temperature and pressure, iv) small half-width of the reflection bands, v) angular dependence of maximum reflection wavelength, vi) anomalous ORD, *but* vii) *no* birefringence, viii) CANO lines in wedge-shaped samples, ix) existence of a critical pitch above which no BP occurs.

The most important problem is to combine the zero birefringence with the other optical properties which are quite similar to that of classical cholesterics and suggest a helical arrangement of molecules also within the BP. The first attempt to solve this problem has been done by SAUPE /9/ who proposed a face-centered cubic structure. Other suggestions have been given assuming helically structured globular assemblies of random orientation /25, 29/. These attempts, however, cannot explain the sharp reflection bands as well as their angular dependence. Generally, one can solve this problem to combine the seemingly contradictious properties by the following way: Nematics can be described by a positive refraction indicatrix of prolate shape. Cholesterics possess a negative indicatrix of oblate shape. Wanted is a spherical indicatrix for the BP state of zero birefringence. One must find out helical molecular arrangements in space which result in such a spherical indicatrix. One can easily imagine that the negative indicatrix of cholesterics changes to a positive one if the tilt angle of the local director against the helical axis decreases from 90° to zero. At a critical angle of 54.74^o the indicatrix becomes spherical /17, 30/. By this "tilt model" (Fig. 17) which has also been proposed by SCHRÖDER /31/ the experimental result of zero birefringence can be understood although the molecules are helically arranged. But no physical reason can be given why the tilt angle should assume this critical value close below T_c.

Other models have been proposed by BRAZOVSKII /32/, SCHRÖDER /33/ and HORNREICH /34 / all basing on the LANDAU-theory but resulting in quite different structures. The body-centered cubic model of HORNREICH /34/ does not seem to be consistent with the occurrence of CANO lines in wedge-shaped BP samples. As this model implies helical axes perpendicular to each

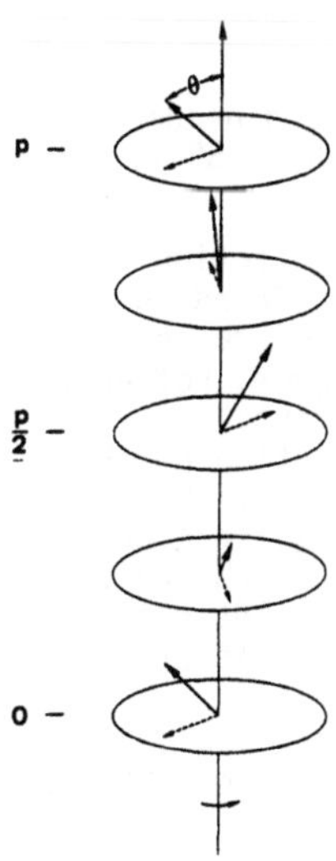

Fig. 17 Tilted helical structure for the BP resulting in a spherical indicatrix. P: helical pitch; Θ: tilt angle

other one should expect two reflection bands at oblique incidence which have not been observed /17,35/. The structural problem of BPs remains unsolved until now and deserves further theoretical work considering the experimental results given in this paper.

10. Historical Comment

In this final section an interesting historical detail should be mentioned. Everybody knows that mesomorphism has been discovered by REINITZER in 1888 in the case of cholesteryl benzoate /37/. The striking observation he described as follows: "On cooling of the molten compound a bright blue-violet colour phenomenon appears which disappears rapidly followed by an uniform turbidity. On further cooling the same colour effect occurs once more but disappears simultaneously on crystallization of the sample."

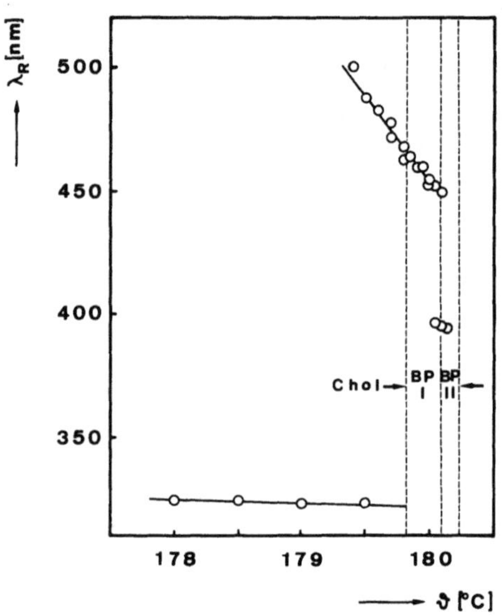

Fig. 18 Temperature dependence of maximum wavelength of SR for CB

We have measured the temperature dependence of the selective reflection of BP of CB which is shown in Fig. 18: Below the clearing point we found an analogous selective reflection of BP II and BP I as discussed before in the other cholesteryl esters. The classical cholesteric state, however, reflects in the u. v. region at temperatures close below the clearing point.

What REINITZER described was nothing but the selective reflection of the BP! Actually, REINITZER detected the mesomorphic behaviour of CB by the reflection property of a *Blue Phase*! That is what he refers to the first occurrence of blue-violet colours. Close above the melting point of CB the reflection of the classical cholesteric CB is shifted up to the violet region (which is not shown in Fig. 18). This explains the second colour effect described by REINITZER.

For my feeling, it is an outstanding fact that the history of liquid crystals began with the observation of the BP whose structure is now - nearly 100 years later - still under discussion.

Acknowledgement. This work has been supported by the Deutsche Forschungsgemeinschaft, the Fonds der Chemischen Industrie, and the Ministerium für Wissenschaft und Forschung des Landes Nordrhein-Westfalen.

References

1 H. Stegemeyer: VDI-Berichte 198, 29 (1973)
2 G.W. Gray, P.A. Winsor: Liquid Crystals and Plastic Crystals, Vol. 1, p. 12 (Ellis Horwood Ltd., Chichester 1974)
3 O. Lehmann: Z. Phys. Chem. 56, 750 (1906)
4 G.W. Gray: J. Chem. Soc. 1956, 3733
5. D. Armitage, F.P. Price: J. Phys. (Paris) Colloq. C 1, 36, C 1-133 (1975)
6 E.M. Barrall, R.S. Porter, J. F. Johnson: Mol Cryst. 3, 103 (1967)
7 H. Arnold, P. Roediger: Z. Phys. Chem. 239, 283 (1968)
8 D. Armitage, F.P. Price: J. Appl. Phys. 47, 2735 (1976)
9 A. Saupe: Mol. Cryst. Liq. Cryst. 7, 59 (1969)
10 G. Pelzl, H. Sackmann: Z. Phys. Chem. 254, 354 (1973)
11 D. Demus, H.G. Hahn, F. Kuschel: Mol. Cryst. Liq. Cryst. 44, 61 (1978)
12 K. Bergmann, H. Stegemeyer: Ber. Bunsenges. Phys. Chem. 82, 1309 (1978)
13 M.F. Vuks: Opt. Spectrosc. 20, 361 (1966)
14 K. Bergmann, H. Stegemeyer: Z. Naturf. 34a, 251 (1979)
15 K. Bergmann: Thesis, Paderborn 1980
16 K. Bergmann, P. Pollmann, G. Scherer, H. Stegemeyer: Z. Naturf. 34a, 253 (1979)
17 K. Bergmann, H. Stegemeyer: Z. Naturf. 34a, 1031 (1979)
18 S. Meiboom, M. Sammon: to be published
19 T.K. Brog, P.J. Collings: to be published
20 H. de Vries: Acta Crystallogr. 4, 219 (1951)
21 J.L. Fergason: Mol. Cryst. 1, 293 (1966)
22 L.S. Goldberg, J.M. Schnur: Radio Electron. Eng. 39, 279 (1970)
23 P. Pollmann, G. Scherer: High Temp. High Pressure, 12, 103 (1980); G. Scherer: Thesis, Paderborn 1979
24 D. Coates, G.W. Gray: Phys. Letters 45 A, 115 (1973)
25 D. Demus, L. Richter: Texture of Liquid Crystals, p. 60, 184 (Verlag Chemie Weinheim, New York 1978)
26 I.G. Chistiakov, L.A. Gusakova: Kristallogr. 14, 153 (1969)
27 D. Coates, G.W. Gray: Phys. Letters 51 A, 335 (1975)
28 H. Finkelmann, H. Stegemeyer: Ber. Bunsenges. Phys. Chem. 82, 1302 (1978)
29 G.W. Gray, P.A. Winsor: Ref. 2, p. 16
30 W. Kuczynski, K. Bergmann, H. Stegemeyer: Mol Cryst. Liq. Cryst. Letters, in press
31 H. Schröder: in The Molecular Properties of Liquid Crystals, Report of the Nato-Conference, Cambridge 1977 (G.R. Luckhurst, G.W. Gray ed., Academic Press, New York, in press)
32 S.A. Brazovskii, S.G. Dmitriev: Sov. Phys. JETP 42, 497 (1976)
33 H. Schröder: to be published; these proceedings, p. 201
34 R.M. Hornreich, S. Shtrikman: J. Phys. (Paris), to be published
35 P.P. Crooker: private communication
36 W. Kuczynski, H. Stegemeyer: Naturwiss., in press
37 F. Reinitzer: Monatsh. Chem. 9, 421 (1888)

Applications of Smectic and Cholesteric Liquid Crystals

E.P. Raynes

Royal Signals and Radar Establishment,
Malvern, Worcestershire, WR14 3PS, United Kingdom

Abstract

The majority of liquid crystal display devices are currently based on the twisted nematic electro-optic effect and although this effect has many advantages its inherent limitations are also becoming known. Attention is therefore being drawn to possible uses of the smectic and cholesteric phases of liquid crystals. These are examined, concentrating exclusively on the electro-optic and thermo-optic applications.

1. Introduction

Liquid crystal displays based on the twisted nematic effect [1] have developed into a large and successful industry; however the large development effort devoted to these displays merely highlights their basic limitations and has generated the feeling that the industry may benefit from an examination of the smectic and cholesteric phases for significant improvements in display devices and other uses. Twisted nematic displays are known to possess several significant advantages:-

(i) High reliability
(ii) Low voltage and power
(iii) High contrast and visibility in high ambient light levels
(iv) Large area

However, many disadvantages are also apparent:-

(i) Low brightness
(ii) Restricted and asymmetric angle of view
(iii) Polarizers degrade
(iv) Restricted multiplexing

The case for seeking alternative display devices and other applications is therefore strong, and the present status of the major electro-optic and thermo-optic effects in cholesterics and smectics will be considered.

2. Electro-optic Effects in Cholesteric Materials

The application of an electric field to a layer of cholesteric material with positive dielectric anisotropy ($\varepsilon_{//} - \varepsilon_{\perp}$) produces the cholesteric-nematic phase change. In this the cholesteric helix progressively distorts and dilates until at a critical field E_C the helix becomes completely unwound, and the layer is a homeotropically aligned pseudo-nematic; this sequence of events is illustrated in Fig.1. For a cholesteric with pitch P_0, twist elastic constant k_{22}, and dielectric anisotropy $(\varepsilon_{//} - \varepsilon_{\perp})$, E_c is given by [2, 3]

$$E_c = \frac{\pi^2}{P_o}\sqrt{\frac{k_{22}}{\varepsilon_o(\varepsilon_{//}-\varepsilon_{\perp})}} \qquad (1)$$

Typically, cholesterics with pitch lengths of $\simeq 3\mu m$ (produced by adding 5% of a normal cholesteric into a nematic material), $(\varepsilon_{//}-\varepsilon_{\perp}) \simeq 10$, and $k_{22} \simeq 10^{-11}N$, possess a threshold voltage of 10 V in a 10μm layer.

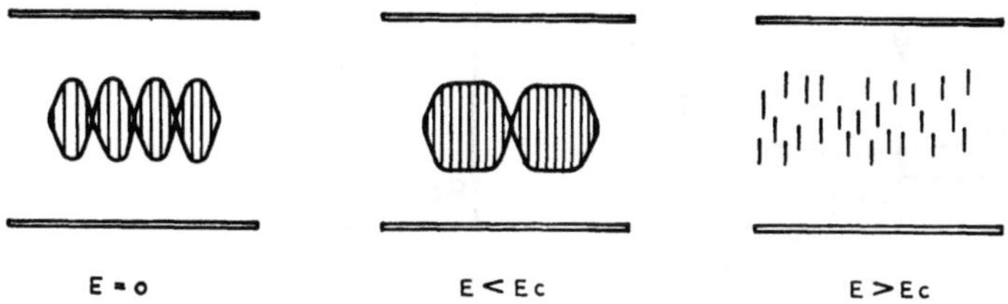

Fig.1 Field-induced unwinding of a cholesteric.

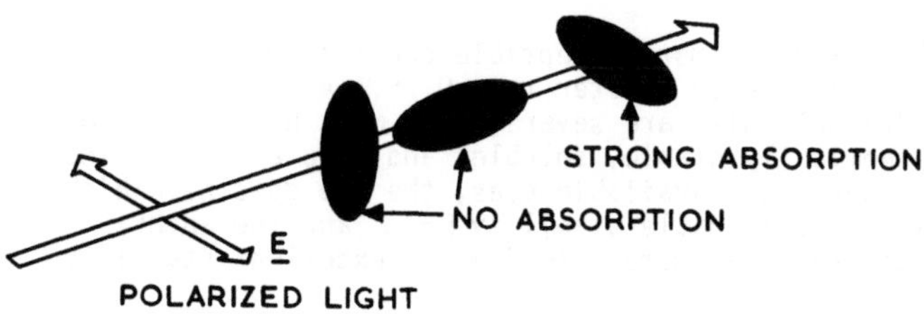

Fig.2 Absorption of light by a pleochroic dye molecule.

With certain boundary conditions the off (E=0) state is a scattering focal conic texture, and the transition between this and the optically clear pseudo-nematic on ($E>E_c$) state forms the basis of a display device [4]. A significant improvement in the optical performance of this display is achieved by changing this scattering effect into one showing absorption of light by the addition of a pleochroic dye [5]. These dyes only absorb light polarized along the long axis of dye molecule (Fig.2), and when they are dissolved in a cholesteric undergoing the phase change it is easy to visualize the strong absorption of all planes of polarization of incident light if a planar layer is considered for the off state, and the weak absorption in the homeotropic pseudo-nematic on state. It is worth noting that the display outlined above only has an acceptable contrast ratio if the optical order parameter of the dye, defined by

$$S = \overline{\frac{1}{2}(3\cos^2\theta - 1} \qquad (2)$$

is higher than the nematic order parameter of the host liquid crystal. For example the order parameter typical of nematic materials (0.65) produces a

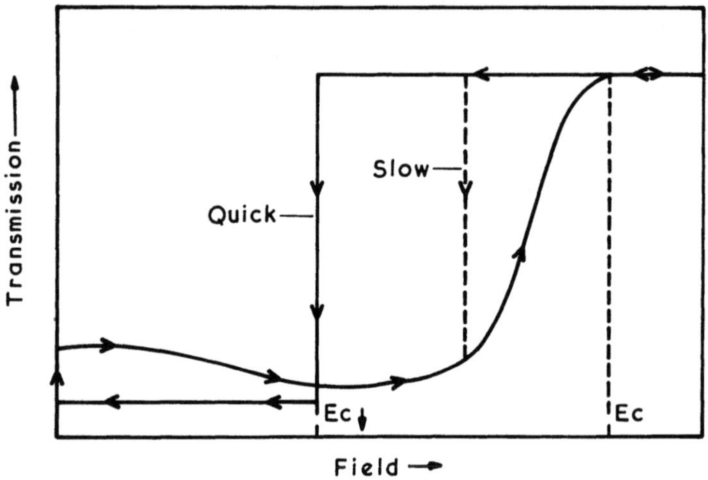

Fig.3 Hysteresis of the cholesteric-nematic phase change.

contrast ratio of only 2:1, and the more acceptable contrast ratio of 5:1 is only achieved with an order parameter greater than 0.75. Altogether the requirements for suitable dye molecules are severe, they must have a high optical order parameter, be chemically stable, soluble, and have a high extinction coefficient. Of the currently available dyes, the azo dyes have high order parameters (0.8) with questionable stability [6], and the anthraquinone dyes have moderate order parameters (0.67) and excellent stability [7].

Although the details of the cholesteric to nematic transition are complex, with the progressive unwinding of the helix being accompanied by removal of twist disclinations from the sample, a few simple conclusions can be drawn about the reverse process as the applied field is reduced and the pseudo-nematic layer reverts to cholesteric. As the field across a layer with homeotropic surface alignment is reduced below E_C, the phase change does not occur until a lower threshold field $E_{C\downarrow}$ is reached (Fig.3). At $E_{C\downarrow}$ a rapid transition occurs in a few msec via a canonical transformation [8], and Fig.4 shows the orientational change involved. If a field between $E_{C\downarrow}$ and E_C is maintained, a slow transformation taking several seconds occurs, and this hysteresis is useful for making displays with up to 100 lines which can be multiplexed. $E_{C\downarrow}$ is given by

$$E_{C\downarrow} = \pi\sqrt{\frac{k_{33}}{\varepsilon_0(\varepsilon_{//}-\varepsilon_{\perp})}\left[\left(\frac{2k_{22}}{P_0 k_{33}}\right)^2 - \frac{1}{d^2}\right]} \qquad (3)$$

and for typical materials the ratio $E_C/E_{C\downarrow}$ is in the range 2 to 3.

The appearance of the dyed phase change display described above is of a clear activated area on a coloured absorbing background. Although the contrast may be reversed by using any one of a number of specialized display construction techniques [9], a simpler method is to use materials with negative dielectric anisotropy ($\varepsilon_{//} < \varepsilon_{\perp}$). Polarizers are not required if a long

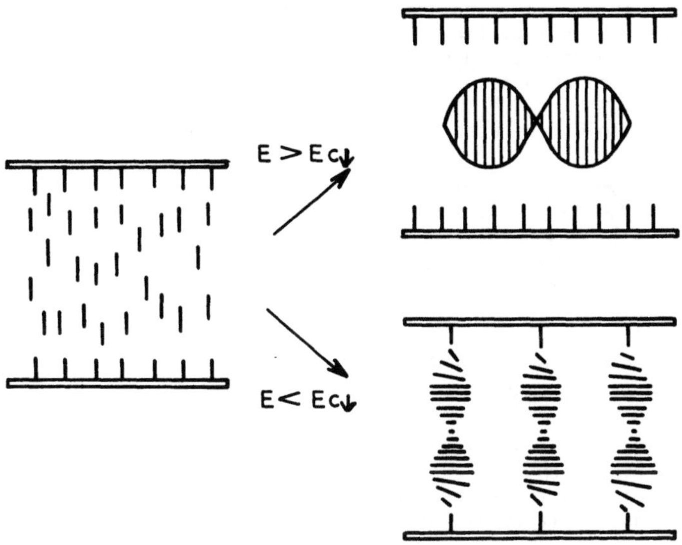

Fig.4 Two mechanisms for relaxing the pseudo-nematic to cholesteric.

pitch cholesteric is used with homeotropic surface alignment and a pleochroic dye is added [10]. Examination of (3) shows that the layer is spontaneously homeotropic with no applied field if

$$\frac{2k_{22}}{k_{33}} < \frac{P_o}{d} \qquad (4)$$

and that it transforms canonically to a cholesteric above a threshold field given by (3) provided $\varepsilon_{//} < \varepsilon_{\perp}$. The inclusion of a pleochroic dye is not a trivial exercise in this display since many nematics with $\varepsilon_{//} < \varepsilon_{\perp}$ seem to significantly reduce the optical order parameter of the dye.

It is appropriate to conclude this section with a summary of the advantages offered by phase change displays over the conventional twisted nematic display:-

(i) Wider and symmetrical angle of view
(ii) Better multiplexing (up to 100 lines by the use of a slow scan mode)
(iii) No polarizers - therefore brighter and more robust
(iv) Parallax can be avoided by the use of internal reflectors
(v) The combination of (iii) and (iv) makes it possible to use active substrates.

3. Electro-optic Effects in Smectic A Materials

The application of a large voltage (100 V) to a thin layer of smectic A produces optical changes in surprisingly short times (10 to 100 msec). The effects can be divided into those based on scattering which seems to have its origins in hydrodynamic motion (like dynamic scattering in nematics), and those based on a reorientation induced by dielectric torques. These electro-optic effects often show memory and can be erased electrically; they therefore offer attractive possibilities for use in multiplexed displays. In most cases no detailed understandings of the phenomena exists and therefore only descriptions based on observations will be given.

Most workers in this field have used the material 8CB because of its convenient temperature range

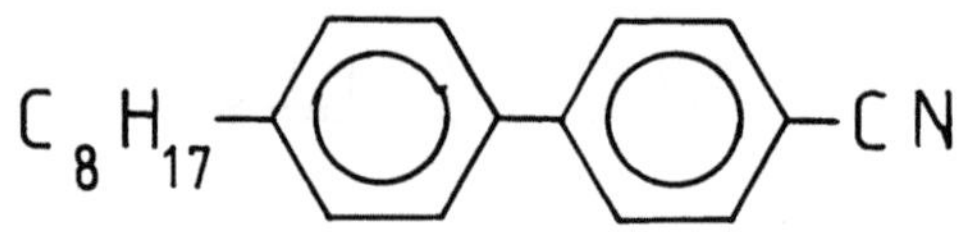

$$K \xrightarrow{21^{\circ}C} S_A \xrightarrow{32^{\circ}C} N \xrightarrow{40^{\circ}C} I$$

Scattering can be induced by the application of a low frequency field to a layer of smectic A material with $\varepsilon_{//} > \varepsilon_{\perp}$ and high electrical conductivity [11,12,13]. An array of defects is the first effect produced in an initially planar layer at a threshold voltage V_{th} which is empirically given by

$$V_{th} \propto d^{1/2} (T_{S-N} - T)^{1/3} \tag{5}$$

where d is the sample thickness, and $(T_{S-N} - T)$ represents the temperature difference below the nematic phase. The $d^{1/2}$ dependence makes it probable that the defects are related to layer undulations [14]. As the voltage is increased above V_{th} a scattering texture is induced. If homeotropic surface alignment is used [13] the effect can be reversed by the application of a high frequency signal, and typical threshold curves are shown in Fig.5.

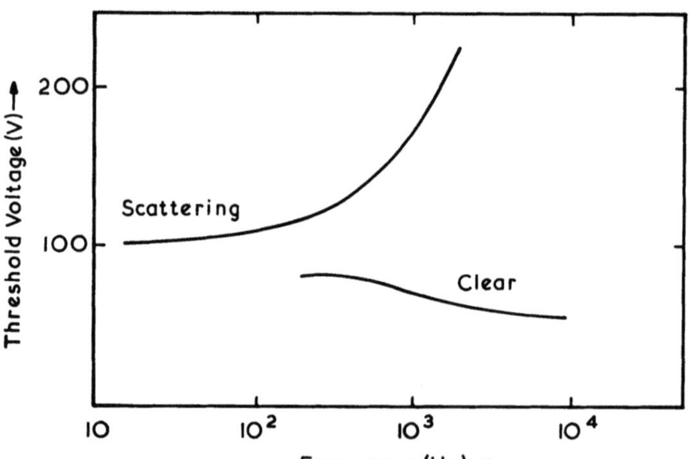

Fig.5 Threshold curves of a smectic A layer.

Therefore by the application of 100 V and switching the frequency from 10^2Hz to 10^3Hz a reversible transition from a clear homeotropic layer to a scattering focal conic texture is induced with response times as short as 10 msec and with both states showing a semi-permanent storage which is useful in multiplexed displays.

The application of even higher voltage to a smectic A with low conductivity and $\varepsilon_{//} > \varepsilon_{\perp}$ induces a planar to homeotropic transition induced by the dielectric torque [15,16], although some of the effects described above still occur at intermediate voltages, and if the smectic has negative dielectric anisotropy ($\varepsilon_{//} < \varepsilon_{\perp}$) an electric field induces a homeotropic to planar transition [16]. Both field transitions occur in 10 to 100 msec when 100 V signal is applied. The two transitions are normally irreversible electrically, but by using a smectic material with a low dielectric relaxation frequency [17], a reversible display has been constructed [18]. The addition of a pleochroic dye imparts contrast to the display [18] which is switched from the strongly absorbing planar texture to the weakly absorbing homeotropic state by a low frequency signal, with the reverse process being induced by a high frequency signal.

4. Thermo-optic Effects in Cholesteric Materials

The temperature sensitive Bragg reflection of light from cholesteric materials, or as it is more commonly known, the thermo-chromic effect, forms the basis of devices which are useful for temperature measurement, surface thermography and thermal imaging. In a well-aligned planar cholesteric texture, reflections occur at normal incidence at a wavelength λ_0 related to the average refractive index $\bar{n}$ and pitch P_0 by

$$\lambda_0 = \bar{n} P_0 \tag{6}$$

The half width $\Delta\lambda$ of the reflection is related to the birefringence Δn by

$$\Delta\lambda = \Delta n P_0 \tag{7}$$

The strongest thermo-chromic effect is shown just above a smectic to cholesteric transition where the pitch diverges rapidly, and the wavelength of the reflected light can be shifted over the whole width of the visible spectrum by a 1°C change in temperature. The present performance of thermo-chromic devices is restricted by the esters of cholesterol normally used. However substantial improvements in performance have recently been made using chiral nematic materials [19]. By substituting the normal alkyl chain of a nematic by a branched optically active alkyl chain, cholesteric materials with pitch lengths down to 0.1μm have been produced. For example the branched pentyl cyanobiphenyl has an intrinsic pitch of 0.15μm. The material is monotropic, so the pitch is estimated from dilute solutions in nematic materials. Di-

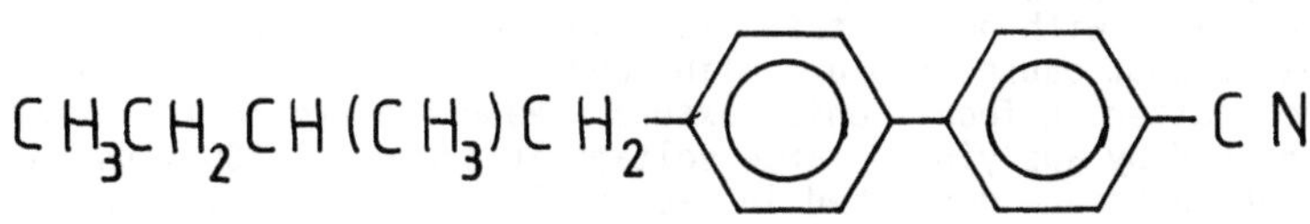

substitution produces even shorter pitch lengths, for example, the phenyl benzoate ester has an intrinsic pitch of 0.1μm. The combination of these systems with nematic cyanobiphenyls yields thermo-chromic mixtures with attractive properties. By comparison with cholesteryl ester mixtures the

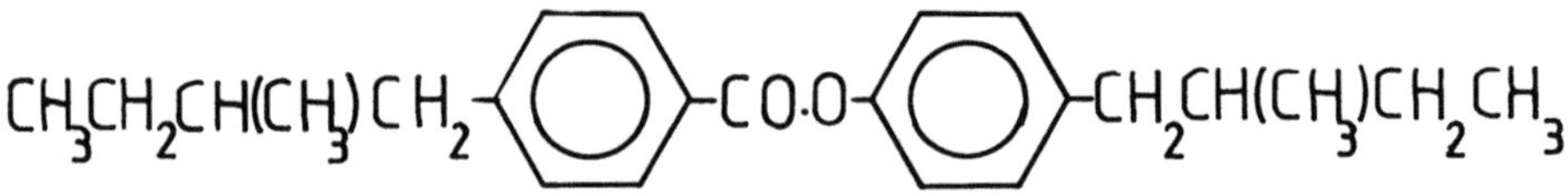

new systems are significantly more stable - both chemically and photochemically, they have a higher birefringence and are therefore brighter, they respond quicker to temperature changes because of their lower viscosity, and can be used at much lower temperatures without crystallization occurring.

5. Thermo-optic Effects in Smectic A Materials

In the basic laser-addressed smectic projection display, a layer of smectic A is written from one side by a laser which changes the texture of the smectic layer by heat, and is read from the other side to produce a projected image [20,21,22]. Although several variations have been described, the essentials of the reflective mode display are shown in Fig.6. The GaAs or argon laser

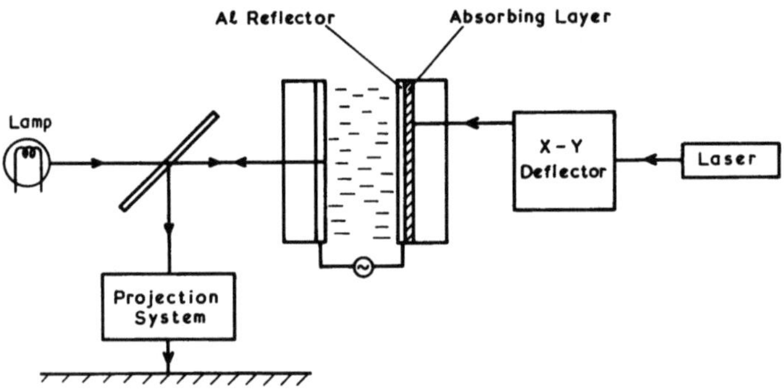

Fig.6 Construction of a reflective laser-addressed smectic projection display.

is modulated acousto-optically and is deflected by galvonometer mirrors onto an absorbing layer in the liquid crystal cell which transfers the laser energy to the liquid crystal layer as heat via an aluminium reflecting electrode. An electric field can be applied between the aluminium and a transparent electrode on the other surface. The image is then read out from the smectic layer by an optical projection system. Writing is accomplished by the application of the laser with no electric field applied, and removal of the laser from the heated spot causes rapid cooling and a transition of the smectic layer to the scattering focal conic texture. Page erasure to the clear state is achieved by applying a large voltage in excess of 100 V using the re-orientation phenomena described in section 3. The selective erasure of information is a more useful attribute than page erasure, and is achieved by the simultaneous application of a lower voltage of 30 V and the laser.

The performance of the system is impressive, with the most recent system [22] being capable of writing a display with over 2 x 10^6 picture elements in only 3 sec. There is almost indefinite storage, a selective erasure capability, and the possibility of gray scale and of two colours. It is perhaps unfortunate that such an impressive system is only achieved by the use of complex and bulky optics which will prevent the system from realizing its full potential.

6. Electrically-addressed Smectic Light Valve

The laser-addressed system described in section 5 has recently been extended by using heat generated not by a laser but by an array of electrical heating elements [23]. A reflective display based on this technique has been used to show restricted complexity TV images on a 100 x 100 matrix array. It has 7 gray scales, the total picture is written in only 10 ms, and it can be stored indefinitely. Although the power consumption is obviously larger than in a normal liquid crystal display, the system is attractive by virtue of its simplicity and novelty.

7. Conclusion

This review has by necessity been limited in scope and has concentrated on the most widespread uses of smectics and cholesterics. Several potential uses have been neglected, and perhaps the most important of these is the work described at this conference on the use of ferroelectricity in a chiral smectic C material to produce a display with memory which can be written in a few μsec by a signal of a few volts. The possible use of re-entrant systems also deserves a mention as these offer either smectics with rather weak smectic properties such as low viscosities, or alternatively, nematics with strong smectic properties over a limited temperature range.

It is clear however that the various optical effects in cholesterics and smectics offer many differences and some advantages over the effects observed in nematic materials, and the research effort being expended on these different phases is more than justified.

Acknowledgements

W A Crossland is thanked for the provision of hitherto unpublished data on reversible field-induced transitions in smectic A materials.

This paper is published by permission of the Controller H.B.M.S.O.

References

1. M. Schadt and W. Helfrich, Appl. Phys. Lett. vol. 18, p. 127, 1971.
2. R.B. Meyer, Appl. Phys. Lett. vol. 12, p. 281, 1968.
3. P.G. deGennes, Solid State Comm. vol. 6, p. 163, 1968.
4. J.J. Wysocki, J. Adams and W. Haas, Phys. Rev. Lett. vol. 20, p. 1024, 1968.
5. D.L. White and G.N. Taylor, J. Appl. Phys. vol. 45, p 4718, 1974.
6. J. Constant, J. Kirton, E.P. Raynes, I.A. Shanks, D. Coates, G.W. Gray and D.G. McDonnell, Electron. Lett. vol. 12, p. 514, 1976.
7. M.G. Pellatt, I.H.C. Roe, and J. Constant, Mol. Cryst. Liq. Cryst., in press.
8. W. Greubel, Appl. Phys. Lett. vol. 25, p. 5, 1974.
9. F. Gharadjedaghi and E. Saurer, IEEE Trans. Electron Devices, in press.
10. F. Gharadjedaghi, German Pat. Appl. 2826440, 1978.
11. C. Tani, Appl. Phys. Lett. vol. 19, p. 241, 1971.

12. M. Steers and A. Mircea-Roussel, J. Physique Coll. No 3, vol. 37, p. 145, 1976.
13. D. Coates, W.A. Crossland, J.H. Morrissy and B. Needham, J. Phys. D: Appl. Phys. vol. 11, p. 1, 1978.
14. W. Helfrich, J. Chem. Phys. vol. 55, p. 839, 1971.
15. M. Hareng, S. Le Berre and J.J. Metzger, Appl. Phys. Lett. vol, 27, p. 575, 1975.
16. M. Goscianski, L. Leger and A. Mircea-Roussel, J. Physique, Lett.vol. 36, p. 313, 1975.
17. D. Coates, Mol. Cryst. Liq. Cryst. Lett. vol. 49, p. 83, 1978.
18. P. Ayliffe and W.A. Crossland, to be published.
19. G.W. Gray and D.G. McDonnell, Mol. Cryst. Liq. Cryst. vol. 48, p. 37, 1978.
20. F.J. Kahn, Appl. Phys. Lett. vol. 22, p. 111, 1973.
21. M. Hareng and S. Le Berre, Electron. Lett. vol. 11, p. 73, 1975.
22. A.G.Dewey, J.T. Jacobs and B. Huth, Proceeding of S.I.D., vol. 19, p. 1, 1978.
23. Electronics, p. 70, April 26, 1979.

Theory of BCC Orientational Order in Chiral Liquids: The Cholesteric Blue Phase

R.M. Hornreich and S. Shtrikman

Department of Electronics, Weizmann Institute of Science, Rehovot, Israel

Abstract

Using Landau theory, it is shown that when the isotropic phase becomes thermodynamically unstable, chiral liquids can undergo a transition to a phase having a body-centered cubic (I432) structure rather than to the usual cholesteric (helicoidal) phase. This will occur when the coefficient of the cubic term in the free energy is sufficiently small with respect to that of the chiral term. It is believed that this bcc structure characterizes the cholesteric blue phase. Theoretical arguments and experimental results strongly supporting this structure are reviewed. It is shown that incorporating higher harmonics into the order-parameter will bring the theoretical predictions into better quantitative agreement with experiment.

1. Introduction

In the past few years, there has been a renewed interest in the properties of the so-called cholesteric "blue phase", which occurs in many cholesterol derivatives [1,2] in a narrow temperature range between the isotropic and usual cholesteric (helicoidal) phases [3-9]. Although this anomalous phase was first observed in 1906 [10], its structure has not as yet been definitely determined. However, as a part of the revived interest in the blue phase, several theoretical models have been proposed, including the hexagonal structure of BRAZOVSKII and DMITRIEV [11], the modified helix structures of BERGMANN and STEGEMEYER [7] and SCHRÖDER [12], and the body-centered cubic (bcc) model of the present authors [13]. Since a bcc structure has received strong support from the very recent experiments of SAMULSKI and LUZ [14] and MEIBOOM and SAMMON [15], we here summarize the theoretical basis for this structure and further analyze its properties. It is shown that these properties are in good agreement with those observed experimentally.

2. Landau Theory of the bcc Phase

As is well-known [16], the order-parameter describing transitions from the isotropic to orientationally ordered states in liquid crystals can be written as a linear combination of the spherical harmonics Y_2^m ($|m| \leq 2$). For the case of cholesterics, which lack inversion symmetry, the order-parameter is also characterized by a set of basic wavevectors, $\{\vec{k}_i\}$. As the structure we wish to describe is bcc, these basic wavevectors, which generate the reciprocal-space unit cell, will lie along <110> directions with all $|k_i| = k \neq 0$. Noting that a bcc structure is invariant under π rotations about the axes $\hat{\zeta}_i = \vec{k}_i/k$, we can take as our order-parameter Φ the linear combination

$$\Phi_{bcc} = \sum_{i=1}^{6} [\mu_i Y_2^2(\theta_i,\phi_i)\exp(i\vec{k}_i \cdot \vec{r}_i) + c.c.], \tag{1}$$

where c.c. denotes complex conjugate.

Several comments on (1) are in order:

(a) The six wavevectors $\vec{k}_i$ associated with the $\ell = m = 2$ spherical harmonics are chosen from <110> so as to form a regular tetrahedron.
(b) The polar and azimuthal angles (θ_i, ϕ_i) are, for each i, defined with respect to a local coordinate system in which $\hat{\zeta}_i$ is the polar axis of a right-handed cartesian coordinate system (ξ_i, η_i, ζ_i), with ξ_i defined in an identical manner for all i.
(c) The amplitudes μ_i are position independent and all $|\mu_i| = \mu$ are equal.
(d) A contribution proportional to Y_2^0, which appears in the order-parameter used to describe the transition from the isotropic to usual helicoidal phase [16], is not included in (1). The reason for this is that such a term cannot couple linearly to products of the $(\mu_i Y_2^2 \exp(i\vec{k}_i \cdot \vec{r}_i) + c.c.)$ functions when it is required that the ordered (i.e., non-isotropic) phase have bcc symmetry.
(e) Our choice of Φ does not include terms involving harmonics of the $\{k_i\}$, which in fact preserve bcc symmetry while coupling linearly to products of the type noted above. In this sense our calculation of the free energy of the bcc phase is restricted (Ritz-type), since including such harmonics would obviously reduce the system's free energy. We shall return to this point in Section 4.

In Landau theory [16,17], the free energy functional associated with a given order-parameter Φ is obtained by expanding the free energy density F in powers of Φ and its spatial derivatives. Each such term contains only those products of the order-parameter which are invariant under the symmetry group of the higher symmetry (here, isotropic) phase. To second-order in μ we also include invariant contributions arising from the first and second-order derivatives of the order-parameter in order to determine the value of k which minimizes the free energy. Note that, in order to calculate the third and fourth-order contributions, it is necessary to transform the $Y_2^{\pm 2}(\theta_i, \phi_i)$ functions in (1) to a common coordinate system and also to recognize that the magnitude of these contributions is dependent upon the relative phases of the amplitudes μ_i. It is straightforward to show that the magnitude of the third-order contribution is maximum when all the μ_i are in phase [18], in which case the free energy density takes the form [13]

$$F_{bcc} = 12(\alpha_0 + dk + ck^2)\mu^2 - (27\sqrt{6}/8)\beta\mu^3 + (1665/8)\gamma\mu^4. \tag{2}$$

Here α_0, d, c, β, γ are phenomenological coefficients which, with the exception of α_0, are regarded as constants in the phase transition region. In Landau theory, α_0 is taken to be linear in the temperature T and can be written as

$$\alpha_0 = \alpha(T/T_0 - 1) = \alpha t, \tag{3}$$

where T_0 is an idealized transition temperature for the case $d=\beta=0$. For stability, it is necessary that c and γ be positive.

A necessary condition for the bcc phase to exist is that, in some temperature region, its free energy be less than that of both the isotropic and usual helicoidal (h) cholesteric phases. For the later the order-parameter and corresponding free energy are

$$\Phi_h = \mu_o Y_2^{o}(\theta,\phi) + [\mu Y_2^{2}(\theta,\phi)\exp(ikz) + c.c.] \ , \qquad (4a)$$

$$F_h = \alpha_o \mu_o^{2} + 2(\alpha_o + dk + ck^2)\mu^2 + \beta(\mu_o^{3} - 6\mu_o\mu^2) + \gamma(\mu_o^{2} + 2\mu^2)^2. \qquad (4b)$$

Note that we have restricted ourselves to considering only the usual transverse helicoidal structure.

For both the bcc and h phases, free energy minima are obtained when $k = k_o = -d/2c$. Consider now (4b) with $k=k_o$. Setting $F_h = \partial F_h/\partial \mu_o = 0$, we find that the thermodynamic phase transition from the isotropic to the cholesteric phase occurs either at

$$t_h = d^2/4\alpha c, \qquad (5a)$$

or at

$$(t_h^{*} - t_h)/t_h = 4(f-1)/9 + \Delta(2f+1)18, \qquad (5b)$$

$$\Delta = (\beta^2 - \beta_c^{2})/\beta_c^{2}, \quad f = [(1+\Delta/4)/(1+\Delta)]^{1/2}, \qquad (5c)$$

$$\beta_c^{2} = (4/9)(d^2/4c)\gamma = (4/9)k_o^{2}\gamma c. \qquad (5d)$$

In the former case the transition is second-order while in the latter it is first-order. The dividing point between the two regions, $t_h^{*} = t_h$, occurs when $\beta^2 = \beta_c^2$. Thus, when $\beta^2 \le \beta_c^2$, as noted by BRAZOVSKII and DMITRIEV[11], the isotropic to helicoidal transition is expected, within Landau theory, to be a continuous one.

On the other hand, inspection of (2) shows that the isotropic-bcc transition is always first-order and occurs at a temperature $t^{*}_{bcc} > t_h$. Setting $F_{bcc} = \partial F_{bcc}/\partial \mu = 0$, we find explicitly that

$$(t^{*}_{bcc} - t_h)/t_h = (9/2960)(1+\Delta), \ (\Delta \ge -1). \qquad (6)$$

The "crossover" from the isotropic-helicoidal to isotropic-bcc transition occurs at $t_h^{*} = t^{*}_{bcc}$ or

$$\beta^2 \simeq 1.15\beta_c^{2}. \qquad (7)$$

In other words, we find that whenever the isotropic-helicoidal transition would be continuous or nearly so, a cholesteric liquid crystal system can lower its free energy by ordering in a bcc phase. This is, of course, intuitively what one would expect due to the cubic term in (2).

One can, naturally, ask if any other phase could order at a temperature $t^{*} > t^{*}_{bcc} > t_h^{*}$? We have elsewhere [13] examined the alternate, hexagonal phase suggested by BRAZOVSKII and DMITRIEV [11] and found that $t^*_{bcc} > t^*_{hex}$ whenever $t^*_{bcc} > t_h^*$. Since these are the only two phases having the "triangular" structures which maximize the magnitude of the cubic free energy invariant, it is reasonable to conclude that it is the bcc structure which is most likely to occur whenever $\beta^2 \lesssim \beta_c^2$.

3. Physical Properties of the bcc Phase

In the previous section we have shown, using Landau theory, that whenever the cubic invariant in the free energy of a cholesteric liquid crystal system is sufficiently small in comparison with the chiral term, an ordered phase having a bcc structure can exist when the isotropic phase becomes thermodynamically unstable. The detailed nature of this phase can be most easily seen by transforming the order parameter, given by (1), into cartesian form in a coordinate system whose axes are parallel to those of the cubic unit cell. One way of doing this is to note that $Y_2^2(\theta_i,\phi_i)$ is proportional to $\sin^2\theta_i \exp(2i\phi_i)$ or $[(\xi_i^2-\eta_i^2) + 2i\xi_i\eta_i]$, and that the latter expression can be written in the equivalent matrix form

$$[\xi_i\ \eta_i\ \zeta_i]\begin{bmatrix}1 & i & 0\\ i & -1 & 0\\ 0 & 0 & 0\end{bmatrix}\begin{bmatrix}\xi_i\\ \eta_i\\ \zeta_i\end{bmatrix} = \rho^T y \rho \ . \tag{8}$$

We can thus use y matrices instead of Y_2^2 spherical harmonics as basis functions, multiply them by appropiate phase factors, $\exp(ik_i\zeta_i)$, and rotate them into the common coordinate frame. Defining Q to be the resulting matrix order-parameter, we obtain [19]

$$Q(\vec{\sigma}) = K\begin{bmatrix}3c_2c_3 - \tau & \sqrt{2}(c_1-c_2)s_3-s_1s_2 & \sqrt{2}(c_3-c_1)s_2-s_1s_3\\ \sqrt{2}(c_1-c_2)s_3-s_1s_2 & 3c_1c_3 - \tau & \sqrt{2}(c_2c_3)s_1-s_2s_3\\ \sqrt{2}(c_3-c_1)s_2-s_1s_3 & \sqrt{2}(c_2-c_3)s_1-s_2s_3 & 3c_1c_2 - \tau\end{bmatrix} \tag{9}$$

where K sets the amplitude of Q, σ_ℓ are reduced cartesian coordinates satisfying $\sigma_\ell = (\sqrt{2}k_0)x_\ell/2\pi$, $c_\ell = \cos 2\pi\sigma_\ell$, $s_\ell = \sin 2\pi\sigma_\ell$ $(\ell = 1, 2, 3)$, and $\tau = c_1c_2 + c_2c_3 + c_3c_1$. (Note that for rod-like molecules, K will be positive).

Inspection of (9) shows that the structure generated by Q is invariant under the operations of the bcc space group I432 (O^5). This space group does not include the inversion operator as a consequence of the chiral nature of cholesteric liquid crystals. Note that Q = 0 at the lattice points. These are therefore defect points at which the system remains unordered or isotropic.

Along an edge of the cubic unit cell we see that the major axes of the quadupolar order-parameter are everywhere parallel to the cubic axes and simply modulated by the factor $\sin^2\pi\sigma_\ell = \sin^2(\sqrt{2}\ k_0x_\ell/2)$. There is neither chirality or biaxiality. Moreover, since the largest (in magnitude) eigenvector, $\lambda_1 = K \sin^2\pi\sigma_\ell$, is positive (except at lattice points), the structure along the edges is "rod-like", as usually found in cholesteric liquid crystal systems.

Along a line parallel to an edge and passing through a face-center we have both chirality and biaxiality. The eigenvalues of e.g., $Q(\sigma_1, 1/2, 0)$ are $\lambda_{1,2} = [1 \pm (9 - s_1^2)^{1/2}]K$ and $\lambda_3 = -2K$. The eigenvalue having the largest magnitude is again positive and nearly constant in magnitude ($3.82 \lesssim \lambda_1/K \leq 4$). Note that the associated eigenvector is perpendicular to $\hat{x}_1$ and that the biaxiality is small, with the asymmetry parameter $\eta = |\lambda_2 - \lambda_3|/\lambda_1 \lesssim 0.045$. Thus the structure, to a first approximation, is simply obtained by rotating a rod of constant length through 180° while advancing along $[\sigma_1, 1/2, 0]$ from $\sigma_1 = 0$ to $\sigma_1 = 1$. At [1/2, 1/2, 0] the rod lies perpendicular to the face of the unit cell.

Along a face diagonal, e.g., $Q(\sigma_1, \sigma_1, 0)$, it is straightforward to see that $[1\bar{1}0]$ is a principal axis (with eigenvalue $\lambda_3 = -K \sin^2(\pi\sigma_1)/2$). The other two eigenvalues, whose eigenvectors are perpendicular to $[1\bar{1}0]$, are $\lambda_{1,2} = \frac{1}{4} \sin^2(\pi\sigma_1)[1 \pm (17 + 8\,c_1)^{\frac{1}{2}}]K$ and the asymmetry parameter $\eta = |\lambda_2 - \lambda_3|/\lambda_1 = [-3 + (17 + 8c_1)]^{\frac{1}{2}} / [1 + (17 + 8c_1)^{\frac{1}{2}}] \to 1/3$ as $c_1 \to 1$. The largest eigenvalue is again positive, thus the structure along [110] is chiral and essentially rod-like, with growing biaxiality and decreasing order as one moves from the face-center toward the lattice points.

Consider finally the body diagonals of the unit cell. It is easily seen that [111] is a principal axis of $Q(\sigma_1, \sigma_1, \sigma_1)$ with eigenvalue $\lambda_1 = -2Ks_1^2$. The eigenvalues with transverse (to [111]) eigenvectors are $\lambda_2 = \lambda_3 = Ks_1^2$, thus $Q(\sigma_1, \sigma_1, \sigma_1)$ exhibits neither chirality or biaxiality. Moreover, the largest (in magnitude) eigenvalue is here <u>negative</u>. Thus the order along [111] is disc-like, rather than rod-like. The average magnitude of the order-parameter along [111] is, however, relatively small.

A schematic view of the bcc structure on a face of the unit cell is illustrated in Fig. 1. Only the magnitude (here everywhere ≥ 0) and direction of the largest eigenvalue is indicated, thus the biaxiality of the structure is not evident. Note the similarity between this view of our structure and that proposed by SAUPE [20] a decade ago.

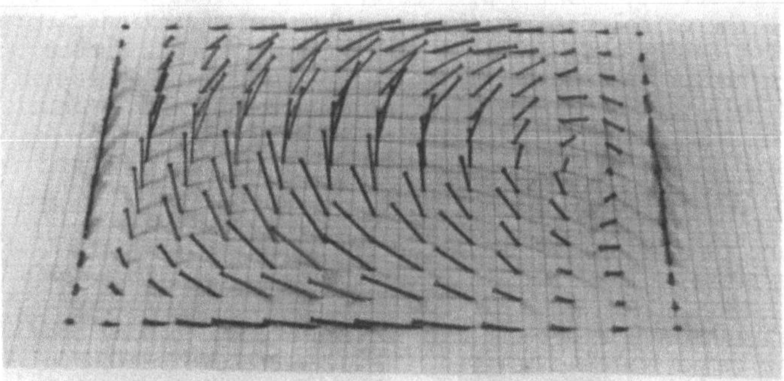

Figure 1. Molecular order on a unit cell face of the I432 structure. Shown is the approximate magnitude and direction of the largest eigenvalue of the order-parameter. The order is rod-like except at the lattice points which are disordered (liquid-like)

Figure 2. Molecular order in the I432 structure. The lines and bands (used to clarify the chiral nature of the structure) give the magnitude and direction of the largest eigenvalue of the order-parameter. The order shown is rod-like exception the body diagonal where it is disc-like. The order in other regions of the unit cell may be estimated by interpolation.

A model of the bcc structure, showing particularly the defect nature of the lattice points and the chiral pattern along $<\sigma_1,1/2,0>$ directions, is shown in Fig. 2. Except on the body diagonal, the lines and bands denote regions in which the structure is rod-like, with their length and direction corresponding respectively to the magnitude and eigenvector of the largest eigenvalue. Along the body-diagonals the order is disc-like. The isotropic or non-ordered nature of the lattice points is clearly evident.

If the cholesteric blue phase is characterized, in some temperature interval, by the bcc structure described above, it would have several attributes which are susceptible to experimental confirmation. These include:

a) The blue phase should appear between the isotropic and ordinary cholesteric phases as the temperature is lowered. The temperature region in which it is thermodynamically stable should be relatively narrow since at lower temperatures the helicoidal structure, in which the magnitude of the order-parameter is constant, is expected to have the lower free energy and to therefore become the stable phase.

b) The blue phase should occur only when the optical pitch of the helicoidal phase $\lambda_0 = \pi/k_0$, satisfies (see (5d) and (7)) $\lambda_0 \lesssim 2.3(\gamma c)^{1/2}/\beta$. That is, it should be characteristic of cholesterics with relatively short pitch.

c) A cubic blue phase would be optically isotropic, i.e., non-birefringent.

d) Since the I432 bcc structure is non-centro-symmetric such effects as optical activity and selective reflection of circularly polarized light are allowed in the blue phase [7,9].

e) Bragg scattering (at optical wavelengths) on polycrystalline blue phase samples should exhibit reflections only for $\{h_1\ h_2\ h_3\}$ satisfying $h_1+h_2+h_3 = 2n$, which is characteristic of all bcc structures. Such a pattern is expected to be observable, even though the bcc "crystal" is relatively "soft" and thus easily distorted since it is basically a liquid phase. Note that it might also be possible to very rapidly quench the material from the blue phase, so as to freeze in its structure. This has been done successfully for the usual cholesteric phase by SACKMANN et al. [21].

f) Quadrupolar NMR spectra for the blue phase can be obtained by averaging $\Sigma_{ij}Q_{ij}\alpha_i\alpha_j$, where α_i are direction cosines, over a unit cell and over all spatial orientations. However, in general, it will be necessary to modify the calculated spectra so as to take diffusion effects into account.

Items (a) to (d) are all in agreement with well-known properties of the blue phase [9] while (e) has been established by the recent work of MEIBOOM and SAMMON [15]. Note, however, that on the basis of our model there is only one reflection which should occur at the same wavelength as the helicoidal phase reflection, i.e. at $\lambda_o = \pi/k_o$. Ref. [15], on the other hand, reports that this reflection actually occurs at approximately $1.4\lambda_o$. We shall return to this point when considering the role harmonics of the basic order-parameter play in the I432 structure. Turning to (f), recent NMR studies by SAMULSKI and LUZ [14] have provided additional support for a cubic structure in the blue phase. In particular, they have taken the molecular diffusion process into account and have shown that 180° molecules reorientation takes place over a length comparable with the unit cell size of the bcc structure.

A schematic phase diagram showing the regions in which the isotropic, bcc, and helicoidal phases are thermodynamically stable according to Landau theory is given in Fig. 3. In Fig. 3a, we use $|\beta/\beta_c|$ and temperature as the thermodynamic parameters while in Fig. 3b we illustrate specifically the effect of varying optical pitch on the phase diagram. Experimental studies of such a phase diagram by e.g., differential thermal analysis or optical techniques [9] could be carried out by mixing two cholestegens with opposite chirality or, alternatively, by diluting a cholestegen having a naturally short pitch [22].

In planning such a study, it would be useful to have an estimate of the maximum optical pitch λ_c for which the intermediate cubic phase can appear. If we assume that the parameters $\beta = \beta_c$, γ, and c are essentially the same in cholesterics as in nematic materials, we can estimate $|\lambda_c| = \pi/|k_o| = 2\pi c/|d|$ from MAIER-SAUPE theory. Using (5d) we have

$$\lambda_c^2 = 4\pi^2 c\gamma/9\beta_c^2 = 4\pi^2 c\gamma/9\beta^2. \tag{10a}$$

From MAIER-SAUPE theory, we obtain [23]

$$\beta^2/\gamma \simeq 10k_B\Delta T^*, \tag{10b}$$

$$c \simeq 10k_B T^*\xi_o^2, \tag{10c}$$

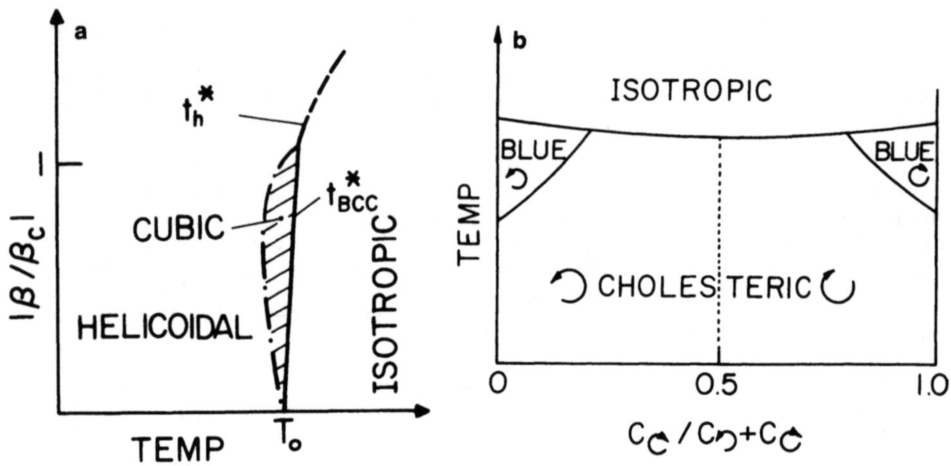

Figure 3. Possible phase diagrams for cholestegens showing schematically the regions in which the isotropic, bcc, and helicoidal phases are stable

and thus, from (11a),

$$\lambda_c(\text{vacuum}) = n\lambda_c(\text{material}) \simeq 3(T^*/\Delta T^*)^{1/2}\xi_o. \qquad (10d)$$

Here T* (~400K) is the extrapolated temperature at which the isotropic to ordered phase transition would occur for $\beta = 0$, ΔT^* is the temperature difference between T* and the actual transition point, ξ_0(~20Å) is the bare coherence length appearing in the expression $\xi = \xi_0|(T/T^*) - 1|^{-\frac{1}{2}}$, and n is the index-of-refraction ($\simeq 1.5$). Clearly λ_c is very sensitive to the value of ΔT^* which, experimentally, is found to be approximately 1K [24]. Substituting this value into (10d) gives λ_c(vacuum) $\simeq$ 4000Å, in surprising agreement with experiment [9].

We now turn to the thermodynamic properties of the isotropic to bcc phase transition. From (2), we find that the magnitude of the order-parameter just below the transition is

$$\bar{\mu}_{bcc} = (3\sqrt{6}/370)|\beta/\gamma| = 0.02|\beta/\gamma|. \qquad (11a)$$

On the other hand, for the isotropic to nematic transition, the corresponding magnitude is [23]

$$\bar{\eta}_n = 0.5|\beta/\gamma|. \qquad (11b)$$

Since the latent heat associated with a first-order phase transitions is proportional to the square of the jump in the order-parameter, (11) would lead us to conclude that the latent heat of the isotropic-bcc transition is less than 1% that of the isotropic-nematic one whereas, experimentally[7,9], they are of the same magnitude.

To summarize, we have shown that our bcc structure provides a basis for satisfactorily explaining the large amount of experimental data on the properties of the blue phase in colestegens. However, on some points, as noted above, quantitative agreement is lacking. We therefore consider ways in which the basic bcc model could be improved.

4. The Role of Higher Harmonics

The basic order-parameter for the bcc structure is given by (1) or, alternately, by (9). As noted earlier, this Φ includes only $\ell = 2$, $m = 2$ spherical harmonics with associated wavevectors $|\vec{k}_i| = k_0$. However, since the phase transition is first-order other terms can contribute to Φ which, while preserving the I432 structure, result in a reduction in the system's free energy. Of particular importance will be terms which, in combination with any two of the basic components, can form third-order invariants. Examples of such terms are

$$\begin{aligned}
&(a)\ Y_2^m(\theta_i,\phi_i)\exp(2i\vec{k}_i\cdot\vec{r}) + c.c. \quad , (m = 0, \pm 2),\\
&(b)\ Y_2^o(\theta_\alpha,\phi_\alpha)\exp(\sqrt{2}ikx) + c.c. \quad ,\\
&(c)\ Y_2^m(\theta_\beta,\phi_\beta)\exp[ik(x + y + 2z)/\sqrt{2}] + c.c. \quad ,\\
&(d)\ Y_2^m(\theta_i,\phi_i)\exp(i\vec{k}_i\cdot\vec{r}) + c.c. \quad , (m = 0, -2)
\end{aligned} \tag{12}$$

Thus we can have contributions with (a) wavevectors <220>, (b) wavevectors <200>, (c) wavevectors <211>, and (d) wavevectors <110>, but associated with $\ell = 2$, $m = 0, -2$ spherical harmonics. Taking all such terms into account is clearly difficult, but it is straightforward to see what the results of such a calculation would be:

a) The equilibrium wavevector k_0' for the modified bcc structure will be less than k_0. This will result in the lowest-order {110} bcc reflection occurring at a longer wavelength than that of the helicoidal phase, in agreement with experiment [15].

b) Higher-order Bragg peaks, such as {211} and {200} [15], will appear for the modified structure.

c) The higher harmonics will effectively reduce the size of the non-ordered (defect) regions associated with the lattice points. This will tend to split the quadrupolar NMR spectrum into two distinct peaks, such as is observed experimentally [14]. This splitting could not be reproduced by us using only the basic order-parameter given in (9).

d) The higher harmonics will shift the isotropic to cubic thermodynamic phase transition to a higher temperature and widen the $|\beta/\beta_c|$ region in which this transition can occur. Also, they will result in a bigger "jump" in the order-parameter at the transition than that given by (11a) and, consequently, a larger latent heat as observed experimentally [7,9].

Thus, we conclude that harmonics of the basic order-parameter play a significant role in determining the quantitative properties of the bcc phase. In principle, they could even result in a bcc structure which belongs to a different space group than I432. Note in this connection that the results of MEIBOOM and SAMMON [15], while conforming the existence of a bcc phase, do not specify the actual space group.

An alternate way of analyzing the I432 phase, which could be particularly useful when higher harmonics are relevant, is to consider the structure as primarily composed of localized point defects. One can then expand the order-parameter as a function of $\vec{r}$ in the neighborhood of each defect, in analogy to the cellular (Wigner-Seitz) method in solids. Due to symmetry constraints, $Q(\vec{r})$ must have the form

$$Q_{ij} = a_o[(x_i{}^2 - r^2/3)\delta_{ij} + a_1 x_i x_j (1-\delta_{ij}) + a_2 \varepsilon_{ijk}(x_i{}^2 - x_j{}^2)x_k + O(r^4)], \quad (13)$$

where a_o, a_1, a_2 are amplitudes and δ_{ij}, ε_{ikj} are the usual Kronecker and antisymmetric tensors. Using (13), a_1 and a_2 would first be fixed by approximately matching the Q_{ij} from adjacent cells at the cell boundaries, then a_o would be determined by minimizing the system's free energy.

5. Summary

In this paper, we have reviewed the body-centered cubic model (space group I432) for the cholesteric blue phase and shown that there are strong reasons, both theoretical and experimental, to believe that this structure is the correct one. We stress that the phase being described by us is that predominantly found between the isotropic and helicoidal phases; there is some evidence [7,9,15] for the existence of a second, distinct blue phase within a very narrow temperature region (~0.1K) just below the clearing point. The structure of this second phase is as yet unknown.

The best method for probing the detailed properties of the blue phase is undoubtably light scattering, using polarized light. Note particularly that by using a laser source it should be possible to study both the phase and amplitude of the coherent (Bragg) scattering. In this way, it would be possible, in principle, to completely determine the tensorial dielectric structure of the blue phase. This is, of course, impossible with conventional incoherent X-ray sources. Thus, in addition to the intrinsic interest in the structure and properties of the blue phase in cholesteric liquid crystals, these materials could provide model systems in which new techniques for studying the properties of ordered phases can be developed.

Acknowledgement

Useful discussions and correspondence with S. Alexander, E. Courtens, P. De Gennes, Z. Luz, P.H. Keyes, S. Meiboom, D. Mukamel, E. Samulski, and H. Thomas are gratefully acknowledged. This work was supported in part by a grant from the U.S.-Israel Binational Science Foundation (BSF), Jerusalem, Israel.

References

1. D. Coates and G.W. Gray: Phys. Lett. 45A, 115 (1973); 51A, 335 (1975); D. Coates, K.J. Harrison, and G.W. Gray: Mol. Cryst.Liquid Cryst. 22, 99 (1973).

2. W. Elser, J.L. Pohlmann, and P.R. Boyd: Mol. Cryst. Liquid Cryst. 20, 77 (1973).

3. D.S. Mahler, P.H. Keyes, and W.B. Daniels: Phys. Rev. Lett. 36, 491 (1976).

4. P.H. Keyes and D.B. Ajgaonkar: Phys. Lett. 64A, 298 (1977).

5. D. Armitage and F.B. Price: J. Chem. Phys. 66, 3414 (1977).

6. P.G. Collings and J.R. McColl: J. Chem. Phys. 69, 3371 (1978)

7. K. Bergmann and H. Stegemeyer: Ber. Bunsenges. Phys. Chem. 82, 1309 (1978); Z. Naturforsch 34a, 251, 1031 (1979); K. Bergmann, P. Pollmann, G. Scherer and H. Stegemeyer: Z. Naturforsch 34a, 253 (1979); P. Pollmann and G. Scherer: Z. Naturforsch 34a, 255 (1979).

8. P.H. Keyes and C.C. Yang: J. de Physique 40, C3-376 (1979).

9. For a review of experimental results on the blue phase, see the paper by H. Stegemeyer and K. Bergmann, these proceedings, p. 161

10. O. Lehmann: Z. Phys. Chem. 56, 750 (1906).

11. S.A. Brazovskii and S.G. Dmitriev: Zh. Eksp. Teor. Fiz. 69, 979 (1979) [trans: Sov. Phys. - JETP 42, 497 (1976)].

12. H. Schröder - to be published; these proceedings, p. 196

13. R.M. Hornreich and S. Shtrikman: Bull. Israel Phys. Soc. 25, 46 (1979); J. de Physique - in press (April 1980 issue).

14. E. Samulski and Z. Luz: J. Chem. Phys. - in press.

15. S. Meiboom and M. Sammon - to be published.

16. See e.g., S.Goshen, D. Mukamel and S. Shtrikman: Solid State Commun. 9, 649 (1971).

17. L. D. Landau and E. M. Lifshitz:"Statistical Physics"(Pergamon Press, Oxford) 1969, Ch.14.

18. Note that, in principle, the phase-dependent part of the fourth-order term could play a role in determining the relative phases of the μ_i for sufficiently large $\gamma/|\beta|$. In such a case, bcc structures different from that presented here might occur.

19. This is in agreement with a result obtained by S. Alexander, who concluded independently that the blue phase has a bcc structure

20. A. Saupe: Mol. Cryst. Liquid Cryst. 7, 59 (1969).

21. E. Sackmann, S. Meiboom, L.C. Snyder, A.E. Meixmer and R.E. Dietz: J. Amer. Chem. Soc. 90, 3567 (1968).

22. G.W. Gray and D.G. McDonnell: Mol. Cryst. Liquid Cryst. 37, 189 (1976) and unpublished results.

23. M.J. Stephen and J.P. Straley: Rev. Mod. Phys. 46, 617 (1974).

24. Y.M. Shih, H.M. Huang, and C.-W. Woo: Mol. Cryst. Liquid Cryst. Lett. 34, 7 (1976); R.G. Prest, Ibid 41, 223 (1978).

Cholesteric Structures and the Role of Phase Biaxiality

H. Schröder

Fakultät für Physik, Universität Konstanz,
7750 Konstanz, Federal Republic of Germany

1. Introduction

In a nematic phase we must in principle distinguish between a uniaxial phase at high temperatures and a biaxial phase at lower temperatures. The two structures are separated by a phase transition, theoretically predicted by FREISER [1]. This spontaneous phase biaxiality is to be understood as a cooperative phenomena due to molecular biaxiality, which is described by a molecular temperature independent parameter. However, it seems that in terms of this parameter the biaxiality of the constituent molecules is not large enough to produce a biaxial phase before the order becomes smectic. The situation is very much different if the constituent molecules are optically active, so that the orientationally ordered phase is chiral. The presence of the helical pitch removes the continuous degeneracy of the nematic free energy with respect to the director, i.e. the cholesteric free energy depends on the relative orientation between the pitch axis and the local nematic director. Therefore the equilibrium fluctuations of the director are anisotropic. This effect is reflected in a non-vanishing phase biaxiality, even if the molecules are uniaxial. Due to this effect the isotropic-cholesteric phase transition is significantly different from the isotropic-nematic transition. Phase biaxiality has been discussed qualitatively by several authors [2-5] previously. In this communication we present an analytical expression for the temperature dependence of the pitch-induced phase biaxiality derived from a Landau expansion of the cholesteric free energy. The result illustrates the different behaviour of the nematic and the cholesteric phase, respectively, in the immediate neighbourhood of the isotropic phase and demonstrates the possibility of an optically isotropic phase with long range helicoidal structure. In addition implications concerning possible structures of the "blue" phase are outlined.

2. The Interaction Model

We are interested in orientational order and therefore make use of the pair potential in form of an invariant expansion in terms of irreducible spherical tensors, as introduced by BLUM and TORRUELLA [7].

$$V_{12} = -\sum_{L,L',L''}\sum_{q,q'} U^{qq'}_{LL'L''}\, S^{qq'}_{LL'L''}(\Omega_1,\Omega_2,\vec{\Omega}_{12}) \qquad (1)$$

$$S^{qq'}_{LL'L''}(\Omega_1,\Omega_2,\vec{\Omega}_{12}) = \sum_{p,p'm} \begin{pmatrix} L & L'' & L' \\ p & m & p' \end{pmatrix} D^{(L)}_{p,q}(\Omega_1) D^{(L')}_{p',q'}(\Omega_2) D^{(L'')}_{m,o}(\vec{\Omega}_{12})$$

Here $\vec{\Omega}_{12} = \frac{\vec{r}_1-\vec{r}_2}{|\vec{r}_1-\vec{r}_2|} = (\sin\psi_{12}\cos\gamma_{12},\sin\psi_{12}\sin\gamma_{12},\cos\psi_{12})$ is the orientation of the intermolecular vector. The orientation of each of the molecular frames with respect to the laboratory frame is described by a set of Euler angles.

$$\Omega_i = (\alpha,\beta,\gamma)_i \qquad i = 1,2 .$$

The expansion, (1), has been discussed previously in detail under the aspect of molecular symmetry in cholesterics [8]. It has been found that in lowest order two different terms in the pair potential may contribute to a chiral interaction, namely a second-rank and a third-rank tensor term, which have different physical meanings and consequences. The second-rank tensor contribution allows a stable cholesteric phase for freely rotating molecules, whereas the third-rank tensor contribution requires biaxial molecules in a biaxial environment in order to stabilize a helicoidal phase. Although this investigation was restricted to simple models, it has shown that it is by no means sufficient to treat mesogenic molecules as structureless rods, a fact which also has been pronounced by other authors of [8]. However, as long as the expansion coefficients in (1) remain unidentified with a known interaction, the situation is still unsatisfactory, because only formal arguments can be applied. For this reason a very detailed discussion of dispersion forces and their contribution to the expansion coefficient $U^{pp'}_{LL'L''}$ has been carried out recently [9]. Without going into detail one can say that the internal molecular structure forbids a representation of realistic molecules as pointlike anisotropically polarizable particles. This is of special importance if one considers the difference between nematogenic and cholesteric molecules. Whereas the anisotropic polarizability of a molecule in lowest order is determined by a dipole-dipole transition matrix element, which is a second-rank cartesian tensor, a chiral molecule possesses at least one asymmetric carbon atom, which accounts for optical activity and gives rise to a dipole-quadrupole transition matrix element, which is a third-rank cartesian tensor. This has been shown earlier by GOOSSENS [10]. On the other hand the reduction of the third-rank cartesian tensor leads to contributions to the chiral interaction containing second-rank as well as third-rank spherical tensors. The latter has to be attributed to the chiral character of the ionic skeleton of the respective optically active molecule, so that this term accounts for form anisotropy [9]. Therefore terms in the chiral part of the pair potential with L,L'=3 become unimportant if the chiral shape effect is negligible. In the present communication we assume that this is the case.

For practical purposes it is convenient to make use of a factorization of the expansion coefficients in the following way:

$$U^{qq'}_{LL'L''} = A^{L''}_{LL'} \; T^{(L,q)}(0) \; T^{(L',q')}(0) \tag{2}$$

The $T^{(L,q)}(0)$ are molecular tensors describing the symmetry properties of the molecules in the individual molecular frame. This factorization is a reasonable approximation for dispersion forces in lowest order [9]. Is has the advantage that the angular functions are spherical tensors, too

$$T^{(L,p)}(\Omega) = \sum_q D^{(L)}_{pq}(\Omega) \; T^{(L,q)}(0) \; . \tag{3}$$

Moreover, with the aid of this representation we can easily illustrate the different behaviour of nematogenic and cholesteric molecules. The pair potential consists of the nematic "background" interaction (L"=0) and the cholesteric correction term (L"=1). In both cases we only deal with second-rank tensors (L=L'=2). For a set of five spherical tensor components $T^{(L,q)}(0)$ one can always find a molecular frame in which the components with $q = \pm 1$ vanish and the others are real. We define for the nematic part of the pair potential

$$T^{(2,0)}(0) = 1 \ , \ T^{(2,\pm 1)}(0) = 0 \ , \ T^{(2,\pm 2)}(0) = \zeta \ , \tag{4}$$

where ζ is the molecular biaxiality parameter identical with the corresponding quantity introduced by LUCKHURST et al. [11]. Now we must remember that for optically active molecules the origine of one set of second-rank tensor components, which we call $\tilde{T}^{(2,q)}(0)$, has nothing in common with the anisotropic polarizability but rather with the distribution and orientation of optically active centers within the molecule. Therefore it is in general not possible to find a molecular frame in which both $\tilde{T}^{(2,\pm 1)}(0)$ and $T^{(2,\pm 1)}(0)$ vanish simultaneously. Consequently the correct expression for the pair potential in the present approximation is:

$$\begin{aligned} V_{12} = &- A^{o}_{22} \sum_{p} \begin{pmatrix} 2 & 0 & 2 \\ p & 0 & -p \end{pmatrix} T^{(2,p)}(\Omega_1)\, T^{(2,-p)}(\Omega_2) \\ &- A^{1}_{22} \sum_{p,p',m} \begin{pmatrix} 2 & 1 & 2 \\ p & m & p' \end{pmatrix} T^{(2,p)}(\Omega_1)\tilde{T}^{(2,p')}(\Omega_2) D^{(1)}_{m,o}(\vec{\Omega}_{12}) \\ &- A^{1}_{22} \sum_{p,p',m} \begin{pmatrix} 2 & 1 & 2 \\ p & m & p' \end{pmatrix} \tilde{T}^{(2,p)}(\Omega_1) T^{(2,p')}(\Omega_2) D^{(1)}_{m,o}(\vec{\Omega}_{12}) \end{aligned} \tag{5}$$

where $T^{(2,p)}(\Omega)$ and $\tilde{T}^{(2,p)}(\Omega)$ are linearly independent. A molecular statistical theory for cholesterics on the basis of this pair potential has not been presented, yet. It is clear that the use of (5) will improve the theoretical understanding of cholesterics, because the only reason for a theoretical prediction of a temperature independent pitch by some authors is the use of

$$T^{(2,p)}(\Omega) = \tilde{T}^{(2,p)}(\Omega) \tag{6}$$

On the other hand, however, the solution of this problem requires some numerical effort and goes beyond the purpose of this short communication, which is to study the effect of pitch-induced phase biaxiality for uniaxial and biaxial molecules on the basis of a Landau expansion of the free energy. For the present purpose it is therefore sufficient to use (6) where a finite ζ is incorporated. But we must bear in mind that the choice of this relation does not describe the symmetry of realistic molecules in the general case, but rather corresponds to a special choice of molecular parameters.

3. The Free Energy and Equilibrium Solutions

After the preparation of the pair potential it is a standard procedure to obtain the corresponding mean field potential and the respective free energy expansion (see for example [8]), where the average molecular rotation about a given direction in the system must be considered, if we want to allow for a stable cholesteric phase. It is convenient to choose the z-axis of the laboratory frame for the direction of the helical pitch. Then the local order parameters at two different positions i and j differ by a phase $\varphi_{ij}=\varkappa(z_i-z_j)$ where $\varkappa$ is the inverse pitch.

$$<T^{(2,p)}(\Omega_i)> = e^{ip\varkappa(z_i-z_j)} <T^{(2,p)}(\Omega_j)> \tag{7}$$

The five components $<T^{(2,p)}(\Omega_i)>$ are not linearly independent; because of L=L'=2 it is possible to find a local frame in which again the components with p = ±1 vanish. This is of course a frame rotating about the laboratory z-axis following the average molecular rotation. The present molecular symmetry allows only for a stable configuration, where the director rotates in planes perpendicular to the pitch. Therefore the components $<T^{(2,\pm1)}(\Omega_i)>$ also vanish in the laboratory frame.

Considering that the length of the pitch is always of the order of several hundred molecular average distances a, and using the following notation for the angular functions

$$\left.\begin{aligned} S_o &= T^{(2,0)}(\Omega) = P_2(\cos\beta) + \sqrt{\tfrac{3}{2}}\,\zeta \sin^2\beta \cos 2\gamma \\ S_2 &= S_{-2} = T^{(2,2)}(\Omega)\sqrt{\tfrac{8}{3}} = \sin^2\beta \cos 2\alpha \\ &\quad + \sqrt{\tfrac{2}{3}}\,\zeta\left[(1+\cos^2\beta)\cos 2\gamma \cos 2\alpha - 2\cos\beta \sin 2\alpha \sin 2\gamma\right], \end{aligned}\right\} \tag{8}$$

we obtain as the effective mean field potential

$$V_o(\Omega) = -\left(<S_o>S_o + \frac{3}{4} <S_2>S_2\right) + \frac{1}{2}(\varkappa a)^2 <S_2>S_2 - \varkappa a\, Q<S_2>S_2 , \tag{9}$$

where $Q = \dfrac{i}{\sqrt{6}} \dfrac{A^1_{22}}{A^o_{22}}$.

The order parameters $<S_o>$ and $<S_2>$ refer to the local, rotating frame and obey the following system of self-consistency equations:

$$<S_p> = \frac{\int d\Omega\, S_p(\Omega)\, e^{-V_o(\Omega)/T}}{\int d\Omega\, e^{-V_o(\Omega)/T}} \qquad p = 0,2 \tag{10}$$

The respective free energy expansion in powers of the order parameters $<S_o>$ and $<S_2>$ can readily be obtained following the procedure given in [8].

$$\begin{aligned} \mathcal{F} = &-\frac{1}{4} Q^2 <S_2>^2 + \frac{1}{2}\left(\frac{5T}{1+2\zeta^2} - 1\right)\left(<S_o>^2 + \frac{3}{4}<S_2>^2\right) \\ &- T\,B(\zeta)<S_o>\left(<S_o>^2 - \frac{9}{4}<S_2>^2\right) + T\,C(\zeta)\left(<S_o>^2 + \frac{3}{4}<S_2>^2\right)^2 \end{aligned}$$

$$- T\,D(\zeta)\langle S_o\rangle(\langle S_o\rangle^2 - \frac{9}{4}\langle S_2\rangle^2)\cdot(\langle S_o\rangle^2 + \frac{3}{4}\langle S_2\rangle^2)$$

$$+ T\,E(\zeta)(\langle S_o\rangle^2 + \frac{3}{4}\langle S_2\rangle^2)^3 + T\,F(\zeta)\langle S_o\rangle^2(\langle S_o\rangle^2 - \frac{9}{4}\langle S_2\rangle^2)^2$$

$$- T\,G(\zeta)\langle S_2\rangle^2(\langle S_o\rangle^2 - \frac{1}{4}\langle S_2\rangle^2)^2 \quad . \tag{11}$$

All coefficients depend on the biaxiality parameter ζ only. The free energy is of the general structure

$$F = U - S\,T \, ,$$

where the internal energy is given by

$$U = -\frac{1}{4} Q^2 \langle S_2\rangle^2 - \frac{1}{2}(\langle S_o\rangle^2 + \frac{3}{4}\langle S_2\rangle^2) \, , \tag{12}$$

so that (11) is essentially an expansion of the entropy in terms of certain invariants. These invariants are members of the irreducible representation of the rotational group and determined by the transformation properties of spherical tensors. It is important to remember that $\langle S_o\rangle$ and $\langle S_2\rangle$ *per definitionem* are referring to a local frame in which the components with p=±1 vanish, where the orientation of the respective local frame itself, relative to the laboratory frame may change in an arbitrary manner with the chosen position. Consequently the present expansion of the entropy is to be used for any other possible structure like a bcc- or hexagonal phase, in the same sense the expansions are identical for the nematic and cholesteric phase, respectively. Because of isotropy of space the cholesteric free energy is continuously degenerate with respect to the orientation of the helical pitch. This circumstance justifies to choose any direction desired for the pitch, which is the z-axis presently. The corresponding symmetry breaking term in the free energy is $-\frac{1}{4}Q^2\langle S_2\rangle^2$, which is the cholesteric correction to the nematic free energy, and because of the magnitude of the pitch very small as compared to the latter. The universal character of the entropy expansion and with it the additivity of cholesteric and nematic contributions reflects the inevitable condition that the "switch-off" of the chiral interaction (Q=0) reproduces the nematic phase. Up to fifth order there are only two invariants in (11), namely $\langle S_o\rangle^2 + \frac{3}{4}\langle S_2\rangle^2$ and $\langle S_o\rangle(\langle S_o\rangle^2 - \frac{9}{4}\langle S_2\rangle^2)$. In sixth order one new invariant,

$$\langle S_2\rangle^2(\langle S_o\rangle^2 - \frac{1}{4}\langle S_2\rangle^2)^2 \, , \tag{13}$$

appears which accounts for phase biaxiality due to molecular biaxiality. In order to understand this we need the transformation presented by $D^{(L)}_{pq}(0,\frac{\pi}{2},0)$.

$$\left.\begin{aligned} \langle S_o\rangle &= \pm\frac{3}{4}\langle S_2\rangle' - \frac{1}{2}\langle S_o\rangle' \\ \pm\langle S_2\rangle &= \pm\frac{1}{2}\langle S_2{}'\rangle + \langle S_o\rangle' \end{aligned}\right\} \tag{14}$$

For Q=0 a stable solution of (11) is the uniaxial nematic phase, i.e. we can find a frame in which the degree of order is described by only one component with p=0, i.e. $\langle S_2\rangle' = 0$. It follows immediately that the absence of phase biaxiality is guaranteed by

$$\langle S_o\rangle = \pm \frac{1}{2} \langle S_2\rangle \quad . \tag{15}$$

Consequently the invariant (13) vanishes if the phase is uniaxial. However, on the other hand one is forced to include sixth order terms into the expansion if one wants to calculate quantitative effects regarding phase biaxiality. As mentioned before, the cholesteric phase is always biaxial, even for uniaxial molecules so that relation (15) does not hold.

The equilibrium solution of (11) are determined by $\frac{\partial F}{\partial \langle S_p\rangle} = 0$ and the corresponding stability limits by the zeros of the determinant

$$\left| \frac{\partial^2 F}{\partial \langle S_p\rangle \partial \langle S_{p'}\rangle} \right| \quad . \qquad (p,p' = 0,2) \tag{16}$$

For $\langle S_o\rangle = \langle S_2\rangle = 0$ we obtain for (16):

$$\frac{\partial^2 F}{\partial \langle S_o\rangle^2} = \frac{5T}{1+2\zeta^2} - 1 \qquad \frac{\partial^2 F}{\partial \langle S_2\rangle^2} = -\frac{1}{2} Q^2 + \frac{3}{4}\left(\frac{5T}{1+2\zeta^2} - 1\right)$$

Therefore the stability limit T^* of the isotropic phase is given by

$$T^* = \frac{1}{5}(1+2\zeta^2)(1+\frac{2}{3}Q^2),$$

which shows that the transition is driven by $\langle S_2\rangle$. Following an idea of FREISER [1], we introduce a set of new variables S and ϑ, where ϑ is an angle in order parameter space

$$\left.\begin{aligned} \langle S_o\rangle &= S\cos\vartheta \\ \langle S_2\rangle &= \frac{2}{\sqrt{3}} S\sin\vartheta \end{aligned}\right\} \tag{17}$$

and the invariant S is the amplitude of the order parameter:

$$S^2 = \langle S_o\rangle^2 + \frac{3}{4}\langle S_2\rangle^2 \quad .$$

In these new variables the free energy is:

$$\begin{aligned} F(S,\vartheta) = &-\frac{1}{3}Q^2 S^2 \sin^2\vartheta + \frac{1}{2}\left(\frac{5T}{1+2\zeta^2} - 1\right) S^2 + T\,C(\zeta)\,S^4 \\ &- T[B(\zeta) + S^2 D(\zeta)]\, S^3 \cos 3\vartheta + T[E(\zeta) + F(\zeta)]\, S^6 \\ &- T[F(\zeta) + \frac{4}{27} G(\zeta)]\, S^6 \sin^2 3\vartheta \end{aligned} \tag{18}$$

$F(S,\vartheta)$ has to be minimized with respect to S and ϑ, where the solution for ϑ is of main interest. We only have to consider the region where spontaneous phase biaxiality in the corresponding nematic phase would be absent, because the molecular biaxiality parameter ζ does not seem to be large enough in the realistic case, so that we can omit fifth and sixth order terms. From

$$\frac{\partial F}{\partial \vartheta} = -\frac{2}{3} Q^2 S^2 \sin\vartheta\cos\vartheta + 3\,T\,B(\zeta)\,S^3 \sin 3\vartheta$$

we obtain as the absolute stable solution

$$\cos\vartheta = \frac{1}{2}\left\{ \frac{Q^2}{18T\,B(\zeta)S} - \sqrt{1 + \left(\frac{Q^2}{18T\,B(\zeta)S} \right)^2} \right\} \quad , \tag{19}$$

where the third order term is explicitly $B(\zeta) = \frac{25}{21}\frac{1-6\,\zeta^2}{(1+2\zeta^2)^3}$. In the low temperature region of the cholesteric phase the quantity $\frac{Q^2}{18TB(\zeta)S}$ is small ($\frac{Q^2}{18TB(\zeta)S} \ll 1$), so that the approximate values for $\langle S_o\rangle$ and $\langle S_2\rangle$ are:

$$\left.\begin{aligned} \langle S_o\rangle &\approx -\frac{1}{2}S + \frac{Q^2}{36T\,B(\zeta)} \\ \langle S_2\rangle &\approx S + \frac{Q^2}{54T\,B(\zeta)} \end{aligned}\right\} \tag{20}$$

According to the transformation (14) this corresponds to

$$\left.\begin{aligned} \langle S_o\rangle' &\approx S \\ \langle S_2\rangle' &\approx \frac{Q^2}{27T\,B(\zeta)} \end{aligned}\right\} \tag{21}$$

This result clearly shows the effect of the pitch-induced phase biaxiality, which is accompanied by a reduced birefringence. This tensorial quantity is determined by

$$\langle \varepsilon_{\shortparallel}\rangle - \langle \varepsilon_{\perp}\rangle = \varepsilon_a \langle S_o\rangle \quad , \tag{22}$$

where ε_a is the anisotropy for Q=0 and $\langle S_o\rangle = 1$. Here the solution (19) has to be inserted instead of $\langle S_o\rangle = -\frac{1}{2}S$. Therefore a comparison of the birefringence in a cholesteric and the respective racemic mixture, which is not biaxial, could serve as an indirect observation of pitch-induced phase biaxiality.

Phase biaxiality can become extremely important for two different physical reasons. The condition

$$\frac{Q^2}{18T\,B(\zeta)S} \gg 1$$

can be fulfilled for a sufficiently small S or $B(\zeta)$. Then the approximate form (19) is

$$\cos\vartheta \approx -\frac{9}{2}\,\frac{T\,B(\zeta)S}{Q^2} \tag{23}$$

Hence: $$\cos 3\vartheta \approx \frac{27}{2}\,\frac{T\,B(\zeta)S}{Q^2} \tag{24}$$

Inserting (24) into the free energy, (18), we obtain up to fourth order:

$$\mathcal{F}(S,T) = \frac{1}{2}\left(\frac{5T}{1+2\zeta^2} - 1 - \frac{2}{3}Q^2\right)S^2 + T\left\{C(\zeta) - \frac{27}{2}T\,\frac{B(\zeta)^2}{Q^2}\right\}S^4 \tag{25}$$

Whereas the isotropic-nematic transition is of second order only, if $B(\zeta)=0$ ($\zeta^2 \approx 1/6$), the isotropic-cholesteric transition can be of second order within a finite region, depending on molecular biaxiality and cholesteric coupling coefficient Q. This possibility has been pointed out earlier in a phenomenological approach by BRAZOVSKII and DMITRIEV [12]. The explicit knowledge of the expansion coefficients enables us to determine the crossover condition which is obtained by

$$C(\zeta) - \frac{27}{2} T \frac{B(\zeta)^2}{Q^2} = 0 .$$

Inserting $C(\zeta) = \frac{25}{196} \frac{7(1+2\zeta^2)^3 + 10(1-6\zeta^2)^2}{(1+2\zeta^2)^5}$ and $T = T^*$ we find:

$$Q^2 = \frac{15}{7} \frac{(1-6\zeta^2)^2}{(1+2\zeta^2)^3} \tag{26}$$

If Q^2 is smaller than the right hand side of (26), a second order transition is not possible. Since $Q = \varkappa_o a = 2\pi \frac{a}{\lambda}$ cannot be made arbitrarily large, the average number of molecules per pitch length λ is always of the order $\sim 10^2$, condition (26) can practically only be satisfied for ζ-values with $\zeta^2 \approx 1/6$. Thus strongly biaxial molecules are needed to produce a second order transition for reasonable values of the cholesteric coupling constant.
All odd-order coefficients are zero for $\zeta^2=1/6$ and the corresponding free energy is explicitly:

$$\mathcal{F}(S,\vartheta,T) = -\frac{1}{3} Q^2 S^2 \sin^2\vartheta + \frac{1}{2} \left(\frac{15}{4} T - 1\right) S^2 + \frac{225}{448} T S^4$$

$$+ \frac{88\ 875}{224\ 224} T S^6 - \left(\frac{25}{8} \right)^3 \frac{6}{1001} T S^6 \sin^2 3\vartheta \tag{27}$$

The absolute stable solution of $S \neq 0$ is $\cos\vartheta = 0$.

Hence: $$\langle S_o \rangle = 0 \qquad \langle S_2 \rangle = \frac{2}{\sqrt{3}} S \tag{28}$$

From this solution it follows that in the laboratory frame on the average all order parameters vanish. Thus this phase has an "isotropic" appearance. It has all optical properties of a cholesteric phase, except for the absence of birefringence.

Discussion

An important result of the present considerations is an analytical expression for the fact that a cholesteric phase cannot exist without phase biaxiality, although the constituent molecules may be uniaxial. This pitch-induced phase biaxiality may be neglected for all practical purposes in the low temperature region, but it becomes very important in the neighbourhood of the isotropic phase. Here it has the consequence that the relevant order parameter is $\langle S_2 \rangle$ instead of $\langle S_o \rangle$ in the case of the nematic phase. Moreover, due to phase biaxiality the isotropic-cholesteric transition may be of second order in a finite region, for which we have given the crossover condition explicitly. In addition we have shown the possibility of a helicoidal phase which is optically isotropic. Despite the fact that the present results are to some extent model dependent (special choice of ζ), we can draw some general conclusions in regards of the blue phase.

One must dispose of molecular parameters which distinguish between cholesterics with and without an optically isotropic modification. These parameters are probably related to a small pitch with strong temperature dependence, i.e. the potential (5) is needed. Since the blue phase is only stable in the neighbourhood of the isotropic phase, phase biaxiality plays a dominant role and must be an inevitable ingredient of any theory. This holds especially for a cubic or hexagonal phase, as, for example, proposed by HORNREICH and SHTRIKMAN [13]. Their conclusions are based on a phenomenological Landau theory for *uniaxial* molecules. The crucial statement is that whenever the isotropic-cholesteric transition would be of second order, a bcc structure should be favored instead. It is clear that the coefficients of the phenomenological Landau expansion are not identical with the present ones, and necessarily do not have to be identical. Nevertheless the phenomenological crossover condition is indeed satisfied with (26). This leads to the conclusion that among other possible properties, especially strongly biaxial molecules would favor a cubic arrangement, which implies that the respective racemic mixture should be nematic biaxial close to the isotropic phase. This can of course be tested. On the other hand, however, one would expect strongly biaxial molecules not to build a cubic structure, because the helical pitch wants to develop in a defined direction with respect to the molecular axes and thus block any but the helicoidal structure. Therefore the problem of the blue phase still must be considered unresolved.

References

1 M.J. Freiser, Phys. Rev. Letters, 24,1041(1970)
2 R.G. Priest, T.C. Lubensky, Phys. Rev. A9,893(1974)
3 A. Wulf, J. Chem. Phys. 59,6596(1973)
4 A. Wulf, Phys. Rev. A8,2017(1973)
5 J.P. Straley, Phys. Rev. A10,1801(1974)
6 W.J. A. Goossens (presented at 7th International Liquid Crystal Conference, Bordeaux 1978)
7 L. Blum, A.J. Torruella, J. Chem. Phys. 56,303(1972)
8 G.R. Luckhurst and G.W. Gray (ed.) "The Molecular Physics of Liquid Crystals", AP, London (1979)
9 H. Schröder, J. Chem. Phys. (1 March 1980)
10 W.J.A. Goossens, Mol. Cryst. Liq. Cryst. 12,237(1971)
11 G.R. Luckhurst, C. Zannoni, P.L. Nordio and V. Segre, Mol. Phys. 30,1345(1975)
12 S.A. Brazovskii, S.G. Dmitriev, Zh. Eksp. Teor. Fiz. 69,979(1975)
13 R.M. Hornreich, S. Shtrikman, these proceedings p. 185

Optical Properties of Cholesteric Liquid Crystals Under a DC Electric Field

F. Simoni, R. Bartolino, N. Scaramuzza
Dipartimento di Fisica, Università di Calabria, Cosenza, Italy

1. Introduction

The high rotatory power of cholesteric liquid crystals (CLC) can be interesting for practical applications when a suitable switching of it is available. It is well known that electric field induces deformations on the CLC structure [1] then we expect variations of the rotatory power versus the applied electric field. The structure deformations modify the apparent pitch of CLC then shift the selective reflection band; the relation between reflection band and rotatory dispersion has been discussed by S. CHANDRASEKHAR [2] whose results will be recalled later.

Different kinds of distortions are possible depending on: the sign of the dielectric anisotropy $\Delta\varepsilon=\varepsilon_{\parallel}-\varepsilon_{\perp}$ and the reciprocal orientation of electric field $\underline{E}$ and helical axis $\underline{z}$. For instance when $\Delta\varepsilon>0$ (the most usual case for CLC) we have the unwinding of helix for $\underline{E}\perp\underline{z}$ [3], while when $\underline{E}\parallel\underline{z}$ in principle several distortions are possible leading to a shorter apparent pitch as discussed in the following. Both the geometries are interesting for the previously recalled applications because they give opposite variation of the rotatory power at a fixed wavelength. Here we report results for the condition $\Delta\varepsilon>0$, $\underline{E}\parallel\underline{z}$.

2. Deformations Under Electric Field and Rotatory Power

As many authors have discussed [1] in our geometry several distortions can take place, i.e. the periodic deformation of the structure as described by W. HELFRICH [4], the focal conics distortion or the sudden 90° tilt of the whole structure. HURAULT [5] showed that HELFRICH's deformation has the lower threshold when the sample thickness L is much greater than the pitch P.

The field threshold value is:

$$E_{th}=2\pi\left(\frac{4\pi(2k_2k_3)^{1/2}}{\Delta\varepsilon}\right)^{1/2}\frac{1}{PL} \qquad (1)$$

The wall's effect may be important for L of the order of P [6].

These deformations act on the reflection band shifting it towards shorter wavelength (the problem is similar to that of multiple reflections in a Fabry-Perot when the light's incidence angle is increased).

The relationship between the shift of the reflection band and the variation of the rotatory power is deductible from CHANDRASEKHAR'S theory. From this theory two different expressions are obtained inside and outside the reflection band for the rotatory power R at normal incidence:

a) inside the reflection band

$$R = -\frac{\pi(n_1 - n_2)^2 P}{4\lambda^2} + \frac{\pi(\lambda-\lambda_o)}{P\lambda} \qquad (2)$$

where $n_1=\varepsilon_2^{1/2}$, $n_2=\varepsilon_1^{1/2}$. This expression is valid in the range $Q^2>G^2$, $|Q|^2$ is the reflection coefficent for a layer and the parameter $G = -2\pi(\lambda-\lambda_o)/\lambda$;

b) outside the reflection band ($Q^2<G^2$)

$$R = -\frac{\pi(n_1 - n_2)^2 P}{4\lambda^2} + \frac{\pi(\lambda-\lambda_o)}{P\lambda}\left[1-\left(1-\frac{Q^2}{G^2}\right)^{1/2}\right] \qquad (3)$$

This theory is valid for infinitely thick specimens, but it remains a good approximation for a sample thickness few times greater than the pitch. Eqs. (2) and (3) are valid for a right handed helical structure. If the structure is left handed the rotatory power changes sign. The most important consequence of this theory is the provision of a large variation in the absolute value of the rotatory power in a narrow band.

3. Experiments

Two kinds of measurements were performed: a) the spectrum of the transmitted light, b) the rotatory power at a fixed wavelength, both versus the applied electric field.

A Cary 17 spectrophotometer was used for the first measurement; while for the rotatory power a standard technique was used with two crossed polarizers and the He-Ne laser (λ=633nm).

The sample was a mixture of cholesteryl chloride, cholesteryl oleate and cholesteryl nonanoate with the reflection peak at λ_o= 675 nm, so that a small shortening of the pitch may induce a large variation of the rotatory power at λ = 633nm.

The sample was prepared in a planar texture by rubbing two optical glasses over which was evaporated a conductive film of SnO_2, so that the electric field resulted parallel to the helix axis. Sample thickness was fixed by means of Mylar spacers, different spacers were used in the range from 6µ to 26µ . The temperature was controlled within 0.1°C. The d.c. electric field reached a maximum value of about $4x10^5$ V/cm .

In Fig. 1 we report the ratio between the actual value of the wavelength of minimum trasmission λ_{oE} and the zero field value λ_o versus the reduced field E/E_{th}, where E_{th} is given by (1).

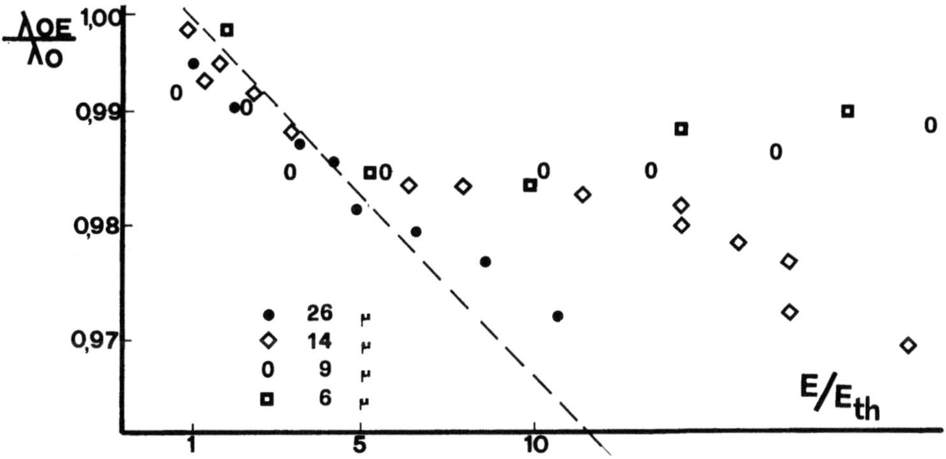

Fig. 1 Ratio λ_{oE}/λ_o of the actual value of the wavelength of the minimum transmission on the zero field value, versus the reduced field E/E_{th} , for different sample thicknesses.

The thickness dependence of the threshold agrees roughly with HELFRICH's theory and for field values a bit higher than E_{th} the behaviour of the ratio is similar for any thickness.

For higher fields the behaviour is different following the sample thickness. λ_{oE} shifts towards lower wavelengths, for the greater thickness, since an irreversible deformation takes place.

For the smaller thicknesses after a lowering of the wavelength of the peak increasing the field again an increase of it is seen.

Looking at the behaviour of the thinner samples, we see that departure from the behaviour of the thickest sample occurs at an electric field which is roughly proportional to the sample thickness.

In all studied range not irreversible processes occur.

Figure 2 shows the results on the variation of the optical rotation versus the reduced field for a sample of 14 μ of thickness. In our experimental conditions (zero-field peaks wavelengths and laser light) few volts under the threshold the rotatory power begins to vary and just after the threshold the variation becomes very important, changing of about $2x10^3$ deg/mm in 10 V of applied field.

For this sample the actual value of the threshold shifts the reflection band near the light wavelength which falls in the large rotatory dispersion range as expected by (2) and (3).

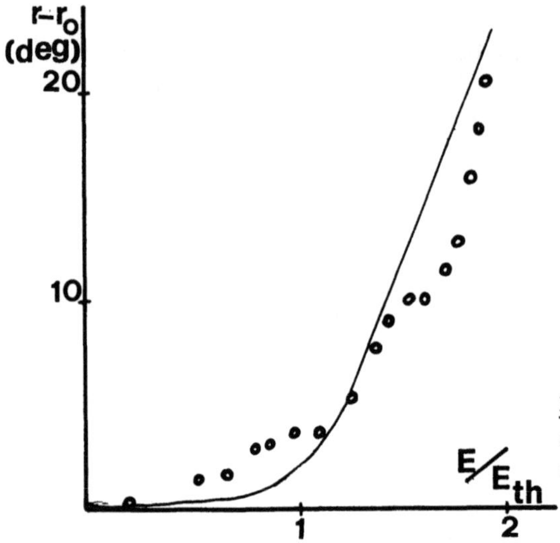

Fig.2 Variation of the optical rotation $r-r_o$ versus the reduced field E/E^o_{th}.

We put the experimental value of λ_{oE}/λ_o in (2) and (3) and fitted the rotatory power results. A quite good agreement is found.

4. Conclusions

The transmission measurements show that for a cholesteric liquid crystal with $\Delta\varepsilon>0$ when $\underline{E}\parallel\underline{z}$ the Helfrich's deformation appears as the first one.

For fields few times higher than the Helfrich's threshold, another kind of deformation takes place (not yet irreversible) which depends strongly on the sample thickness.

In the future we want clear out this point by more accurate selective reflection measurements, coupled with dielectric measurements.

The rotatory power measurements confirm the expected behaviour as pointed out by the Chandrasekhar's theory, without taking care of the occurred kind of deformation. This result allows to consider this mechanism as suitable for future applications. In fact for small variation of field values around the threshold we see a spectacular change of the rotatory power of about 2×10^3 deg/mm. Experiments are under way also with a.c. fields with the spinning analyzer technique [7] allowing dynamical measurements.

Aknowledgements

Discussion on this paper, during the Garmisch Conference, are greatly acknowledged with:
G.Durand, P.Martinot-Lagarde, P.G.de Gennes, F.Rondelez and W. Helfrich.

REFERENCES

1 For a general review see P.G.de Gennes "The physics of the Liquid Crystals" Oxford University Press 1973, and references therein.

2 S.Chandrasekhar, K.N.Srinivasa Rao Acta Cryst.A24,445 (1968)

3 R.B.Meyer Appl.Phys.Lett.12,281(1968)

4 W.Helfrich Appl.Phys.Lett.17,531(1970)
W.Helfrich Jour.Chem.Ph.55,839(1971)

5 J.P.Hurault Jour.Chem.Ph.59,2068(1973)

6 F.Rondelez Thesis,Univ.Paris IX(1973)

7 P.E.Sokol, J.T.Ho Appl.Phys.Lett.31,487(1977)

New Simple Model of a Liquid Crystal Light Valve

B. Kerllenevich and A. Coche
Centre de Recherches Nucléaires, F-67037 Strasbourg Cédex, France

The device described consists of two transparent electrodes (an In_2O_3 film and a conducting glass electrode) and between them a photosensitive element in series with a liquid crystal. It does not require an optical blocking layer to protect the photosensitive material from the "readout" light". It is based on the electric field induced cholesteric-nematic phase transition. The photosensitive element is an indium oxide (In_2O_3)/silicon heterojunction. A 13 μm thick layer of a mixture of 1132 TNC Merck nematic product (92%) + cholesteryl nonanoate (8%) is sandwiched between the silicon wafer (10 000 Ω·cm) and the conducting glass electrode. An ac voltage is applied between the In_2O_3 film and the conducting glass electrode. When the In_2O_3 side is illuminated with the "writing light", the diode impedance decreases and the electric field in the liquid crystal increases becoming sufficient to produce the cholesteric-nematic transition. The applied voltage necessary to induce this transition has been found to decrease for a given frequency when the writing light power P_{Wr} increases and to be independent of P_{Wr} for its higher values. With a convenient choice of the cell voltage, a sensitivity better than 10 $\mu W \cdot cm^{-2}$ can be easily obtained. The sensitivity is maximum for a writing light wavelength of 0.8 μm, therefore such a structure can be used in the near infrared. Rise times of a few tens of ms can be obtained.

Restabilized Planar Texture in Homogeneously Aligned Cholesterics and Its Application to a Color Display Device

Y. Ebina and H. Miike

Department of Electrical Engineering, Yamaguchi University, Tokiwadai 2557, Ube, Japan

This report describes the dissipative structure[1,2], and the possibility of color device(CPT-cell) by utilizing the discrete change of rotatory dispersions under the applied a.c. electric field V[3].

The sample is the mixture of cholesteryl nonanoate(CN) and MBBA with negative dielectric anisotropy. The concentration of CN is within 10 wt%.

The dissipative electric power W is computed from the electric current and its phase angle, in order to see another aspect of induced patterns when V is increased. W is not proportional to V^2 above the first threshold V_c. The nondimensional quantity $\Delta Ne=(W-W_0)/W_0$ is calculated, where W_0 is the power extrapolated from ohmic region($V<V_c$). ΔNe has a minimum point b'. Around this point the plain part of the pattern occupies the major part of the nonuniformly distorted patterns. The dissipative structure around b' can be attained reversibly by applying V.

The cholesteric phase occupying dominant part has the optical rotatory dispersion and displays coloring when we insert the cell between two polarizers with a suitable angle. The rotatory dispersion characteristics are examined in CPT-cells made by various concentrations of CN. Favorable dispersion ones are obtained on the mixtures of 3-3.5 wt% or L/P=2.5 or 3, in which L is the film thickness with the nominal value of 9 μm and P is the realized pitch length. Typical CPT-cell of 3.15 wt% is examined in details yielding dominant wavelengths of 571, 488 and 472 nm for yellow, green and blue colors, respectively. This gives good color purity(0.43, 0.42 and 0.774). The transient times of color change are computed from the transmittance change. If the strength and frequency of V are suitable, about 30 msec switching time is obtained. The proposed two color display device is superior in blue color. The CPT-cell has some favorable characteristics as color display device.

1. T. Kohno, H. Miike and Y. Ebina, J. Phys. Soc. Japan 44(1978) 1678.
2. H. Miike, T. Okazaki, T. Kohno and Y. Ebina, J. Phys. Soc. Japan 45(1978) 1174.
3. H. Miike, T. Yamada and Y. Ebina, Japan J. of Applied. Phys. 19, No.4(in press).

Orientation of the Chiral Solutes in Induced Cholesteric Solutions

E.H. Korte, P. Chingduang
Institut für Spektrochemie,
D-4600 Dortmund 1, Federal Republic of Germany

1. Introduction

In nematic liquid crystals a cholesteric molecular arrangement is induced by dissolving a small quantity of a chiral compound [1]. The induction is independent of whether the solute is mesogenic or not and of the actual type of its chirality (centre, axis, plane or chirality; helical molecule) [2]. The induced cholesteric solution exhibits cholesteric features, in particular selective reflexion and an anomaly of the optical rotation around the same centre wavenumber $\tilde{\nu}_R$, which is related to the pitch z of the structure by the well known equation

$$\tilde{\nu}_R = (\bar{n}z)^{-1} \qquad (1)$$

where $\bar{n}$ denotes the mean refractive index. The sense of circular polarisation of the selectively reflected light and the sign of the rotatory anomaly indicate the induced handedness. Due to the low solubility of non-mesogenic chiral compounds in nematic liquid crystals, the induced twist is quite weak so that these phenomena occur at *infrared* wavelengths.

Since enantiomers lead to countercurrently coiled structures, we use the induced handedness together with the helical twisting power as an analytical tool for the dissolved chiral compounds [2,3]. The rotatory anomaly at the selective reflexion band we refer to as *R-Cotton effect* (R meaning reflexion), offers additionally information on the mesophase: its amplitude is related to the orientational order parameter [4]. This is mainly determined by the nematic solvent. Its temperature dependence in the induced cholesteric state is similar to that in the nematic state even though the clearing temperature is effected by the type and concentration of the chiral solute, in particular, if this were non-mesogenic as usual for most applications. Such molecules often are less distinctly prolonged so that their orientation within the mesomorphic solvent might a priori be unpredictable.

However, the orientation determines the solvent-solute interaction which in turn, governs the cholesteric induction and the temperature dependence of the pitch. To contribute to this problem we measured the infrared rotatory dispersion spectra of induced cholesteric solutions and evaluated the anomalies resulting from the chiral solute molecules with respect to their orientation.

2. Absorption Induced Rotatory Anomalies

Scanning the infrared rotatory dispersion of an induced cholesteric solution one will find the R-Cotton effect provided the pitch is appropriate. In any case, several anomalies are displayed the position of which on the wavenumber scale is essentially independent of the pitch. In these cases no matching of wavelength and pitch takes place but these are caused by absorption bands of either the solvent or the solute: polarized bands of orientated molecules exhibit *linear* dichroism which, for a twisted pile, results in a *circular* dichroism band correlated with a rotatory anomaly [5]. The sign of each of these anomalies which we refer to as *A-Cotton effects* (A meaning absorption), depends [6] on the handedness and the pitch of the induced cholesteric structure, but also on the angle (α) between the transition moment involved and the local director as given by (2) [2]

$$\text{sign (RCE)} = \text{sign (ACE)}\ \text{sign}\ (\tilde{\nu}_R - \tilde{\nu}_A)\ \text{sign}\ (\cos 2\alpha) \qquad (2)$$

where sign (RCE) and sign (ACE) denote the signs of the R-Cotton effect and a given A-Cotton effect, respectively, while the difference between the centre wavenumber $\tilde{\nu}_R$ of the R-Cotton effect and the wavenumber $\tilde{\nu}_A$ of the maximum of the absorption band accounts for the pitch. If the pitch and the handedness have been obtained from the R-Cotton effect or from other experiments, by means of (2) it can be derived from the observed sign of an A-Cotton effect whether the transition moment involved is orientated parallel or perpendicular to the local director of the mesomorphic solution. Knowing the orientation of the transition moment within the molecule, information on the orientation of the molecule with respect to the director is obtained in this way.

Following the calculations by HOLZWARTH [5] the shape of an A-Cotton effect can be described in terms of the absorption band profile and its Kramers-Kronig transform. Contributions proportional to the band profile and the square of its transform distort the sigmoidal shape to various extents depending on the angle α and shift the centre of the anomaly from the peak of the absorption band. The amplitude of the A-Cotton effect depends additionally on the ratio $\tilde{\nu}_R/\tilde{\nu}_A$.

Due to the excess of solvent molecules, their absorption bands and, consequently, their A-Cotton effects will be predominant in the spectra. However, in many cases the infrared range provides A-Cotton effects based on absorption bands of the solute, which are isolated from those of the solvent, characteristic for well defined vibrations and strong enough to be observed in spite of the low concentration attainable with non-mesogenic solutes. Particularly suitable are the valence vibrations of C=O, C≡N, and N=C=O as well as the out-of-plane ring deformation of monosubstituted benzene.

3. Experimental Method

The rotatory dispersion spectra were measured using a modified IR-spectrometer Perkin-Elmer model 180 [3]. The sample and the analyzer are positioned in a part of the optical path common to sample and reference beam, thus compensating the absorption by the sample. The vector of the polarizer in the sample beam forms an angle of 45° with that of the analyzer. The recorded output of the spectrometer is given by (3)

$$I_S/I_R = T_P(1 + \sin 2\rho)/2 \qquad (3)$$

if I_S and I_R denote the intensities of the sample and reference beam, respectively, T_P is the transmission of the polarizer and ρ the optical rotation. The sinusoidal scale of ρ distorts the shape of the Cotton effects; nevertheless, their sign is unambiguous in all cases. The signal-to-noise ratio is decreased compared to normal usage of the spectrometer due to the fact that both beams are affected by the sample and analyzer.

4. Results and Discussion

The upper diagram of Fig. 1 shows the transmission spectrum of an induced cholesteric solution of menthone in Nematic Phase IV Licristal® (E. Merck) at 30° C (molefraction x = 0.17; sample thickness s = 25 µm) while the correspondent rotation spectrum is given in the lower part. This consists of a number of A-Cotton effects superimposed by a broad R-Cotton effect as indicated by the broken line. This is positive, indicating that a lefthanded structure has been induced. The related selective reflexion band causes the obvious decrease of the transmission in this wavenumber range. All pronounced A-Cotton effects are due to absorption bands of the solvent.

The A-Cotton effects at 1600 cm^{-1} and 1500 cm^{-1} are negative; since they occur at larger wavenumbers than the R-Cotton effect ($\tilde{\nu}_R$), the difference ($\tilde{\nu}_R$-$\tilde{\nu}_A$) is negative also. Therefore, it follows from (2) that cos 2α = 1, i.e. the involved transition moments are expected to be parallel to the local director. The same result is obtained for the 1155 cm^{-1} A-Cotton effect since its sign as well as ($\tilde{\nu}_R$-$\tilde{\nu}_A$) are positive. However, the A-Cotton effect at 837 cm^{-1} is negative even though it occurs at wavenumbers smaller than $\tilde{\nu}_R$, consequently, cos 2α must be negative and the transition moment is expected to be perpendicular to the director. Assuming the molecular long axis to be aligned parallel to the director these results is in accordance with the assignments of the bands [7].

The calculations by HOLZWARTH [5] explain the shape of the A-Cotton effect around 1250 cm^{-1} as being a consequence of the close neighbourhood to $\tilde{\nu}_R$ for a vibration parallel to the director, this orientation is confirmed by the assignment [7].

4.1 Menthone

In the transmission spectrum (Fig.1, top) three resolved bands are indicated by arrows which are caused by the dissolved menthone. The band at 1712 cm^{-1} due to the C=O vibration is isolated from the bands of the solvent and strong enough to cause a measurable A-Cotton effect. Fig.2a shows this on an enlarged scale together with the line of zero rotation obtained by using the same

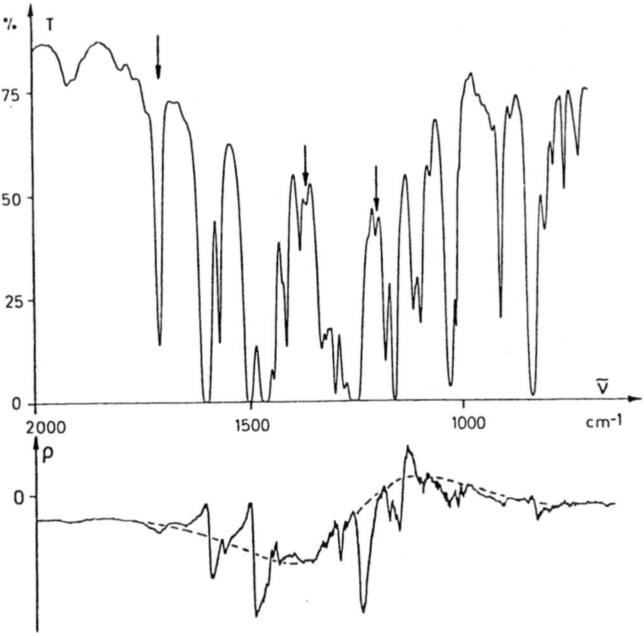

Fig.1 Infrared transmission (T) and optical rotation (ρ) of an induced cholesteric solution of menthone in Nematic Phase IV ($x=0.17$; $s=25\mu m$), broken line indicates positive R-Cotton effect

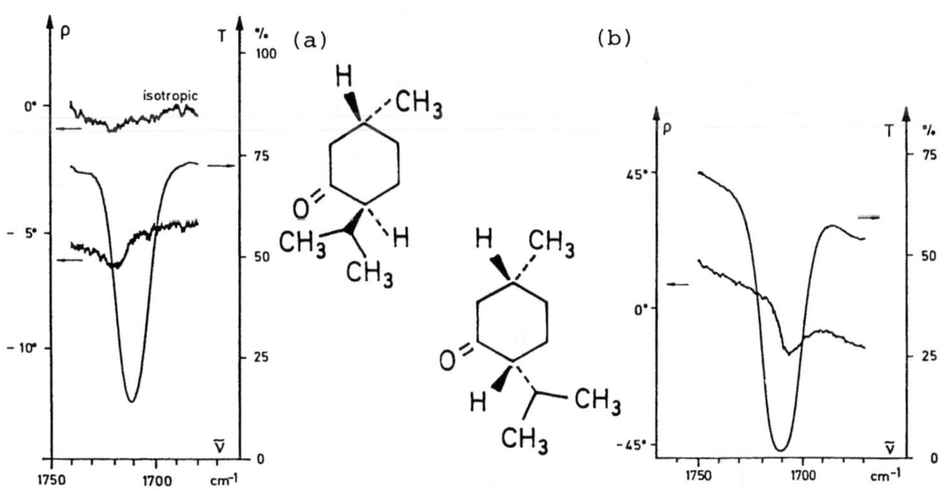

Fig.2 Absorption band and A-Cotton effect due to C=O vibration of (a) menthone, (b) isomenthone ($x=0.13$; $s=50\ \mu m$; $\tilde{\nu}_R \approx 20\ cm^{-1}$) in Nematic Phase IV

sample in the isotropic state. Comparison with the A-Cotton effects due to parallel vibrations of the solvent molecules shows that in first order the C=O vibration and, consequently, the group itself must be orientated perpendicular to the director. However, from the distorted shape it must be concluded that the angle differs from 90° to some extend. This is in accordance with the assumption that the para-axis of menthone is aligned parallel to the director, the C=O bond thus being oblique to it. Additionally, the influence of thermal motion has to be taken into consideration.

Isomenthone, on the other hand, is less distinctly prolonged, so that the preferred orientation seems to be questionable. Nevertheless, the A-Cotton effect due to its C=O band indicates that the transition moment is oriented in a similar way as for menthone. The A-Cotton effect is shown in Fig.2b. Its sign is obviously negative, however, contrary to menthone, isomenthone induces a righthanded cholesteric structure exhibiting a negative R-Cotton effect. The evaluation by using (2) for both diastereomers is shown in Table 1.

Table 1 Evaluation using (2)

Compound	sign(RCE)	sign(ACE)	sign($\tilde{\nu}_R-\tilde{\nu}_R$)	=	sign(cos2α)	orientation
menthone	(+1)	(+1)	(-1)	=	(-1)	⊥
isomenthone	(-1)	(-1)	(-1)	=	(-1)	⊥

4.2 Androstenone

Since the skeleton of 17β-acetoxy-5α-androst-1-en-3-one is similar to the ones of some cholestogenic molecules, one is tempted to assume an orientation parallel to the nematic solvent molecules (Nematic Phase IV). The observed A-Cotton effects support this. The compound owns two C=O groups which lead to clearly separated absorption bands: one appertains to ring A, the other is part of the substituent of C-17. While the A-Cotton effect at 1680 cm^{-1} indicates that the keto group is orientated mainly parallel to

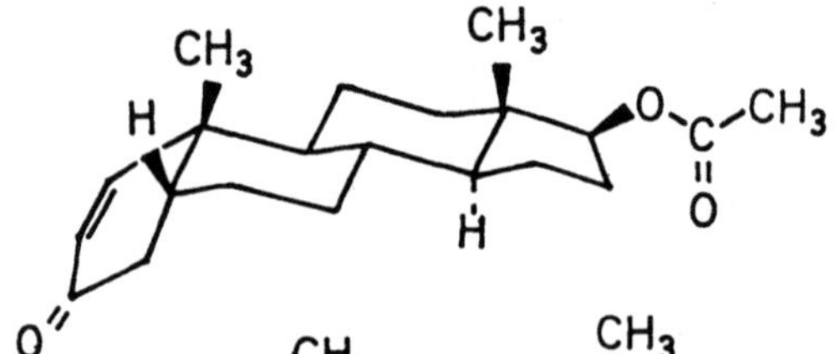

17β-acetoxy-
5β-androst-1-en-3-one

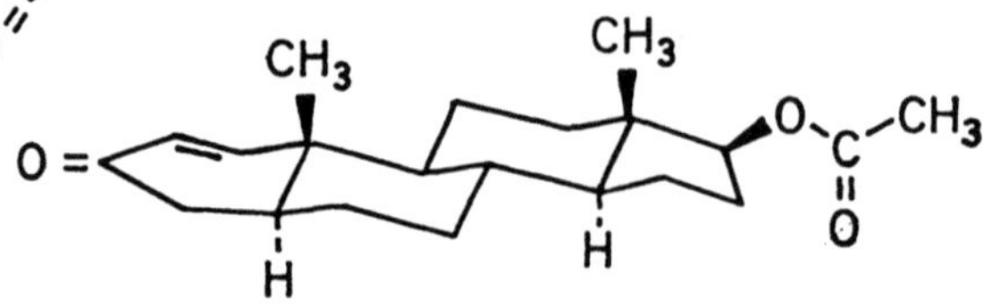

17β-acetoxy-
5α-androst-1-en-3-one

the director, the A-Cotton effect around 1740 cm^{-1} confirms that the carboxyl group is perpendicular to the director.

For the 5β isomer (showing greater helical twisting power) ring A and, consequently, the C-3 keto group is oblique to the plane of the molecule. As a consequence we found no measurable A-Cotton effect for this absorption band, while the carboxyl band causes a similar A-Cotton effect as observed when using the 5α isomer.

4.3 1-Phenylethylisocyanate

Here two substituents directly bonded to the chirality centre (C^*) are suitable labels for the orientation: the N=C=O group absorbs at 2260 cm^{-1} in a definitely empty part of the solvent spectrum and the out-of-plane ring deformation of the monosubstituted benzene leads to an isolated absorption band at about 700 cm^{-1}. The transition moment of the latter vibration is perpendicular to the plane of the ring. From the signs of the related A-Cotton effects it must be concluded that the transition moment of the 700 cm^{-1} vibration is perpendicular to the director, while that of the 2260 cm^{-1} vibration is parallel to the director. This would be consistent with an alignment of the C^*-C_6H_5 and N=C=O bonds almost parallel to the long axes of the solvent molecules. The sign of the linear dichroism exhibited by a solution of the racemic mixture of the enantiomers confirms this result. Additionally, the linear dichroism measurement allows to determine the orientational order parameter which comes out to be for the N=C=O transition moment one third only of that of the ring deformation.

Treating the cell windows with lecithin and choosing the sample thickness smaller than the pitch enforces a nematic order in homeotropic orientation [8]. Comparison of the measured absorbance with that of the same sample in the isotropic state shows orientation and order parameter for the different vibrations [9]. On the first glance the results we obtained in this way are contradictory to the ones stated before: *both* transition moments are indicated to be perpendicular to the director. The order parameter of the N=C=O vibration comes out to be considerably smaller than the one of the ring deformation: 0.12 and 0.52, respectively.

In spite of the pretended discrepancy of the N=C=O orientation the two results are consistent. From the definition of the order parameter

$$S = 1 - \frac{3}{2}\,\overline{\sin^2\theta} \qquad (4)$$

an expectation value $\langle\theta\rangle = 50^{\circ}$ is obtained for $S = 0.12$.

On the other hand, the linear dichroism exhibited by a solution of the racemic mixture is measured using light incidenting from a direction perpendicular to the director. The same holds for the A-Cotton effect exhibited by an induced cholesteric solution. The linear dichroism caused by a valence vibration (absorption coefficient ε) the transition moment of which forms an angle ϕ

with the plane of polarization of one of the two beams is given by

$$\Delta\varepsilon = \varepsilon\ (\sin^2\phi - \cos^2\phi) = \varepsilon \cos 2\phi \qquad (5)$$

Starting from $\phi = 0$, $\Delta\varepsilon$ changes sign for $\phi = 45^{\circ}$ corresponding to $S = 0.25$. Therefore, for $\phi = \langle\theta\rangle$ one obtains here $\Delta\varepsilon = -0.17\ \varepsilon$, so that the sign of the A-Cotton effect indicates an orientation of the transition moment parallel to the director. $\langle\theta\rangle$ being close to 45° explains the observed distortion of the sigmoidal shape [5].

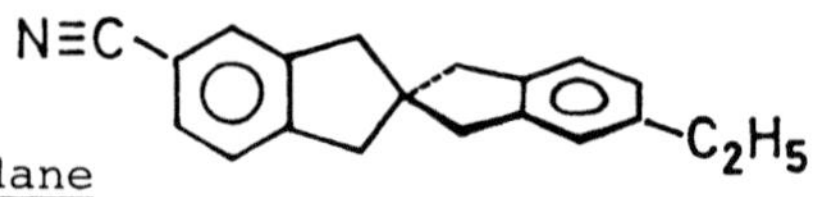

4.4 5-Cyano-5'-ethyl-2,2'-spirobiindane

The C≡N group producing a strong and completely isolated absorption band at 2220 cm^{-1} is rigidly linked to the spirobiindane skeleton. Therefore, the orientation of the transition moment parallel to the director indicated by the A-Cotton effect confirms the alignment of the whole molecule in this way.

Other 5,5' derivatives of spirobiindane as well as different compounds substituted by more flexible groups like $COOCH_3$ or $COCH_3$ are subject to further studies including the linear dichroism of the homeotropic state and calculations of the A-Cotton effect shape [5] in dependence on the angle α between transition moment and director, and taking into account the thermal motion.

Acknowledgements

We would like to thank Prof. Dr. H. Stegemeyer for helpful discussion and Prof. Dr. K. Schlögl for making available the spirobiindane derivatives. The financial support by the Minister für Wissenschaft und Forschung des Landes Nordrhein-Westfalen, the Bundesminister für Forschung und Technologie, and the Deutsche Forschungsgemeinschaft is gratefully acknowledged.

Literature

1 G. Friedel, Ann. Phys. (Paris) 18, 273 (1922)
2 E.H. Korte, B. Schrader, S. Bualek, J. Chem. Research (S) 1978, 236; (M) 1978, 3001
3 E.H. Korte, Appl. Spectrosc. 32, 568 (1978)
4 E.H. Korte, Mol. Cryst. Liq. Cryst. 44, 151 (1978)
5 G. Holzwarth, N.A.W. Holzwarth, J. Opt. Soc. Amer. 63, 324 (1973)
6 E. Sackmann, J. Voss, Chem. Phys. Letters 14, 528 (1972); G. Holzwarth, I. Chabay, N.A.W. Holzwarth, J. Chem. Phys. 58, 4816 (1973)
7 E.B. Wilson, Jr., J.C. Decius, P.C. Cross, "Molecular Vibrations", McGraw-Hill, New York 1945
8 M. Brehm, H. Finkelmann, H. Stegemeyer, Ber. Bunsenges. phys. Chem. 78, 883 (1974)
9 H. Kelker, R. Hatz, G. Wirzing, Z. Anal. Chem. 267, 161 (1973)

Ultrasound Effects on Cholesterics

F. Scudieri, M. Bertolotti, L. Sbrolli

Istituto di Fisica-Facoltà di Ingegneria, Università Rome, Italy
and Gruppo Nazionale Elettronica Quantistica Plasmi, of CNR, Roma, Italy

1. Introduction

To have large effects in cholesteric submitted to an ultrasound field it is necessary to work with materials with a large pitch value. Such materials can be obtained by doping a nematic with a small amount of cholesteric.

As it is well known in this case the pitch of the resulting structure follows a law of the kind p(μm) = 0,3/C where c is the weight percent of the cholesteric material into the nematic one.

In the following a mixture of MBBA with 0.5% in weight of cholesteric oleate is used, and the effects of a ultrasound field are studied, considering both the changes of the structure as a function of the ultrasound frequency and the time response under pulsed operation.

2. Experimental set-up

The mixture of MBBA and cholesteryl oleate was sandwiched between two optical glasses, treated for homeotropic alignment. The sample, whose thickness was 20μm, exhibited usually two different regions where the homeotropic and finger structures [1] were present (s.Fig.1) one cell window was cemented to a cylindrical piezoelectric transducer radially poled. The observation was performed with a polarizer microscope to study the effect of changing frequency of the ultrasound field on

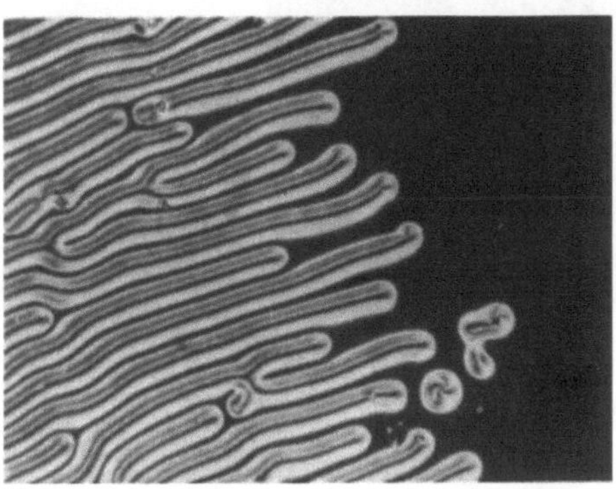

Fig.1 Finger structure

the sample structure. The time response was also studied of the Bragg diffracted light by the periodic structure under pulsed applied stress and of the light directly transmitted through it.

3. Pattern changes under mechanical stress

Attention was given to the finger structure behavior under the mechanical stress. By changing frequency starting from very low frequencies up to some tens KHz the finger structure was subjected to several modifications. At very low frequencies (from 0.1 up to 1 Hz) it was clearly visible, and was modulated at the same frequency of the applied stress. Increasing frequency in the region from 1 to 130 Hz it disappeared and the material exhibited a uniform darkness. The pattern reappeared with an orthogonal orientation with respect to the previous one when frequency was increased in the region from 130 Hz to 14 KHz. At higher frequencies it reappeared with the initial orientation and disappeared only at the resonance frequencies of the mechanical mounting.

4. Time response under pulsed operation

By using a pulsed square wave the time response of both Bragg diffracted light by the finger structure and of the light passed through the material can be determined (s.Fig.2).
Fig.2a shows the response by observing between crossed polarizers of the first order Bragg diffracted light and the compressional square wave pulse.

As a consequence of the compression the diffracted light presents a sort of oscillating behavior interrupted by the decompression at the end of the pulse (lasting 20 msec) after which the light reaches a maximum value and subsequently decays exponentially with a time constant of about 24 msec.

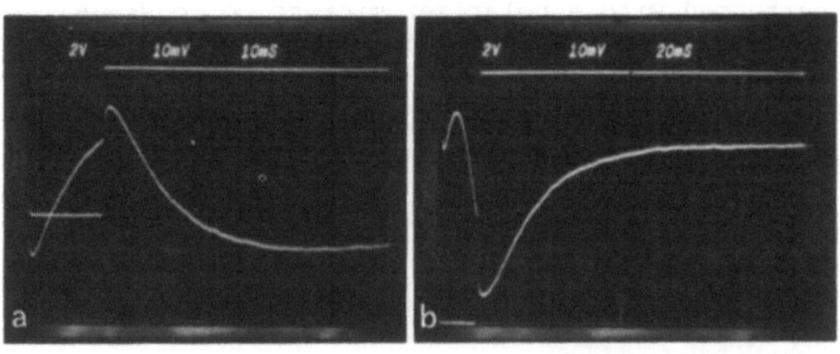

Fig.2a *Fig.2b*

Fig.2 Time behavior for transmitted (2b) and diffracted (2a) light for finger structure.

In Fig.2b the transmitted light still observed between crossed polarizers shows a kind of complementary behavior with a time constant in the slow exponential relaxation of about 15 msec.

5. Discussion

The behavior of the finger pattern at various frequencies shows different regions in which different processes play a role. At very low frequencies the whole structure is able to follow the field and the pattern is therefore modulated at the same frequency of the mechanical stress.

This can happen untill the time characteristic of the mechanical oscillation is larger than the structural relaxation time.

Once these two times are of the same order or the structural relaxation time is somehow larger due to a cumulative effect of the applied stress the finger pattern disappears and a new TIC structure [1] takes its place.

At higher frequencies when the structural relaxation time is much larger than the inverse frequency and so does the molecular relaxation time, finger pattern reappears as suggested by Press and Arrott [2] with an orthogonal orientation with respect to the previous one. At still higher frequencies the finger is turned to the original direction because no transverse macroscopic fluxes can be present at these frequencies except at some resonance frequencies where it is observed a reversible disappearing of any structure.

The relaxation times derived by the time response behavior described in § 4 are the molecular visco-elastic relaxation times that do not differ greatly by those of a pure nematic. The oscillatory processes of the light intensity are connected to tunable birefringence processes that take place any time the molecular director is rotating away from the equilibrium position.

REFERENCES

1. M.J.Press and A.S.Arrott - J.Physique 37, 387, 1976

2. M.J.Press and A.S.Arrott - J.Physique 39, 750, 1978

A Microsecond-Speed, Bistable, Threshold-Sensitive Liquid Crystal Device

N.A. Clark[1] and S.T. Lagerwall

Chalmers University of Technology, Department of Physics,
S-412 96 Göteborg, Sweden

Current liquid crystal devices (LCD) are based on dielectric alignment effects in nematic or cholesteric phases in which, by virtue of the dielectric anisotropy, the average molecular long axis takes up a preferred orientation in an applied electric field. Since the coupling of an applied electric field by this mechanism is rather weak, the electro-optical response time for these devices is too slow for many potential applications. The slow response and the insufficient non-linearity in LCD's have been the serious limitations. The lack of speed becomes especially important in proportion to the number of elements that have to be addressed in a device. This leads to increasingly impractical production costs for flat-panel displays with potential use in computer terminals, oscilloscopes, radar and T.V. screens.

A promising way for overcoming these difficulties is to use chiral smectic C liquid crystals in a carefully chosen geometry. These media are ferroelectric and thus permit a very direct action by the external field. As reported below the resulting electro-optic device will show the following characteristic properties:

1) High speed even at very low voltages. The electro-optical response is as much as 1,000 to 10,000 times faster (for the smectic C) than currently available electro-optical devices using liquid crystal.

2) Bistability. The electro-optical response is characterized by two stable states, either of which may be selected by an appropriate electric field and either of which is stable in the absence of the field.

3) Threshold behavior. The change from no switching response to full switching response is made over a very small range in the amplitude and duration of the applied field.

[1] Permanent address: Department of Physics and Astrophysics, University of Colorado, Boulder, Co 80309, USA

4) Large electro-optical response. The optical change induced by the electric field corresponds to a rotation through a 20^0 to 60^0 angle of a uniaxial material having a refractive index and anisotropy Δ n of greater than 0.2. This response is 10 to 100 times larger than that attainable in other bistable electro-optical devices.

In addition, gray-scale control is possible by pulse amplitude and width modulation. Switching between two birefringence colours (or between several, with superposed samples) is also possible.

Physical principles

As was discovered by R.B. Meyer [1, 2], any tilted smectic phase built up by chiral molecules ought to have an intrinsic ferroelectric property in the sense that every smectic layer possesses an electric dipole density, $\vec{P}$, which is perpendicular to the molecular tilt direction, $\hat{n}$, and parallel to the smectic layer plane. The presence of the electric dipole in these chiral smectics provides a much stronger coupling of the molecular orientation to the applied electric field, $\vec{E}$, than is available via the dielectric anistropy. Furthermore, the coupling is polar in that the preferred orientation of $\vec{P}$ is parallel to $\vec{E}$ so that reversing the polarity of the applied electric field reverses the preferred orientation of $\vec{P}$, meaning that field reversal can be effectively used to control molecular orientation. When estimating the possible speed for the director reorientation one finds, from

$$\tau^{-1} \approx P \cdot E/\eta$$

times of the order of microseconds, or even less, for easily achievable fields and current values of P. However, as an additional result of the molecule chirality, in a bulk ferroelectric smectic C or H liquid crystal, the unit vector $\hat{n}$ and polarization $\vec{P}$ spiral about the axis normal to the layers from layer to layer through the sample. The spiralling cancels the macroscopic dipole moment and corresponds to macroscopic cancelling of polarization by domain formation in crystalline ferroelectrics.

As a result of this,the bulk behaviour has been found as slow in these media as in other LCD materials, and a dielectric rather than ferroelectric response above the 100 kHz frequency regime has been reported [1, 3].

In order to achieve high speeds the director spiralling must be suppressed, and in order to achieve bistable operation the domains must be re-established. Both can be done in the same step. Figure 1 shows the appropriate geometry. By making the sample thin enough, a couple of microns, and orienting the smectic

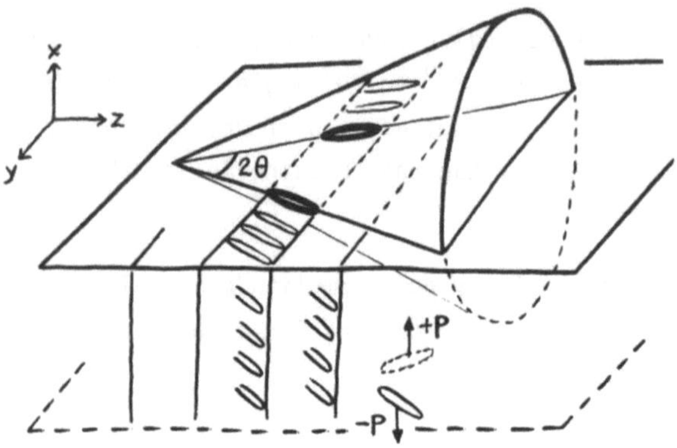

Fig. 1

layers perpendicular to the surrounding glass walls, the helix, which wants to be along the Z direction, is prevented from forming. The boundary condition should imply that the director is parallel to the glass surface but free to move in the surface plane. The conical degeneracy of the director is then replaced by the two directions formed by the cut of the plane and the cone. These $\hat{n}$ directions define the two possible, and energetically equivalent, domains of the sample, with their corresponding P pointing in the +X and -X direction, respectively. The director can thus be switched back and forth between these two directions by an electric field E applied across the sample after coating the bounding surfaces with a conductive layer. On switching, the optic axis of the sample changes direction, with an angular difference of 2Θ, Θ being the tilt and also the aperture angle of the cone. The tilt is a function of temperature, but typical values ($\Theta \approx 15\text{-}20^{0}$) will give an effective change of the optic axis by $30\text{-}40^{0}$ in space direction.

Operation

The simplest geometry has the sample between crossed polarizers with $\hat{n}$ parallel to the polarization direction in the DOWN state leading to extinction of light passing through the polarizers and sample (DOWN = OFF). In the UP state the polarization will make an angle 2θ with the optic axis and a fraction of the incident optical power, I, will be transmitted, with $I = I_0 \cdot \left\{ \sin(4\theta) \cdot \sin(\pi\, \Delta n\, d/\lambda)\right\}^2$. Here Δn is the refractive index anisotropy, λ the vacuum optical wavelength, and I_0 the parallel polarizer transmission. $I = I_0$ can be achieved for $\theta \gtrsim 23^0$ (this condition is met for DOBAMBC for $T \lesssim 85^0$C, and $d > \lambda/2\, \Delta n$, implying $d > 2.5\lambda \sim 1.2\ \mu m$ ($\Delta n \sim 0.2$

for DOBAMBC). Hence the electro-optic effect is large, being equivalent to the rotation of a uniaxial material with $\Delta n \sim 0.2$ through 45°.

The surrounding plates used were microscope coverglasses coated with semi-transparent conductive (100 Ω/cm^2) SnO_2 layers. The SnO_2 surfaces were cleaned of contaminants and dust with spectrographic acetone and placed together without spacers (overlap area = 6mm X 6mm). The sample material was introduced between them by capillary suction from the isotropic phase, resulting in samples which were slightly wedged, typically varying from 0,5 μm to 3 μm in thickness. The compounds used in this study were optically active DOBAMBC and HOBACPC. The desired smectic alignment was obtained upon cooling from the isotropic and subsequent gentle shearing in the A phase. The overall behaviour of the two compounds was qualitatively similar except as noted below.

Bistability and speed were studied by applying pairs of opposite polarity rectangular voltage pulses of selectable amplitude, V, width, t, and time separation, to the sample. To give an example, for a 1.5 μm thick HOBACPC sample at T = 88^oC a rectangular voltage pulse of duration t = 5 μs and amplitude V = 10 volts will switch the liquid crystal orientation field in about 5 μs. As t or V is reduced the switching threshold is approached such that a 10 volts, 4 μs or 8 volts, 5 μs pulse will not actuate switching. The sample will remain in the switched state until an above threshold pulse of the opposite polarity is applied. For Vt sufficiently large, the optical response is bistable, with the (+, -) pulse latching the monitored area (200 μm X 200 μm) into the (ON, OFF) state. The bistable latching exhibits a relatively sharp threshold, going from zero to saturated memory response for a less than 25% change in Vt. The dynamic behaviour of the optical response to a pulse is characterized by a risetime, τ_r, which depends on pulse amplitude, increasing from a minimum of 1μs for V = 20 volts to 4 ms at V = 0.2 volts. This general trend and fast response is expected from the simplest theoretical estimate referred to above: $V\tau_r \sim \eta\, d/P \sim 10^{-4}$ volts-sec, where η is an orientational viscosity, although the predicted $\tau_r \sim V^{-1}$ dependence is not obeyed. The minimum response time (≈ 1μs) is comparable to the RC time constant of the sample sandwich.

In general, for latching to occur, V and t must be such that the saturated optical response is achieved during the applied pulse. The full response, once attained, was stable over periods of at least several hours. The dynamic response to fast risetime pulses was homogenous (i.e. independent of the size, ℓ, of the sample area monitored for ℓ > 5 μm), reflecting the nucleation and motion of many domain walls. Results in HOBACPC and DOBAMBC were similar, with risetimes and requisite pulse widths two to three times longer in DOBAMBC, presumably a result of the smaller value of P in this material.

The contrast ratio critically depends on the quality of the surface treatment, the polarizers employed and their orientation, and on the pulse height and width. With 5 μs, 10 volts pulses, and with unsophisticated surface treatment, a contrast ratio of better than 20:1 was easily obtained in laboratory samples.

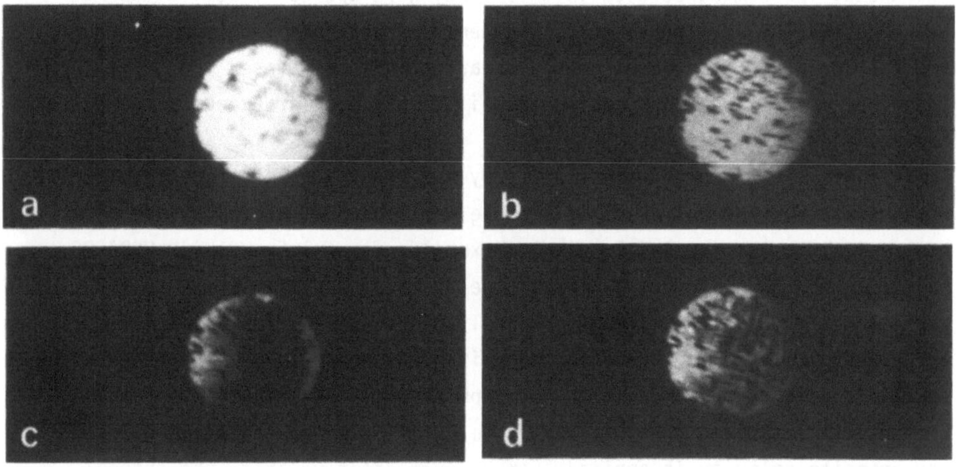

Fig. 2

To illustrate the operation, a very primitive device was made, in which the glass plates had undergone no surface treatment whatsoever, apart from simple washing - and which hence exhibits very inhomogenous alignment and multi-domain behaviour. The features are shown in Fig. 2 in the succession: turn-on pulse (up), field off, turn-off pulse (down), field off. The bistability is clearly demonstrated, the memory effect less pronounced (sample: DOBAMBC in H phase, ±12 volts).

The bistable characteristic of the present device is advantageous in situations requiring a large number of individual electro-optical devices, such as graphic or pictorial planar displays. The bistability obviates the need for an external electronic or other memory to maintain an image. The technique may therefore be employed in applications like matrix-addressed video displays. The response of the individual elements is rapid enough to permit the required rate of frame change and its threshold sensitive is such that pulses applied to alter a particular element do not alter others in the same row or column. Also, the optical change associated with the electro-optic effect is large enough to allow for its convenient use. In contrast, for bistable crystalline ferroelectrics, operated by considerably higher voltages, the optical anisotropy is very small and light control effects can be made operative only over a range of viewing angle which is too small to be useful in display applications.

This work was supported by the Swedish Natural Science Research Council and the Swedish Board of Technical Development.

References

1. R.B. Meyer, Mol. Cryst. Liq. Cryst. 40, 33 (1977).
2. R.B. Meyer, L. Liébert, L. Strzelecki, P. Keller, J. de Phys. Lett. 36, 69 (1975).
3. Ph. Martinot-Lagarde, J. de Phys. 37, C-129 (1976).

Part VI

Liquid-Crystalline Polymers

Weak Nematic Gels [1]

P.G. de Gennes

Collège de France, F-75231 Paris Cédex 05, France

Abstract : We give some theoretical predictions concerning the mechanical, optical and electrical properties of weak networks made with nematic polymers. The network may be permanently crosslinked, or may exist only in transient states, and be due only to physical entanglements. We construct an elastic theory involving both the deformations of the gel and the distortions of the director field. We conclude that weak gels may show some interesting mechanooptic properties.

1. Introduction

A nematic liquid combines the anisotropic properties of uniaxial solids and the strong deformability of a fluid [1]. If we freeze the liquid, the deformability is lost. For instance, if we add a few chiral impurities to a nematic, we get a helical structure. But in a crystal, the addition of chiral impurities has no long range effects.

In the present note, we wish to discuss an intermediate situation, corresponding to weak solidification of a nematic phase. This can occur in various instances :

1.a starting from a nematic polymer (with the nematogen in the backbone [2]) and crosslinking, chemically, an anisotropic gel can be formed. (The condition of weak crosslinking is absolutely necessary here to avoid destruction of the nematic order by the reticulation groups).

1.b a similar (and more flexible) situation could be obtained with a gel of nematogenic chains swollen by an adequate solvent. Recent theoretical work [3] shows that the solvent itself must be nematogenic to ensure both a significant swelling and the preservation of nematic order. The simplest case corresponds to a solvent identical to the monomer.

1.c with a melt, of nematic polymers, even in the absence of chemical crosslinks, we expect to find a behavior of the gel type if the chains are strongly entangled, and if the experiments are performed at frequencies ω larger than a certain inverse terminal time T_t^{-1} (T_t is the time required to disentangle the chains [4] [5] [6]). An Essential parameter here is the average number of monomers N_e between consecutive entanglement points. For conventional (non

---

[1] The following text expands on one particular point of the oral presentation, which covered various features of polymers confined in sheets, or tubes. A general review on confined polymers can be found in [5] below.

nematic) polymers N_e is of order 100 - 300. For chains with a nematogenic backbone, the existence of a preferred orientation probably increases N_e. The elastic moduli (measured at $\omega T_t > 1$) are of order

$$C = \frac{kT}{vN_e} \tag{1.1}$$

where T is the temperature and v the monomer volume.

The above list is not exhaustive : for instance, at temperatures slightly above a glass transition point T_G , a conventional nematic, studied at finite frequencies, will begin to show a solid like response. But this case is less interesting, because the elastic moduli are large. In the present paper, we restrict our attention to very weak solids. We also simplify our discussion by omitting case (b) : the swollen gel problem (b) is interesting, but is somewhat more complex than (a) or (c) because two different velocity fields must be introduced : one for the network, and one for the solvent. We consider cases (a) and (c) here, and hope to come back to (b) in later work. Finally we restrict our attention to small deformations of the solid. The opposite limit of strong deformations (and their effect on the nematic - isotropic transition) has been studied previously [7] [8].

2. Coupling Between the Director and the Deformations

Let us assume that our nematic gel at rest has its director $\underset{\sim}{n}_o$ lying in the z direction. In a distorted state we shall observe displacements $\underset{\sim}{u}(xyz)$ giving deformations and rotations

$$e_{ij} = \frac{1}{2}\left(\frac{\partial u_i}{\partial x_j} + \frac{\partial u_j}{\partial x_i}\right) \tag{2.1}$$

$$\omega_x = \frac{1}{2}\left(\frac{\partial u_z}{\partial y} - \frac{\partial u_y}{\partial z}\right) \qquad \text{etc.} \tag{2.2}$$

To describe the distortions of the molecular alignement we should, in principle, introduce a second rank tensor $\delta\varepsilon_{ij}$ giving the changes of dielectric properties. Here we use a simplified picture, where the medium is treated as uniaxial, with a certain director $\underset{\sim}{n} = \underset{\sim}{n}_o + \delta\underset{\sim}{n}$. This is not rigorous, but avoids the introduction of a very large number of mechanooptic constants. We put

$$\underset{\sim}{n} - \underset{\sim}{n}_o = \delta\underset{\sim}{n} = \underset{\sim}{\Omega} \wedge \underset{\sim}{n} \tag{2.3}$$

The energy density is the sum of three parts $E_{tot} = E_e + E_n + E_{en}$

2.a a conventional elastic energy for a uniaxial solid

$$E_e = \frac{1}{2} c_{11}(e_{xx}^2 + e_{yy}^2) + c_{12}\, e_{xx}\, e_{yy} + c_{13}\, e_{zz}\,(e_{xx} + e_{yy})$$
$$+ \frac{1}{2} c_{33}\, e_{zz}^2 + 2c_{44}\,(e_{yz}^2 + e_{xy}^2 + e_{xz}^2) + 2(c_{11} - c_{12})\, e_{xy}^2 \tag{2.4}$$

2.b the Frank elastic terms for the director [1]

$$E_n = \frac{1}{2} K_1 \left(\frac{\partial n_x}{\partial x} + \frac{\partial n_y}{\partial y}\right)^2 + \frac{1}{2} K_2 \left(\frac{\partial n_x}{\partial y} - \frac{\partial n_y}{\partial x}\right)^2 + \frac{1}{2} K_3 \left[\left(\frac{\partial n_x}{\partial z}\right)^2 + \left(\frac{\partial n_y}{\partial z}\right)^2\right] \tag{2.5}$$

2.c a sum of two coupling terms :

$$E_{en} = \frac{1}{2} D_1 \left[(\underset{\sim}{\Omega} - \underset{\sim}{\omega}) \wedge n \right]^2 + D_2 (\underset{\sim}{\Omega} - \underset{\sim}{\omega}) \; \underset{\sim}{n} \cdot \hat{e} \; \underset{\sim}{n} \tag{2.6}$$

The structure of (2.6) is geometrically reminiscent of the Leslie equations for viscous flow of conventional nematics [1]. Instead of dealing with velocities, we are dealing with displacements ($\underset{\sim}{u}$ and $\delta \underset{\sim}{n}$). The energy density (2.6) is invariant by the change $\underset{\sim}{n} \to -\underset{\sim}{n}$. It is also invariant by a simultaneous rotation of the solid and of the director. With $\underset{\sim}{n}_o$ along z the form (2.6) may be written as

$$E_{en} = \frac{1}{2} D_1 \left[(\Omega_x - \omega_x)^2 + (\Omega_y - \omega_y)^2 \right]$$

$$+ D_2 \left[(\Omega_y - \omega_y) \, e_{xz} - (\Omega_x - \omega_x) \, e_{yz} \right] \tag{2.7}$$

Stability requires $D_1 > 0$ and also

$$D_2^2 < 4 D_1 \, c_{44} \tag{2.8}$$

But the sign of D_2 is not prescribed. If D_2 was equal to zero, pure strains would not affect the director : $\underset{\sim}{n}$ would simply rotate with the solid ($\underset{\sim}{\Omega} = \underset{\sim}{\omega}$).

3. Stress Optical Coefficients

The role of the coupling coefficient D_2 can be understood on a simple case, such as a pure strain in the xz plane :

$$\left.\begin{aligned} u_x &= \frac{1}{2} \varepsilon z \\ u_z &= \frac{1}{2} \varepsilon x \end{aligned}\right\} \to e_{xz} = \varepsilon \tag{3.1}$$

Then optimisation of E_{ne} at fixed ε leads to

$$\delta n_x = - \frac{D_2 \varepsilon}{2 D_1} \tag{3.2}$$

We may usually expect D_2 and D_1 to be comparable in magnitude, and δn_x to be of order ε. The mechanical stresses associated with (3.1) are

$$\sigma_{xy} = \tilde{c}_{44} \, 2 e_{xy} \tag{3.3}$$

with a renormalised shear modulus

$$\tilde{c}_{44} = c_{44} - D_2^2/(4D_1) \tag{3.4}$$

In the uniaxial approximation used here, the birefringence tensor is

$$\hat{\varepsilon} = \hat{1}\,\varepsilon_\perp + (\varepsilon_{/\!/} - \varepsilon_\perp)\,\underset{\sim}{n} : \underset{\sim}{n} \tag{3.5}$$

where $\varepsilon_{/\!/}$ and $\varepsilon_\perp$ are the principal dielectric constants in the rest state. This gives a stress optical coefficient

$$\frac{\varepsilon_{xz}}{\sigma_{xz}} = -\frac{1}{2}\,\frac{\varepsilon_{/\!/} - \varepsilon_\perp}{\tilde{c}_{44}}\,\frac{D_2}{D_1} \tag{3.6}$$

The major interest of (3.6) lies in the smallness of the shear modulus $\tilde{c}_{44}$: using the estimate (1.1) either for a gel (N_e : distance between reticulation points) or for a melt (N_e : distance between entanglement points) we see that

$$\left|\frac{\varepsilon_{xz}}{\sigma_{xz}}\right| \cong \frac{vN_e}{kT} \tag{3.7}$$

and this may be unusually large. Of course, for practical applications, one often works not with pure strains, but with a mixed displacement field : for instance in simple shear ($u_x = \varepsilon z$) normal to the unperturbed axis, (3.2) must be replaced by

$$\delta n_x = \frac{\varepsilon}{2}\,(1 - D_2/D_1) \tag{3.8}$$

and the added term ($\varepsilon/2$) corresponds to the rotational component. But the order of magnitude estimate (3.7) is maintained.

4. Characteristic Lengths

We want to find out now under what conditions a weakly crosslinked anisotropic gel can still behave like a nematic fluid. We shall see that the answer depends critically on the sample size L : when L is larger than a certain characteristic thickness λ the behavior is solid like, while for $L < \lambda$ the effects of crosslinking are weak.

The precise definition of λ depends somewhat on the type of director distortion which is studied. Consider for instance a situation where the gel is fixed ($\underset{\sim}{u} = 0$) and where a small deviation (n_x) is imposed at the sample surface. Then from the elastic energies (2.4,5,6) we arrive at a local equilibrium condition

$$-K_i\,\frac{\partial^2 n_x}{\partial x_i^2} + D_1\,n_x = 0 \tag{4.1}$$

giving an exponential decay with depth (x_i = x, y, or z being the direction normal to the sample surface)

$$n_x = n_{xo} \exp - (x_i/\lambda_i) \tag{4.2}$$

$$\lambda_i = \sqrt{\frac{K_i}{D_1}} \tag{4.3}$$

Thus there are three lengths λ_i associated with splay, twist, or bend deformations. If the sample dimension L (along x_i) is smaller than λ_i the Frank elasticity is dominant, and the gel structure is not essential. But if $L > \lambda_i$ the director is more effectively locked to the gel.

The scaling properties of the various lengths λ_i are non trivial, and may depend on the index i. Consider for instance the case of a nematic polymer melt. The bend elastic constant is expected to be essentially independent of molecular weight, and of order $kT\, a^2/v$ (where a is the monomer length) for a polymer with flexible "spacers" between nematogenic units [2]. Then, using (1.1) we are led to the conjecture :

$$\lambda_3 \cong a\, N_e^{1/2} \tag{4.4}$$

However, if we switch to splay deformations, the laws are different, since the splay elastic constant K_1 is expected to be anomalously large : from [9] we predict

$$K_1 \cong \frac{kT}{v} (Na)^2 \tag{4.5}$$

where N is the total number of monomers per chain (N must be larger than N_e to achieve an entangled melt).

Then the characteristic length λ_1 should have the form

$$\lambda_1 \cong a\, N_e^{1/2}\, N \tag{4.6}$$

If $a = 10$ Å, $N_e = 10^2$ and $N = 10^3$, we expect $\lambda_1 \cong 10$ microns.

5. An Example : The Helfrich Instability Under Electric Fields

We now consider a weak nematic gel in the form of a thin slab (planar texture) under an electric field E, (refer to Fig.1). Our analysis follows the notation of [1], Eqs. 5(85-90). For simplicity we neglect the dielectric anisotropy ($\varepsilon_{/\!/} = \varepsilon_\perp = \varepsilon$) while retaining a conductance anisotropy ($\sigma_H = \sigma_{/\!/} - \sigma_\perp > 0$). The charge balance equation is not modified by gelation and reads

$$\mathring{q} + \frac{q}{\tau} + \sigma_H \psi E = u \tag{5.1}$$

where $\psi = \frac{\partial n_x}{\partial z}$ is the curvature of the director lines and τ a dielectric relaxation time. Following the simple Helfrich approach, the gel displacement is taken in the form

$$u_x(z) = u \cos kz \tag{5.2}$$

(i.e. the detailed boundary conditions at the walls are not included). The force balance is

$$qE - c_{44} \frac{\partial^2 u_x}{\partial z^2} = 0 \tag{5.3}$$

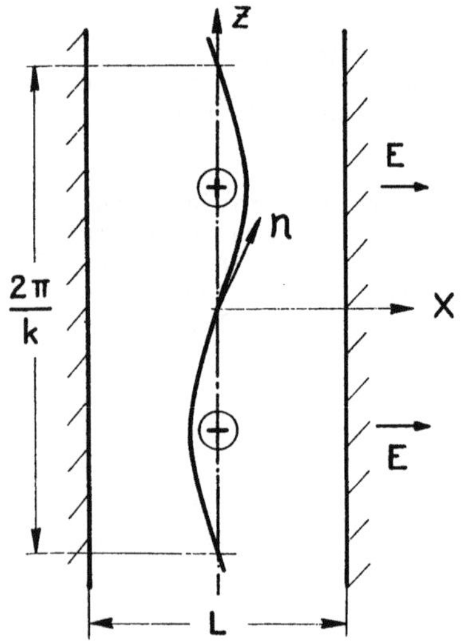

Fig.1 Helfrich instability in a weak nematic gel. A small deviation $n_x(z)$ from the planar texture is assumed. Charges q pile up as shown because of the conductance anisotropy. The electric force on these charges distorts the gel and amplifies the perturbation.

The director orientation is given by an extension of (2.8)

$$D_1 \left(n_x - \frac{1}{2} \frac{\partial u_x}{\partial z} \right) + D_2 \frac{1}{2} \frac{\partial u_x}{\partial z} - K_3 \frac{\partial^2 n_x}{\partial z^2} = 0 \tag{5.4}$$

giving :

$$\psi - \lambda_3^2 \frac{\partial^2 \psi}{\partial z^2} = \frac{1}{2} \frac{\partial^2 u_x}{\partial z^2} (1 - D_2/D_1)$$

$$\psi = - \frac{qE}{2C_{44}} (1 - D_2/D_1) \frac{1}{1 + k^2 \lambda_3^2} \tag{5.5}$$

Inserting this into (5.1) and looking for a static (or nearly static) threshold, we put $\dot{q} = 0$, and find :

$$E^2 = \frac{2C_{44}(1 + k^2 \lambda_3^2)}{\tau\sigma_H(1 - D_2/D_1)} \tag{5.6}$$

In a rough approximation [10] we expect πk^{-1} to scale like the sample thickness L. We find that when $L \lesssim \pi\lambda_3$ the k^2 term dominates and (5.6) predicts a critical voltage EL = V independent of L. But in the (more common) opposite limit ($L > \pi\lambda_3$) it is the field which is independent of L.

We conclude that there may exist electrical instabilities in nematic gels : but the magnitude of the critical field increases rapidly with the gel rigidity (i.e. with C_{44}). The denominator in (5.6) is of order unity. If we had a strong solid, the threshold field E would be comparable to a local field inside a molecule (10^7 - 10^8 volts/cm). For the cases at hand, C_{44} may be reduced by $\sim 10^{-4}$ put this still gives a threshold $\gtrsim 10^5$ volts/cm. Thus the instabilities will be hard to see. On the other hand, if weak nematic gels do become available in the future, they may give us a very good way of studying the vicinity of a sol-gel transition [5] : here the elastic moduli are very small, and difficult to measure by conventional means. However they are interesting (and involve some delicate critical exponents). They could be monitored during gelation by measurements of an electric (or magnetic) instability threshold.

NOTES AND REFERENCES

1. See for instance, P.G. de Gennes, The Physics of Liquid Crystals, Oxford, 3d printing (1980).
2. H. Finkelmann, these proceedings p. 238
3. F. Brochard, J. de Phys. 40, 1049 (1979).
4. J. Ferry, Viscoelastic Properties of Polymers, Willy, N.Y. (1970).
5. W. Graessley, Adv. Pol. Sci. 16 (1974).
6. P.G. de Gennes, Scaling Concepts in Polymer Physics, Cornell U.P., Ithaca, N.Y. (1979).
7. P.G. de Gennes, Comptes Rendu Acad. Sci. Paris, 281 B, 101 (1975).
8. J.P. Jarry, Thèse Université Paris 6 (1978).
9. P.G. de Gennes, Molecular Crystals Letters 34, 177 (1977).
10. In the present case the usual relation between k^{-1} and L might be upset because the splay elastic constant K_1 is much larger than the bend constant K_3.

Thermotropic Liquid Crystalline Polymers

H. Finkelmann

Physikalisch-Chemisches Institut der Technischen Universität Clausthal, D-3392 Clausthal Zellerfeld, Federal Republic of Germany

1. Introduction

Studies concerning liquid crystalline polymers have become of increasing interest because of their theoretical and technological aspects. Looking for literature with the key words "liquid crystal" and "polymer" a mushrooming number of papers is available. In most cases, however, polymers are described, which exhibit an anisotropic structure in the glassy state but not the liquid crystalline state above the glass transition temperature. Only some papers remain, which deal with thermotropic liquid crystalline polymers, where the macromolecules can be obtained in the liquid crystalline state like a low molecular weight liquid crystal. These polymers could be realized just in the last few years.

For the synthesis of macromolecules which are expected to be liquid crystalline a simple concept can be followed up by using suitable mesogenic monomers, which are able to build up a liquid crystalline phase. If the mesogenic group of these molecules is substituted with appropriate functional groups A and B, macromolecules can be synthesized (Fig.1).

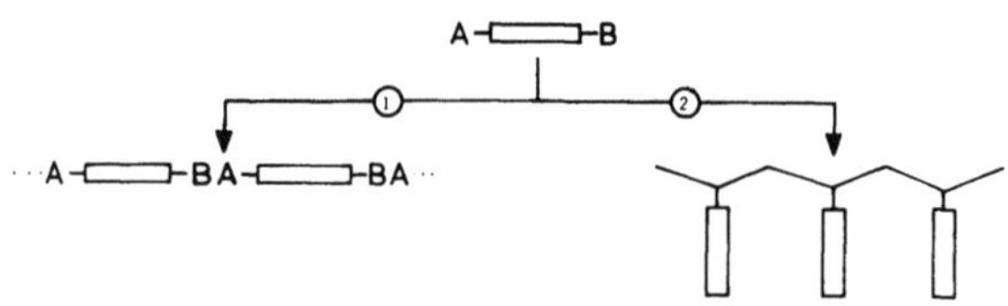

Fig.1 Polymerisation of a mesogenic monomer (symbol ▭)

1. If A and B are able of performing a condensation reaction, the mesogenic groups build up the polymer main chain. These polymers will be named "Liquid Crystalline (LC) Main Chain Polymers".
2. If A is capable of performing an addition polymerisation reaction, the polymer main chain is built up by A and the mesogenic groups are fixed like side chains to the polymer main chain. These polymers will be called "LC Side Chain Polymers".

For both polymers the original mesogenic moieties of the low molecular weight compound has been preserved, which suggests the idea, that the ability of forming a liquid crystalline phase can be preserved.

2. LC Main Chain Polymers

In principle for the LC main chain polymers two different structures can be obtained, which depend on the chemical nature of the functional groups A and B (Fig.2). If A and B are directly linked to the rigid mesogenic moiety, a rigid rod like structure is also produced for the resulting polymer back bone. If on the other hand A and B are substituted via flexible spacers (e.g. flexible alkyl chains) to the mesogenic moieties, the resulting polymer main chain consists of alternating rigid segments (mesogenic groups) and flexible segments (e.g. alkyl chains). In contrast to the previous polymers, the flexibility of the over-all polymer main chain can be regulated by using different chemical structures or different length of the flexible segments.

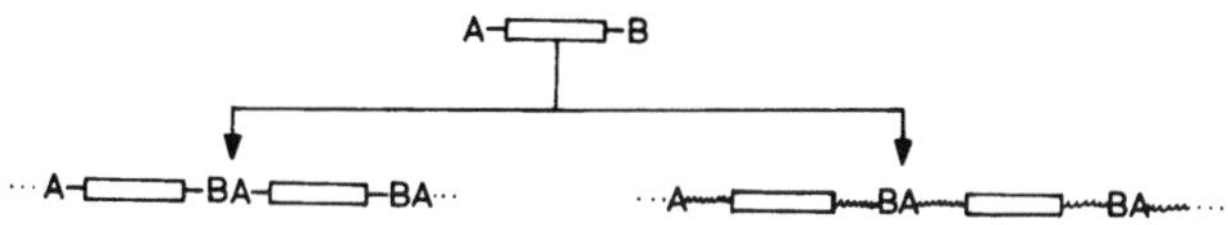

Fig.2 Structures of liquid crystalline main chain polymers

2.1. Theories

Already in 1949 Onsager (1) and in 1951 Isihara (2) treated theoretically the packing of rigid rod like molecules. They calculated the excluded volume as a function of the orientation of the molecules. In 1956 Flory (3) has overcome the restriction to dilute solutions using the lattice model as convenient method, and calculated the statistics of packing of monodisperse macromolecules and solvent molecules without considering any dispersion interactions. If a critical concentration is reached for the polymer in solution a lyotropic liquid crystalline phase is observed. This critical concentration depends on the molecular weight and the flexibility of the macromolecule. In case of low molecular weight a high concentration is needed to obtain the liquid crystalline phase. With increasing flexibility of the polymer back bone the critical concentration increases up to a limit, where even in bulk no liquid crystalline phase can exist. Liquid crystalline polymers basing on these principles have been reviewed by Papkov recently (4). Following these theories thermotropic liquid crystalline polymers are obtained in the limiting case of no solvent being present. In 1960 Di Marzio (5) pointed out in his calculations, that in the case of the stable lyotropic phase of rigid rod like molecules the solvent molecules can be withdrawn and replaced by flexible polymers. The theories should still be valid, even if the flexible polymers are tied to the ends of the rigid rod like molecules, which results in thermotropic liquid crystalline polymers with flexible and rigid segments.

2.2. Examples and Properties

Some typical features of the liquid crystalline main chain polymers will be marked out by some arbitrarily chosen examples. The very first polymers were published by Roviello and Sirigu (6). Looking at the structure, the polymers consist of rigid mesogenic segments and flexible segments formed by the alkyl chains(Table 1).

Table 1 Phase transitions of liquid crystalline polymers (6)

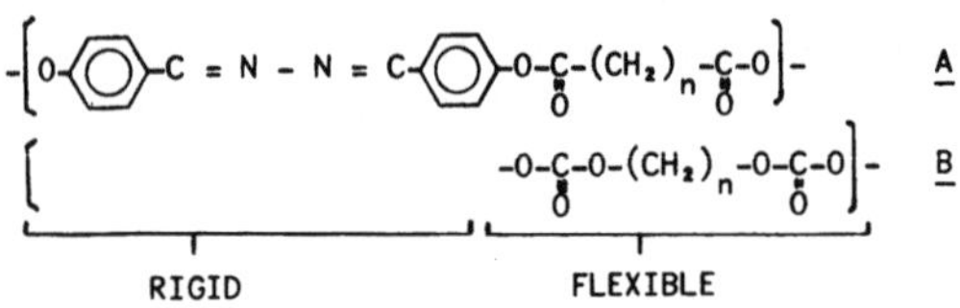

A:

n	T_c(K)
8	568
10	529
12	514

B:

n-4	T_c
10	530
12	508
14	486
16	467

According to the theories the stability of the liquid crystalline phase decreases with increasing over-all flexibility of the polymer main chain for constant molecular weight. This is confirmed by those polymers, where the length n of the flexible segments is varied. With increasing length n of the alkyl chain the examples show a decreasing phase transition temperature liquid crystalline - isotropic for the esters A as well as for the carbonates B. Recent papers indicate (7), that smectic and nematic phases can be prepared depending on the chemical structure of the mesogenic moieties. Using chiral elements in the polymer back bone even cholesteric polymers are described by Blumstein (8).

Another method of realizing liquid crystalline main chain polymers was applied by Jackson and Kuhfuss (9). Starting with the easy accessible poly(ethylen therephthalate) (PET), rigid mesogenic elements are inserted into the polymer back bone by transesterification with p-hydroxybenzoic acid (PHB). If a sufficient stiffness of the polymer main chain is reached, the anisotropic phase becomes stable. The authors also investigated the mechanical properties of these polymers, which point out the very important features of these materials. In Fig.3·the tensile strength of PET is plotted over the amount of PHB, which was inserted into the main chain by transesterification.

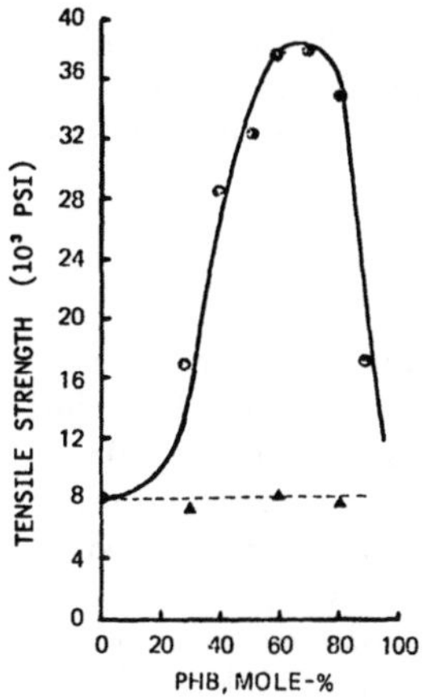

Fig.3 Tensile strength of PET trans-esterified with PHB (9)

Starting with pure PET the tensile strength increases rapidly, if the amount of PHB is larger than 35 mole-%, because of the influence of the liquid crystalline order. The mechanical behaviour exceeds the mechanical behaviour of conventional materials and gives reason for their increasing technological applications.

In contrast to low molecular weight liquid crystals the polymers can be hardly oriented in an external magnetic field. To induce a measurable molecular orientation of the polymers described above, high temperatures and a high magnetic field were used (9), furthermore this process required a long time (60 minutes). In contrast to these measurements recent results by Liebert et al.(10) indicate a good orientation of the macromolecules in the external field. For these investigations the crucial point seems to be the molecular weight of the polymer, which will determine the mobility. Careful measurements have to be made to clear up the orientation effects as a function of the molecular weight of the macromolecules.

3. Liquid Crystalline Side Chain Polymers

Several groups were engaged with the synthesis of polymerizable liquid crystals and their polymerisation reactions since years. Although numerous polymers were prepared, which are summarized by the reviews of Blumstein (11) and Shibaev (12), polymers exhibiting the liquid crystalline state could be received only in some exceptional cases (13). In most cases only frozen in liquid crystalline structures were obtained, which are irreversibly lost after having heated the polymer above the glass transition temperature. The few obtained liquid crystalline polymers, however, confirmed the possibility of realizing this material. But no concept was discernible concerning the structural elements producing the liquid crystalline state.

Subsequently a simple model will be described (14), which enabled the systematic synthesis of liquid crystalline side chain polymers. According to these model considerations some aspects of the properties of these polymers will be described.

3.1. Model Considerations

In the liquid crystalline state of the polymer two aspects have to be considered. The polymer main chain exhibits a high mobility of the chain segments and a tendency towards a statistical chain conformation. On the other hand the mesogenic groups tend towards an anisotropic orientation. Both tendencies are conflicting and it will depend on the molecular structure which tendency predominates. The crucial point, whether a liquid crystalline polymer or an isotropic polymer melt will appear, is the chemical structure and the property of the linkage between polymer main chain and the rigid mesogenic group. Two extreme limiting conditions for this linkage can be considered (Fig.4).

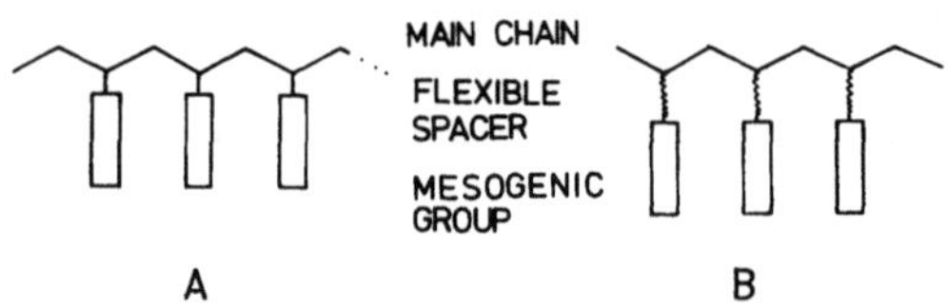

Fig.4 Linkage of the mesogenic group to the polymer main chain A directly, B via the flexible spacer

A) The rigid rod like mesogenic group is directly coupled to the polymer main chain (Fig.4A). The tendency towards a statistical chain conformation hinders the anisotropic orientation of the mesogenic side groups. Furthermore due to their voluminous molecular structure a steric hindrance suppresses the formation of the liquid crystalline order. Under these conditions above the glass transition temperature only an isotropic polymer melt will be observed.
B) The rigid rod like mesogenic group is decoupled by the polymer main chain (Fig.4B). This fictitious condition can be approximately achieved by a flexible spacer between polymer main chain and mesogenic side chain. Due to this indirect linkage the motions of the polymer main chain do not effect the anisotropic orientation of the mesogenic groups. The polymer will exhibit the liquid crystalline state.

The actual conditions for the linkage will be between the extremes of the complete decoupling and the direct coupling of the mesogenic group to the polymer main chain. Therefore the main chain will influence more or less the motions of translation and rotation of the mesogenic groups and vice versa, compared with a low molecular weight molecule in the liquid crystalline state. The mechanism of these interactions will depend on the chemical nature of the polymer main chain, the spacer and the mesogenic group. To get detailed information about these interactions the liquid crystalline properties can act as indicator to investigate changes of polymer characteristic properties.

3.2. Phase Behaviour

For low molecular compounds the liquid crystalline state is limited by first order transitions to the crystalline state at lower temperatures and the isotropic melt at higher temperatures. The typical P-V-T behaviour for a nematic compound is illustrated in Fig.5a (15). In

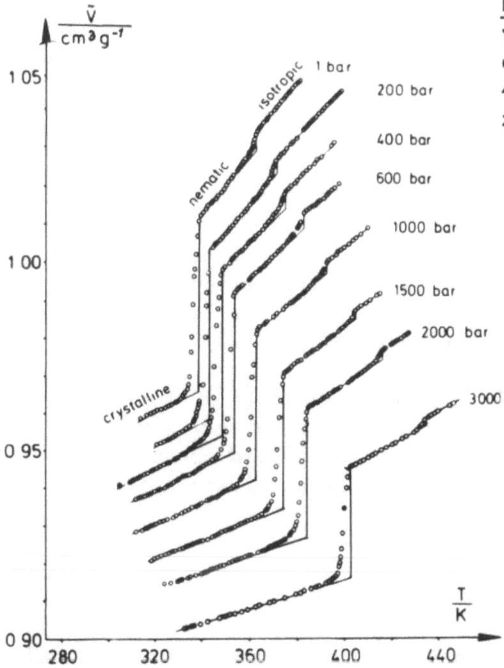

Fig.5a Isobaric specific volume vs. temperature for the low molecular weight liquid crystal 4-hexyloxy-phenyl 4-hexyloxybenzoate

comparison to the conventional liquid crystal Fig.5b shows the P-V-T behaviour of a nematic side chain polymer, where the slightly modified mesogenic group of the low molecular weight compound is fixed to a poly(methacrylate) main chain. The polymer exhibits a wide nematic phase, which also changes to an isotropic polymer melt with a first order phase transition like the conventional material in Fig.5a. This transition obeys the Clausius Clapeyron equation.

Going to low temperatures a remarkable difference can be found in comparison to conventional liquid crystals. Due to the polymer main chain no crystallisation is observed, but the polymer is transformed into a polymer glass, indicated by the bend in the V-T curve. This characteristic feature of the material shows the combination of liquid

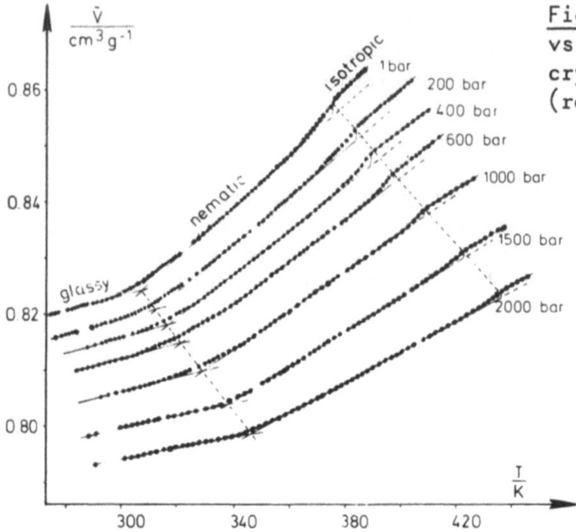

Fig.5b Isobaric specific volume vs. temperature for the liquid crystalline polymer No. 1 (refer to Table 2, p. 8)

crystalline and polymer specific properties. In view of technological applications it has to be pointed out, that the transition into the polymer glass does not influence the liquid crystalline texture, which can be frozen in by this process without any changes (23).

It could also be proved by the P-V-T measurements (15), that the liquid crystalline polymer phase is a homogeneous phase macroscopically like conventional liquid crystals. With this the liquid crystalline polymers clearly differ from partially crystalline polymer systems.

3.3. Influence of the Mesogenic Group

3.3.1. Nematic and Smectic Polymers

Following the model considerations, the formation of the liquid crystalline phase has to be assumed to be more or less influenced by the polymer main chain, if the mesogenic group is fixed via the flexible spacer to the polymer back bone. Therefore starting from a defined mesogenic group a nematic and a smectic polymer phase should be determined by the substituents of the mesogenic moiety. In Table 2 some typical polymers are summarized with the benzoic acid phenylester as mesogenic group. For the low molar mass derivatives it is well known, that with increasing length of the alkyl substituents a smectic phase becomes stable (16). This tendency is also obtained for the polymers. For the poly(methacrylates) the short $-OCH_3$ substituent most distant from the polymer main chain is exchanged by a long $-OC_6H_{13}$ substituent (polymer 2) and the nematic phase of polymer 1 converts to the smectic phase of polymer 2. This principle is also observed if the poly(methacrylate) main chain is exchanged by a siloxane main chain. Although their chemical and

Table 2 Nematic and smectic liquid crystalline side chain polymers

No.	Polymer	Phase transitions (K)	Lit.
1	$-[CH_2 - C(CH_3)]-$ with side chain $COO - (CH_2)_6 - O - C_6H_4 - COO - C_6H_4 - OCH_3$	g 309 n 374 i	17, 18
2	$- OC_6H_{13}$	g 303 s 374 i	
3	$-[Si(CH_3) - O]-$ with side chain $(CH_2)_3 - O - C_6H_4 - COO - C_6H_4 - OCH_3$	g 288 n 334 i	19
4	$- OC_6H_{13}$	g 288 s 385 i	
5	$-[Si(CH_3) - O]-$ with side chain $(CH_2)_3 - O - C_6H_4 - COO - C_6H_4 - OCH_3$	g 288 n 334 i	19
6	$(CH_2)_6 -$	g 278 s 319 n 391 i	20

physical properties strongly differ, the order of the mesogenic side chain also changes from nematic to smectic with increasing length of the substituent (polymer 3 and 4). But not only the substituent, which is most distant from the polymer back bone , influences the structure of the liquid crystalline phase. This is indicated with polymer 5 and 6. In this case the substituent most distant from the main chain remains constant but not the length of the spacer. Here again with increasing length of the alkyl substituent of the mesogenic moiety a low temperature smectic phase becomes stable for polymer 6.

Presuming that the polymer main chain does not strongly influence the mesogenic side chain, which requires a sufficient length of the spacer (refer to 3.4.), these principles were also found for different mesogenic moieties (21). The results indicate, that the formation of the liquid crystalline phase of the polymers essentially follows the principles known for conventional liquid crystals, which indicates the validity of the model considerations.

3.3.2. Cholesteric Polymers

Chiral molecules have to be present to obtain a cholesteric (chiral nematic) phase (22). Therefore the substitution of chiral substituents to the mesogenic groups described above should result in cholesteric homopolymers. Until now, however, only smectic phases were obtained when chiral molecules were polymerized (11, 12).

Cholesteric phases can also be achieved using the well known concept of "induced cholesteric phases" (23). Assuming a direct analogy between polymers and low molecular weight liquid crystals, the cholesteric phase has to be induced if chiral comonomers are added to the nematic polymer host phase (Fig.6).

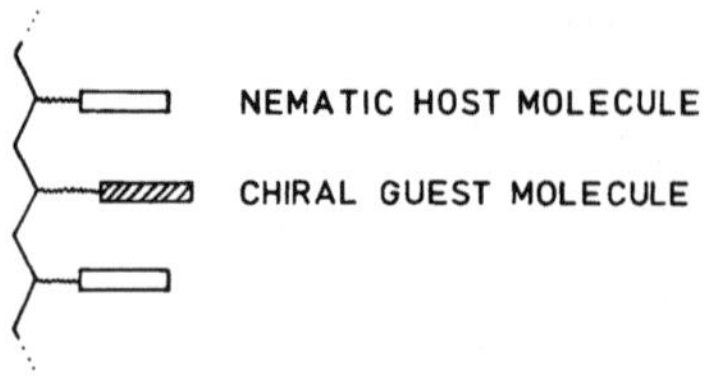

Fig.6 Schematic representation of an "induced cholesteric polymer phase"

The validity of this concept has been proved by different systems, which are listed in Table 3. The temperature dependence of the wave-

Table 3 Induced cholesteric polymer systems

Polymer Main Chain	$-CH_2-C(CH_3)(COO-R_{1,2})-$ (20)	$-Si(CH_3)(R_{1,2})-O-$ (23)
Nematic Group	$-(CH_2)_6-O-C_6H_4-COO-C_6H_4-C_6H_4-OCH_3$	$-(CH_2)_n-O-C_6H_4-COO-C_6H_4-OCH_3$
Chiral Group	$-(CH_2)_2-O-C_6H_4-COO-C_6H_4-CH=N-CH(CH_3)-C_6H_5$	$-(CH_2)_3$-COO-cholesteryl

length λ_R of reflected circularly polarized light is shown in Fig.7 for the methacrylate, which is listed in Table 3. In analogy to low molecular weight induced cholesteric phases only slight slopes of λ_R with the temperature are observed. As the temperature increases all the polymers show a blue shift of λ_R, the positive value $d(1/\lambda_R)/dT^+$ increasing with increasing amount of the chiral comonomer. Furthermore these systems confirm the linear dependence of $1/\lambda_R$ with the mole fraction x_{chiral} of the chiral part of the sidegroups of the macromolecule (Fig.8) in the case $x_{chiral} < 1$ (24).

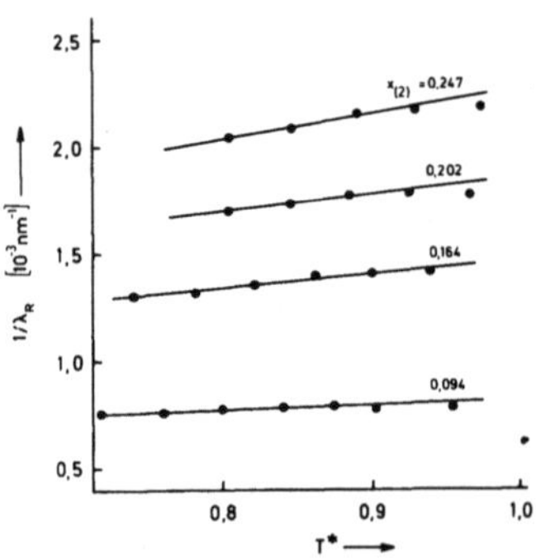

Fig.7 Inverse wavelength of reflection $1/\lambda_R$ as a function of the temperature for the copolymer (20)

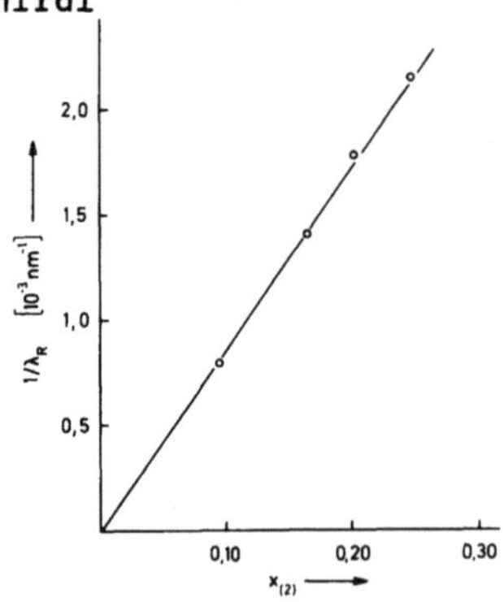

Fig.8 Inverse wavelength of reflection $1/\lambda_R$ as a function of the mole fraction of the chiral comonomer (20)

The investigations of the influence of the mesogenic group on the phase behaviour of the polymers confirm the model considerations. Although the mesogenic moieties are covalently fixed via the flexible spacer to the polymer main chain, the variation of their substituents changes the character of the liquid crystalline phase.

3.4. Influence of the Polymer Main Chain and the Spacer Length

The orientation of the mesogenic side groups, which are decoupled by flexible spacers, follow the principles similar to that of low molar mass liquid crystals, forming nematic smectic and cholesteric phases. It could be supposed, that the main and side chain are essentially decoupled as expected by the model considerations. Actually, however, the tendency towards a higher ordered phase is always observed, if the monomer is converted to the polymer (11, 12). If nematic polymers are obtained, in most cases the corresponding monomers exhibit none or only a monotropic nematic phase. Starting with nematic monomers mainly smectic polymer phases are obtained. Only very few examples exist, where a nematic monomer becomes a nematic polymer at the same temperature (18). For the polymers, however, the phase transition nematic - isotropic are always observed at higher temperatures than for the corresponding monomers. This tendency towards a stabilisation of the liquid crystalline state has to be perceived, because the conditions for the motions of translation and rotation of the mesogenic groups had been changed. Detailed measurements have to be made, to get a quantitative insight into these mechanisms.

Corresponding to the model considerations a limiting case can be supposed. The mesogenic groups are just able to build up an anisotropic aggregation, whereas on the other hand the polymer main chain disturbs a macroscopical order, which is known for conventional liquid crystals. This has to be expected for little or no decoupling by a flexible spacer. In Table 4 some polymers are listed with decreasing length of

Table 4 Nematic polymers with different spacer length

No.	Spacer Length	Polymer	Phase	Δn	Lit.
7	9	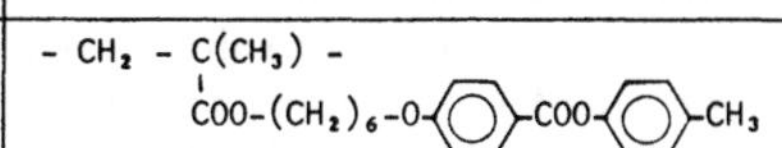	nem	+	18
3	5	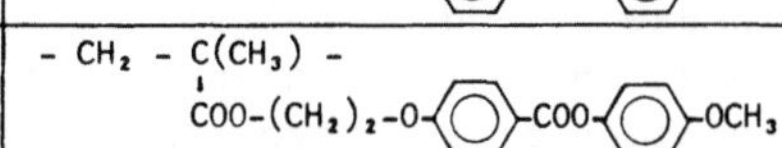	nem	-	17 25
8	2	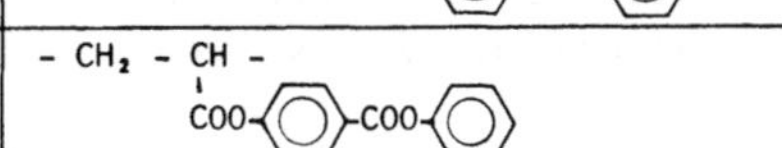	nem	-	27
5	4	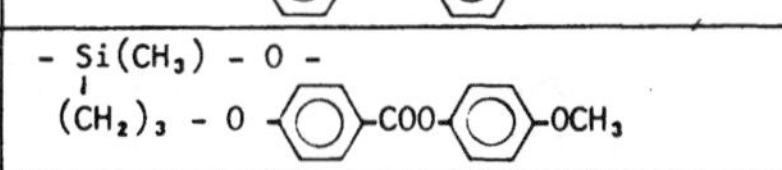	nem	+	19

the number of atoms between the main chain and the rigid mesogenic moiety. They were classified to be nematic following x-ray investigations and DSC measurements. Looking at the texture of these polymers using the polarizing microscope some characteristic differences can be stated. Polymers 5 and 7 have the texture of conventional nematic phases if they are annealed for some time near the clearing temperature. By conoscopic observation a positive uniaxial texture can be identified. For polymer 7 the mesogenic group is fixed via a long flexible spacer of nine atoms to a poly(methacrylate) main chain. Polymer 5 has only a short spacer of four atoms, but the mesogenic groups are fixed to the very flexible poly(methyl siloxane) main chain, causing the low glass transition temperature.

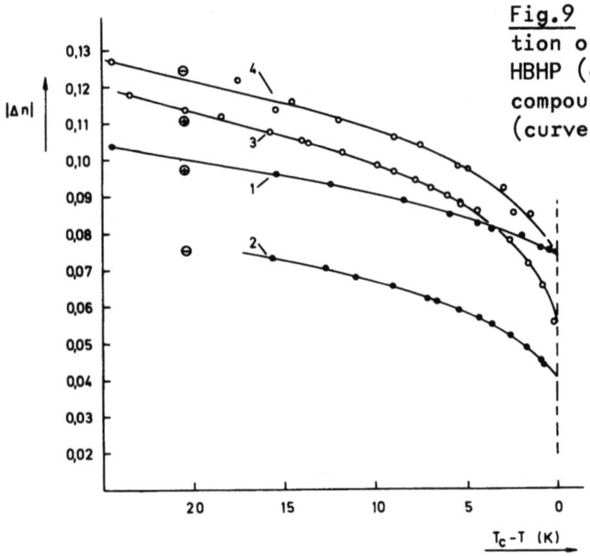

Fig.9 Birefringence as a function of the temperature for HBHP (curve 1), HBHP + chiral compound (curve 2), polymer 7 (curve 3), polymer 3 (curve 4)

In contrast to these polymers for polymer 3 and 8 no texture can be found, which is equal to conventional liquid crystals (25). Conoscopic observations indicate a negative uniaxial texture. Detailed investigations with polymer 3 indicate a strong negative birefringence (26) shown in Fig.9 curve 4. In comparison curve 3 shows the positive uniaxial polymer 7 and curve 1 the low molecular weight nematic 4-hexyloxybenzoic acid hexyloxy phenylester (HBHP). If HBHP is converted into the negative uniaxial cholesteric phase by adding chiral molecules, the birefringence is shown in curve 2. The results clearly show, that a polymer structure with an orientation of the mesogenic groups perpendicular to the optical axis can be definitely excluded, which was supposed by Cser (27).

Using a dye probe no preferred macroscopic orientation could be found for the mesogenic groups, although polarizing microscopy indicated a macroscopic uniaxial negative order (26). A possible explanation for this unconventional structure might be the disturbing effect of the polymer main chain as described above. The macroscopic orientation of the mesogenic groups is suppressed by the motions of the main chain, which might result in clusters of very small size. Under these conditions the birefringence has to be interpreted carefully. Further investigations have to clarify the detailed structure of these negative uniaxial polymers.

The investigations indicate, that not only a sufficient length for the spacer is necessary. While polymer 7 (spacer length 9 atoms) still exhibits the conventional texture, polymers 3 and 8 show disturbed structures. If, however, the poly(methacrylate) chain is exchanged by the very flexible poly (methyl siloxane) chain (polymer 4) a conventional liquid crystalline structure is obtained, although compared with polymer 3 the spacer length is even shorter by one atom. In order to obtain a liquid crystalline polymer the flexible spacer as well as the flexibility of the polymer main chain have to be considered.

Investigations on liquid crystalline polymers originally aimed at the question, whether the liquid crystalline order influences the tacticity of the polymer main chain, or if the polymerisation in the liquid crystalline state favours the formation of a stereospecific main chain. Following the results above, the main chain and the mesogenic side chain are more or less decoupled by the spacer. Consequently the tacticity has little or no influence on the structure and behaviour of the liquid crystalline phase. This is actually confirmed by several authors (28).

3.5. Applications

In the last few years the application of low molecular weight liquid crystals has become of increasing interest in the display technology. This poses the question whether the liquid crystalline side chain polymers can be oriented under the influence of an electric field and whether the orientation of the mesogenic moieties is restricted by the polymer main chain. In Fig.10 the very first qualitative investigations (29) are shown for a copolymer, which contains the polar component R_1 to achieve high dielectric anisotropy. For this copolymer with increasing voltage the response time decreases down to a value smaller than 200 ms for 10 V. Although no defined boundary conditions were used, these investigations indicate the high mobility of the mesogenic groups. Quantitative measurements of the orientation of the polymers in the electric field will give detailed insight into their elastic properties and the mechanism of the interaction of polymer main chain and mesogenic side chain.

On the other hand the orientation of the polymers in the electric or magnetic field opens the possibility to freeze in the obtained in-

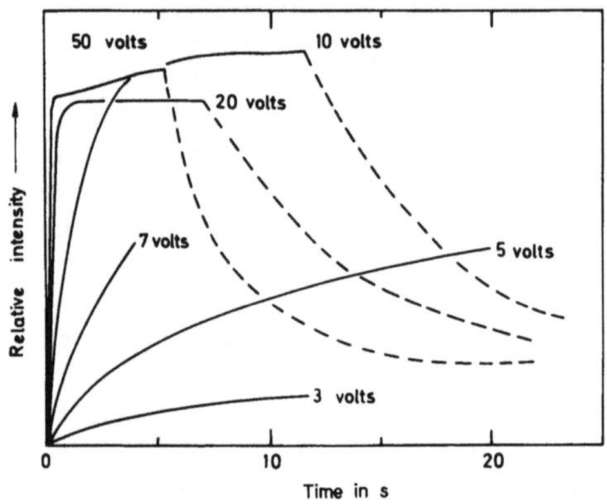

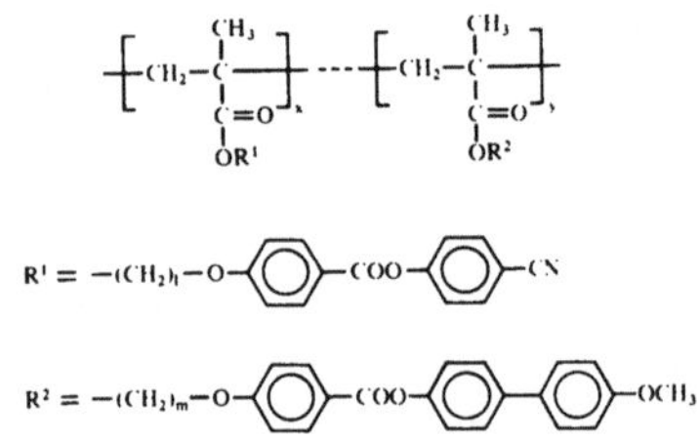

Fig. 10 Orientation of a nematic copolymer in the electric field (29)

formation. In contrast to the known starage effects, here the information is locked in the glassy state of the polymer. A very obvious application in this way is to fix the circularly polarized light reflecting Grandjean texture. In this way foils can easily be prepared, which can be used as filters or reflectors.

4. Literature

1 L. Onsager, New York Acad. Sci. 51, 627 (1949)
2 A. Isihara, J. Chem. Phys. 19, 1142 (1951)
3 P.J. Flory, Proc. Roy. Soc. A234, 60 (1956)
4 S.P. Papkov, Polymer Sci. USSR 19, 1 (1977)
5 E.A. DiMarzio, J. Chem. Phys. 35, 658 (1961)
6 A. Roviello, A. Sirigu J. Polym. Sci., Polym. Let. Ed. 13, 455 (1975)
A. Roviello, A. Sirigu, European Polym. J. 15, 61 (1979)
7 A. Blumstein, K.N. Sivaramakrishnan, S.B. Clough, R.B. Blumstein, Mol. Cryst. Liq. Cryst. 49, 119 (1978)
8 A. Blumstein, K.N. Sivaramakrishnan, S. Vilasagar, R.B. Blumstein, S.B. Clough, these proceedings, p. 252
9 W.J. Jackson, H.F. Kuhfuss, J. Polym. Sci. 14, 2043 (1976)
10 L. Liebert, L. Strzelecki, D. van Luyen, A.M. Levelut, these proceedings, p. 262
11 A. Blumstein, E.C. Hsu, Liquid Crystalline Order in Polymers, Academic Press, Inc., New York (1978)
12 V.P. Shibaev, N.A. Plate, Vysokomol. soyed. A19, 923 (1977)
13 E. Perplies, H. Ringsdorf, J.H. Wendorff, J. Polym. Sci., Polym. Lett. Ed. 13, 243 (1976)
F. Cser, K. Nyitrai, Magy. Chem. Foly. 82, 207 (1976)
V.P. Shibaev, J.S. Freidzon, N.A. Plate, Dokl. Akad. Nauk, SSSR 227, 1412 (1976)

14 H. Finkelmann, H. Ringsdorf, J.H. Wendorff, Makrom. Chem. 179, 273 (1978)
15 J. Frenzel, G. Rehage, Makromol. Chem. 1, 129 (1980)
16 R. Steinsträsser, Z. Naturforschg., Teil B: 27, 774 (1972)
17 H. Finkelmann, H. Ringsdorf, W. Siol, J.H. Wendorff, Mesomorphic Order in Polymers, A. Blumstein Ed., ACS Symposium Series 74 (1978)
18 H. Finkelmann, M. Portugall, H. Ringsdorf, ACS Polymer Preprints 19, 183 (1978)
19 H. Finkelmann, G. Rehage, Makromol. Chem., Rapid Commun. 1, 31 (1980)
20 H. Finkelmann, G. Rehage, to be published
21 H. Finkelmann, M. Happ, M. Portugall, H. Ringsdorf, Makromol. Chem. 179, 2541 (1978)
22 H. Stegemeyer, K.J. Mainusch, Chem. Phys. Lett. 6 5 (1970)
23 H. Finkelmann, J. Koldehoff, H. Ringsdorf, Angew. Chem. Int. Ed. Engl. 17 No 12, 935 (1978)
24 H. Finkelmann, H. Stegemeyer, Ber. Bunsenges. Phys. Chem. 82, 1302 (1978)
25 H. Kelker, U.G. Wirzing, Mol. Cryst. Liq. Cryst. Lett. 49, 175 (1979)
26 H. Finkelmann, D. Day, Makromol. Chem. 180, 2269 (1979)
27 F. Cser, J. Phys. (Paris) 4, C3, 459 (1979)
28 A. Blumstein, y. Osada, S.B. Clough, E.C. Hsu, R.B. Blumstein, Mesomorphic Order in Polymers, ACS Symposium Series 74 (1978)
B.A. Newman, V. Frosini, P.L. Maganini, Mesomorphic Order in Polymers, ACS Symposium Series 74, (1978)
E. Perplies, H. Ringsdorf, J.H. Wendorff, Ber. Bunsenges. Phys. Chem. 78, 921 (1974)
29 H. Finkelmann, D. Naegele, H. Ringsdorf, Makromol. Chem. 180, 803 (1979)

Thermotropic Polymeric Liquid Crystals: Polymers with Mesogenic Elements and Flexible Spacers in the Main Chain

A. Blumstein, K.N. Sivaramakrishnan, S. Vilasagar, R.B. Blumstein, and S.B. Clough

Department of Chemistry, Polymer Program, University of Lowell, Lowell, MA 01854, USA

It is known since the work of ROVIELLO and SIRIGU (1) that polymers with mesogenic moieties and flexible sequences of methylene groups in the main chain display thermotropic liquid crystalline behavior. It was assumed that nematic phases are strongly favored. More recently we have studied poly (ester) derivatives of 4,4'-dihydroxyhyphenyl and 4,4'-dihydroxystilbene, condensed with sebacic acid and shown that smectic mesophases appear in these systems (2).

In this paper the influence of structural factors is discussed. It is shown that the crystal to smectic transition temperature decreases with the length of the flexible spacer. For example spacers based on succinic acid or 3-methyl adipic acid give a highly crystalline polymer with transition temperatures beyond 280°C while spacers based on sebacic acid give transitions in the range of 200°C. The lengthening of the spacer through the addition of an ethylene oxide group on each side of the corresponding diphenol (such as for example the poly(4,4'-bis-2-hydroxyethoxydiphenyl sebacate) results in a family of polymers characterized by mesophases with lower transition temperatures. These mesophases display much less order than the smectic mesophases of the poly(4,4'-diphenyl sebacate) (3).

It is shown in addition that the mesogenic group has a determining influence on the nature of the polymeric mesophase. For example the substitution of the 4,4'-diphenol or 4,4'-dihydroxystilbene moiety with the well known nematogene 4,4'-nitrosodiphenol leads to thermotropic nematic polymers.

Linear cholesteric (twisted nematic) polymeric mesophases can be obtained by the incorporation into the flexible spacer of a chiral group {(+3)-Methyladipic acid}. The cholesteric mesophases display irridescence of planar Grandjean textures. The pitch of the cholesteric helix can be varied as a function of the chiral content of the spacer (4).

Thus all three basic liquid crystalline phases : nematic, cholesteric and smectic are obtained with this type of polymers.

(1) A. ROVIELLO and A. SIRIGU, J. Polymer Sci. (Letters), 13, 455 (1975)
(2) A. BLUMSTEIN, K.N. SIVARAMAKRISHNAN, S.B. CLOUGH and R.B. BLUMSTEIN, Mol. Cryst. Liq. Cryst., 49 (Letters) 255 (1979)
(3) A. BLUMSTEIN, K.N. SIVARAMAKRISHNAN, R.B. BLUMSTEIN and S.C. CLOUGH Polymer (submitted for publication).
(4) S. VILASAGAR and A. BLUMSTEIN. Mol.Cryst.Liq.Cryst.,(Letters)(in print)

Polymerization of Lipid and Lysolipid Like Diacetylenes in Monolayers and Liposomes

H.H. Hub, B. Hupfer, and H. Ringsdorf

Institut of Organic Chemistry, University of Mainz,
D-6500 Mainz, Federal Republic of Germany

1. Introduction

The bilayer type lipid membrane has a variety of important functions that are necessary in order to sustain life. A lot of these functions have been studied with artificial model membranes such as black lipid membranes(BLM) or liposomes. A great disadvantage of all these models is their instability - BLM for instance can only exist for minutes or hours and only under the presence of water. Therefore,it was the aim of this work to produce simple model membranes that can retain their structure under a variety of test conditions. The route chosen to obtain such membranes was through the polymerization of lipid and lysolipid like molecules in monolayers at the gas-water interface or in liposomes. After polymerization the lipid molecules are bound to each other by covalent bonds and a much greater stability of BLM or liposomes can be expected.

As initial model compounds we investigated diacetylene carbonic acids with different alkyl chain lengths(1) [1] . These monomers polymerize in the monolayer upon exposure to UV-light as it was observed in the solid state [2] and in multilayers [3] resulting in polymers containing a conjugated backbone (2).

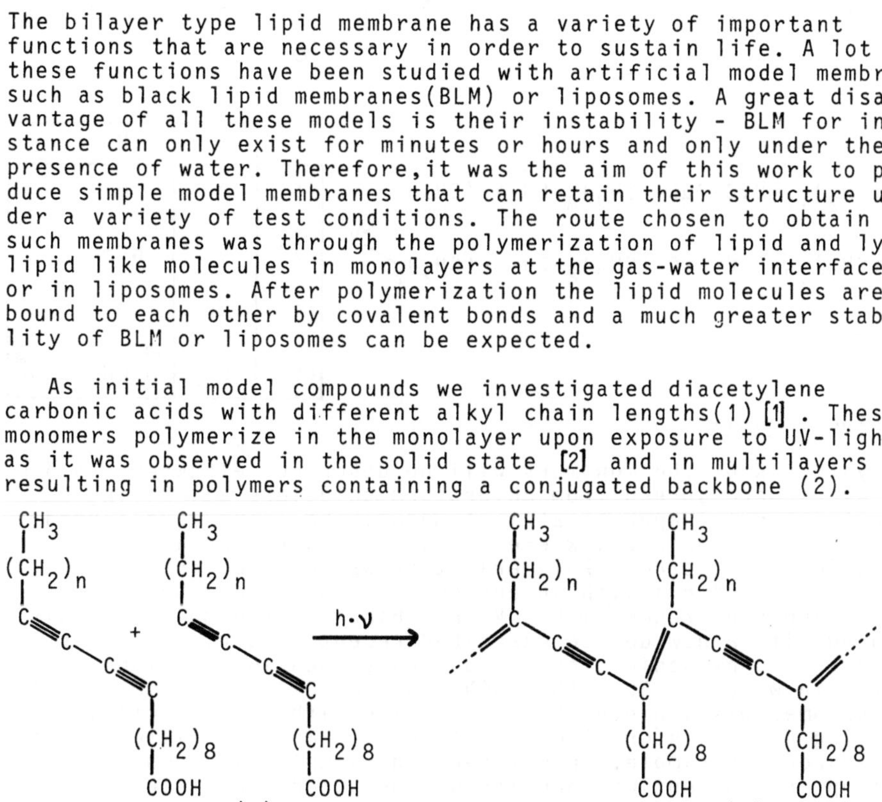

(1) (2)

These monolayer polymer films have also been found to form air-stable bilayers if deposited onto a porous substrate [1] . These bimolecular layers, which can be viewed as inverse model cell membranes are approximately 6 nm thick, stable in air for months and can span holes as large as 0.5 mm in diameter. This very simple model demonstrates the great advantage of polymer model membranes over conventional BLM.

The next step was the preparation and investigation of polymerizable lysolipid like compounds (3,4,5) resembling the naturally occurring lipids in the head group. Since natural lipids contain two alkyl chains, as model compounds molecules were prepared with simple hydrophilic head groups (according to KUNITAKE [7]) and two chains containing the diacetylene functionality (6,7,8,9).

$H_3C-(CH_2)_{12}-C{\equiv}C-C{\equiv}C-(CH_2)_9-OR$

(3): $R = PO_3H_2$

(4): $R = \underset{\displaystyle O^{\ominus}}{\underset{|}{P}}O-O(CH_2)_2\overset{\oplus}{N}H_3$

(5): $R = \underset{\displaystyle O^{\ominus}}{\underset{|}{P}}O-O(CH_2)_2\overset{\oplus}{N}(CH_3)_3$

$$\left.\begin{array}{l} H_3C-(CH_2)_{12}-C{\equiv}C-C{\equiv}C-(CH_2)_8-CO-Y-(CH_2)_2 \\ H_3C-(CH_2)_{12}-C{\equiv}C-C{\equiv}C-(CH_2)_8-CO-Y-(CH_2)_2 \end{array}\right\rangle X$$

	Y	X
(6)	O	O
(7)	O	NCH_3
(8)	NH	O
(9)	O	$\overset{\oplus}{N}(CH_3)_2\ Br^{\ominus}$

2. Experimental

The synthesis of the monomers will be described elsewhere [4,5]. Monolayers were spread from chloroform solutions all having a concentration of approximately 1 mg/ml. The films were spread on a LAUDA film balance, where surface pressure and area were automatically recorded. Polymerization was carried out via UV-irradiation (254 nm) with an energy of 5 mW/cm^2 at the water surface under nitrogen [1]. UV-absorbance in the monolayer was measured with a new device described recently [6]. Liposomes were prepared by sonication of aqueous suspensions of monomers (6)-(9) in water at 50 °C (BRANSON sonifier B 15). Polymerization of liposomes was achieved by irradiation with multichromatic light (Hg high pressure lamp) at 18 °C with a water filter between beam source and sample. In between, the sample was removed from the lamp and the optical density was measured. The techniques for electron microscopy were freeze fracture, freeze etching, and negative staining (Uranylacetate).

3. Spreading Behaviour of the Monomers

The surface pressure-area diagrams of compounds (3)-(5) (Fig.1) exhibit significant differences due to the different head groups.

The phosphatidic acid analogue (3) at temperatures below 35°C shows a single condensed phase with a collapse point at 0.25 nm^2/molecule, whereas the cephaline analogue (4) already at 20°C forms both a condensed and an expanded film. In the diagram of compound (5) in the whole investigated temperature region (2°C to 50°C) only a liquid expanded phase occurs. This is due to the great volume of the trimethylammonium group, which prevents the crystallization of the alkyl chains in the monolayer.

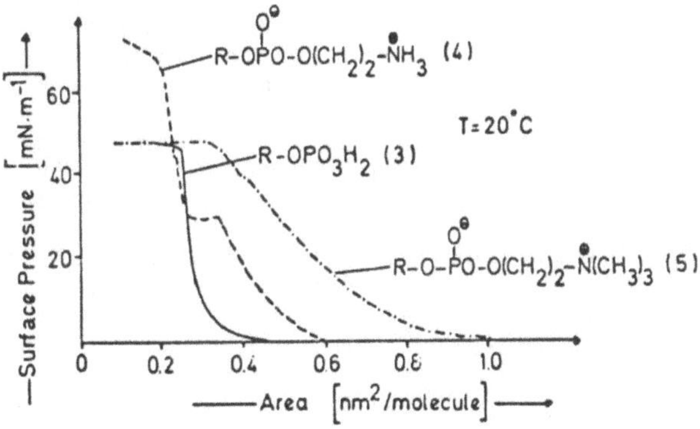

Fig.1 Surface pressure-area diagrams of lysolipid analogues ($R= CH_3-(CH_2)_{12}-C{\equiv}C-C{\equiv}C-(CH_2)_9$)

In contrast to monomers (3)-(5) the lipid analogues (6)-(9) show collapse areas greater than o.40 nm^2/molecule, as to be expected from the two long chains per molecule (see Fig.2).

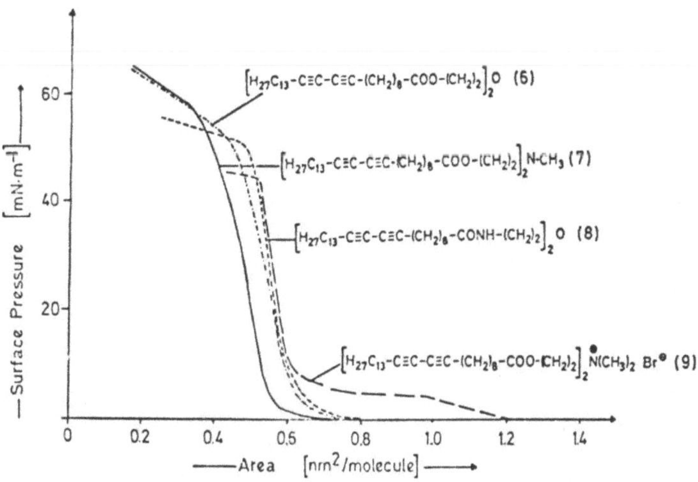

Fig.2 Surface pressure-area diagrams of lipid analogues at 20°C

Monomers (6) and (7) exhibit collapse areas of 0.42 and 0.43 nm^2/ molecule respectively without a real sharp collapse point, whereas (8) and (9) show collapse areas of 0.48 and 0.51 nm^2/ molecule with sharp collapse points similar to monomer (3). The increase in collapse areas for compounds (8) and (9) may be explained by hydrogen bonding between the molecules of (8) and a steric and charge repulsion for the ammonium compound (9).

4. Polymerization Behaviour

Polymerization only occurs, if the temperature is below the phase transition temperature, i.e. in the condensed phase. Monomers (3) and (4) at 20 °C under contraction of the film exhibit a complete conversion after less than 15 min and constant surface pressures of 10 and 35 mN/m respectively. Monomer (5) did not polymerize in the monolayer: caused by the great head group a close packing of the chains necessary for topochemical polymerization is not possible. The behaviour of the lipid molecules (6)-(9) is comparable to the acids (1) [1] only in the cases of (7) and (9): under a contraction of the films of about 10% the polymerization is finished after 5 min. Monomer (8) exhibits a two step contraction and shows no further reaction after 10 min of UV-irradiation, whereas (7) expands during polymerization.

5. Monolayer Absorbance

As a result of the conjugated nature of the polymer backbone the final polymerized monolayers of all the monomers have high extinction coefficient in the visible spectrum and can be seen as a faint reddish tint on the water surface. Because the initial monomer is colorless, the degree of absorbance of the

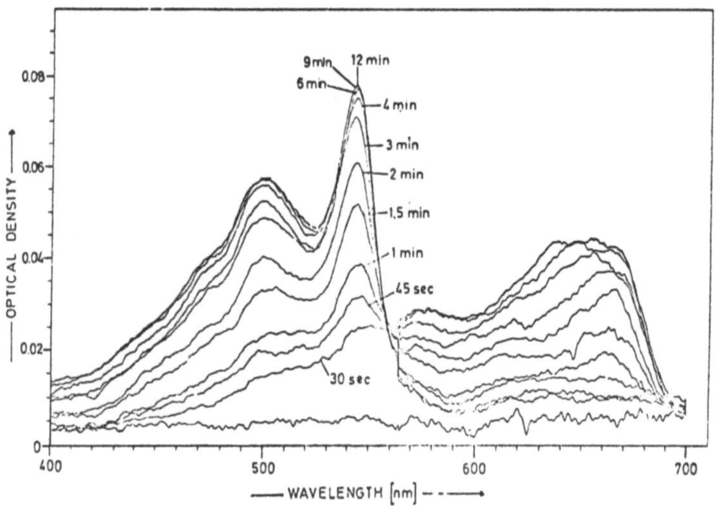

Fig.3 UV-Multiplot of absorbance of a monolayer of (8) vs. polymerization time (constant pressure 10 mN/m, N_2-atmosphere)

monolayer offers another method of following the polymerization mechanism. For this purpose an improved special device was used, which had already been useful for the investigation of the polymerization kinetics of the acids (1) [6]. With this apparatus monolayer UV-spectra of compounds (7)-(9) were recorded during the polymerization. The reaction rate of these bifunctional monomers is higher than those of the diacetylene carbonic acids. The blue species of the polymer can only be observed in the time interval from 0 up to 2 min, after that time the conversion to the red polymer takes place. The maximum absorbances are slightly different from those of the monofunctional compounds. Mono- and bifunctional monomers show a decrease of the maximum absorbance on further UV-irradiation eventually due to a degradation of the polymer backbone by radicals. Fig.3 summarizes the original absorbance curves of (8) at various irradiation times.

6. Polymer Liposomes

Sonication of suspensions of the phospholipid analogues (7) and (9) results in the formation of clear colorless solutions of monomer liposomes. Investigation by electron microscopy proves the formation of bi- and multilayer vesicles of a defined spherical shape and of different diameters (range: 100 nm up to several µm). The diameter and number of bilayers strongly depend on sonication time and intensity. After 30 min of sonication only bilayer liposomes of a diameter of 100 nm are formed. Further sonication does not alter the size. Diameter and number of bilayers are also affected by the concentration of the monomer suspensions. A ten fold increase in concentration results in a two fold increase in diameter at a given sonication time and intensity.

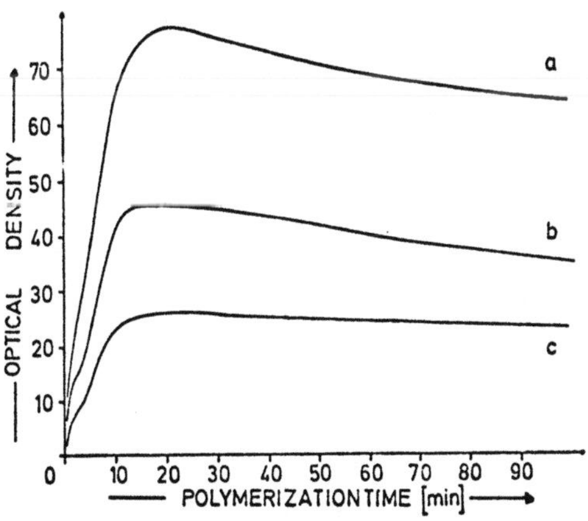

Fig.4 Dependence of polymerization rate on sonication time measured by increase of optical density, monomer (9), sonication times: a: 5 min; b: 15 min; c: 30 min

UV-irradiation of such a monomer liposome solution of (9) results in the formation of polymer liposomes indicated by a color change from colorless via blue to red. The polymerization rate is followed by the change of optical density and exhibits a strong dependence on sonication time (Fig.4). Electron microscopy shows that the shape of the liposomes remains unchanged during polymerization, only the diameter of the spheres is smaller than those of the monomer liposomes. The polymer as well as the monomer liposome solutions are stable for several months. Precipitation can be achieved by the addition of salts, due either to a "salting out" or a destruction of the polymer liposomes. Investigations on leakage and ruptures in the shell of the polymer vesicles as well as attempts to "co-sonicate" polymerizable and natural lipids under the presence of proteins in order to obtain stable cell models after polymerization are in progress.

7. References

[1] D. Day, and H. Ringsdorf, J.Polym.Sci.,Polym.Lett.Ed., 16, 205(1978)

[2] G. Wegner, Makromol.Chem., 154, 35(1972)

[3] B. Tieke, H.-J. Graf, G. Wegner, D. Naegele, H. Ringsdorf, A. Banerjie, D. Day, and J.B. Lando, Colloid Polym.Sci., 255, 521(1977)

[4] H.H. Hub, and H. Ringsdorf, Angew.Chem., to be published

[5] B. Hupfer, and H. Ringsdorf, in preparation

[6] D.R. Day, and H. Ringsdorf, Makromol.Chem., 180, 1059(1979)

[7] T. Kunitake, J.Macromol.Sci.-Chem. A13, 587(1979)

Spin Probe Studies of Oriented Liquid-Crystalline Polymers

G. Kothe, K.-H. Waßmer, E. Ohmes, M. Portugall, and H. Ringsdorf

Institute of Physical Chemistry, University of Freiburg
D-7800 Freiburg, Federal Republic of Germany, and
Institute of Organic Chemistry, University of Mainz
D-6500 Mainz, Federal Republic of Germany

The spin probe technique is employed to study liquid-crystalline side chain polymers [1], oriented in a high frequency electric field [2]. Temperature and angular dependent electron spin resonance spectra are analyzed, using a comprehensive lineshape model [3]. Computer simulations provide the order parameters $\overline{P2}$ and rotational correlation times τ of the nitroxide probes [4]. They are related to the structure and dynamics of the nematic polymer:

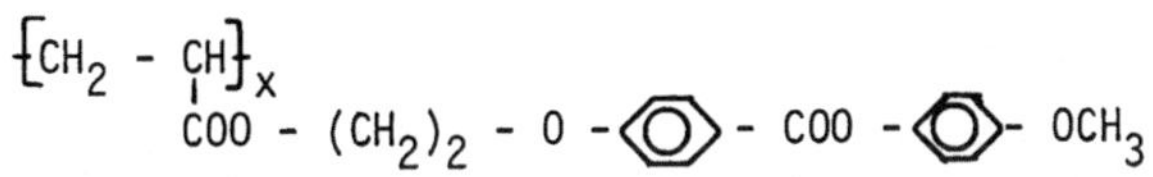

A plot of τ versus 1/T shows two breaks, which correspond to the isotropic-nematic and to the glass transition temperature Tg, respectively. Within a particular phase the plot is linear, yielding rotational activation energies of E = 44 [kJ/mol] (T > Tg) and E = 23 [kJ/mol] (T < Tg). Apparently molecular rotation is determined by different processes.

The order parameters in the isotropic phase are $\overline{P2} = 0$, indicating a random orientation of the side chains. At the isotropic-nematic transition $\overline{P2}$ jumps to a finite value and then increases with decreasing temperature, approaching a high limiting value of $\overline{P2} = 0.65$. It is essentially maintained when the polymer is cooled below the glass transition temperature. This result clearly shows, that nematic order can be frozen in.

References

1 H. Finkelmann, H. Ringsdorf, J.H.Wendorff: Makromol.Chem. 179, 273 (1978)
2 G. Kothe, T. Berthold, E. Ohmes: Molec.Phys. (1980, in press)
3 G. Kothe: Molec.Phys. 33, 147 (1977)
4 J.F.W. Keana, S.B.Keana, D. Beetham: J.Am.chem.Soc. 89, 3055 (1967)

Photochromic Polymers in Two Dimensions

F. Rondelez, H. Gruler and R. Vilanove

Collège de France, Physique de la Matière Condensée,
11 Place Marcelin-Berthelot, F-75231 Paris Cédex 05, France

We have investigated the properties of monomolecular films of photochromic polymers spread at an air-water interface. A statistical copolymer of polymethyl methacrylate and spirobenzopyran derivatives was used in all experiments (molecular weight 215.000 - 5.1 mole % of spiropyran). Under ultra-violet excitation, the pyran ring undergoes an heterolytic scission followed by a rotation of one part of the molecule so as to approach coplanarity. This large conformational change of the chromophores attached as side groups on the polymeric backbone induces an increase in the coil radius of gyration. Consequently, large changes in the lateral surface pressure of the two-dimensional film are observed. The static conformation of the macromolecules in both the excited and unexcited state can be readily derived from these pressure measurements. Under visible light excitation, the chromophores can be returned to their initial state, which makes the mechanical process fully reversible. A full account of this work will be reported in the Physical Review Letters (March 1980). Mixed films of photochromic polymers with lipids could serve as model systems for photoregulated biological processes. Indeed membrane proteins as the rhodopsin in the eye rod outer segment and the phytochrome in plant cells undergo photo-isomerisation as the primary step to trigger the biological activity. Photo-induced surface pressure changes can also be used in dynamic studies of mono and bilayers to induce instantaneous mechanical stresses without spurious hydrodynamic velocity gradients.

Nematic Phases of Polymers

A. Thierry, B. Millaud, A. Skoulios

Centre de Recherches sur les Macromolécules,
6, rue Boussingault, F-67083 Strasbourg-Cédex, France

In a rough attempt to answer the question of, *whether a polymer of any given flexibility is able to yield a nematic lyotropic phase,* we have calculated the extension of the domain of stability of the nematic phase in some typical binary polymer/solvent mixtures, these being *assumed ideal* (1) ; we thus could have an idea of the tendencies of polymers to yield lyotropic phases :

- when both components are nematic in the same range of temperatures, the nematic phase observed extends over the whole range of concentration in the binary mixture
- when only one component is nematic, the liquid-crystalline phase is restricted to a small range of concentrations in the vicinity of the corresponding pure component
- when none of the components is nematic, no nematic phase usually appears

Qualitatively these results hold well in the case of binary mixtures of Schiff-base polymers of various degrees of condensation (2). Quantitatively however they lead to some discrepancies which increase with molecular weight and which can be attributed to the non ideality of the mixtures for entropic reasons.

The lyotropic nematic phases observed in the presence of a very polar solvent (ex. : aromatic polyamides + sulphonic acid (3)) are evidently to be related with the existence of strong enthalpic interactions among the polymer and solvent molecules.

References

1. B. Millaud, A. Thierry, A. Skoulios, Colloid and Polymer Sci., 257, 247, (1979)
2. B. Millaud, A. Thierry, A. Skoulios, Mol.Cryst.Liq.Cryst. (Lett.), 41, 263 (1978)
3. Du Pont de Nemours, E.I., B.F. N°21 134 582 (1972)

Nematic Thermotropic Polyester[1]

L. Liebert[2], L. Strzelecki[2], D. Van Luyen[3], and A.M. Levelut[2]

[2]Laboratoire de Physique des Solides associé au CNRS n°2, Bât. 510, F-91405 Orsay, France

[3]Ecole Supérieure Polytechnique, Hanoi, Vietnam

Synthetized by us[1] [2] [3] the polymer :

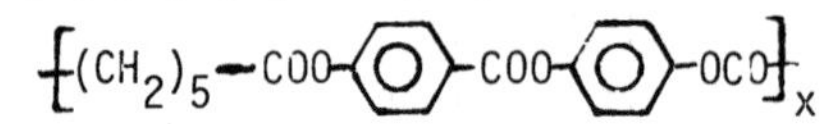

has the following transition temperatures :

$$\text{Solid} \xrightarrow[145°C]{T_1} \text{Anisotropic Liquid} \xrightarrow[300°C\text{-}350°C]{T_2} \text{Isotropic Liquid}$$

These temperatures are determined by optical microscopic observations and checked by D.T.A.. The anisotropic liquid presents the threaded texture of nematic phase but it is more viscous.

We have studied the structure of the solid and nematic phases by X-Rays measurements. The X Ray scattering patterns of the solid phase of the polymer (purified by reprecipitation in aceton from a solution in the dichloroacetic acid) correspond of unoriented assemblies of lamellar crystals. In the nematic phase, without magnetic field, we have obtained a typical pattern of an unoriented nematic. The nematic phase of the polymer is oriented in a magnetic field of 0,3 T and the pattern presents two crescents; these crescents are characteristic of a liquid of parallel polymer chains and their position is related to the mean distance between two adjacent chains. By decreasing the temperature the polymer becomes solid, conserves the orientation and in this case the obtained pattern is characteristic of a fiber with an axis parallel to the magnetic field.

In conclusion, a relatively low magnetic field orients the polymer in the nematic phase and this orientation is preserved in the solid phase. The order parameter determined by densitometry measurements [4] depends of the molecular weight of the polymer estimated by inherent viscosity from a solution in dichloroacetic acid.

BIBLIOGRAPHY

1) L. Strzelecki and D. Van Luyen, European Polymer Journal (to be published)
2) D. Van Luyen and L. Strzelecki " " " "
3) D. Van Luyen, L. Liébert and L. Strzelecki " " "
4) A.J. Leadbetter and P.G. Wrighton, Coll. C3, Suppl. n°4, Tome 40, C3-234 (1979).

[1] will be published in European Polymer Journal

Part VII

Lyotropic Liquid Crystals

Lyotropic Nematic Phases of Amphiphilic Compounds

J. Charvolin and Y. Hendrikx

Laboratoire de Physique des Solides, Université Paris-Sud,
F-91405 Orsay, France

1. Introduction

In 1967 Lawson and Flautt created new lyotropic phases from a classical water-surfactant (sodium decyl sulfate, or SDS) mixture by adding to it small amounts of long chain alcohol (decanol) and salt (sodium sulfate) [1]. They observed fluid anisotropic phases whose textures and spontaneous orientation in a magnetic field, uncommon to lyotropic systems, appeared quite similar to those observed with thermotropic systems, hence the denomination of "lyotropic nematics". The approximative location of the nematic phases in the phase diagram is shown in Fig.1, [2].

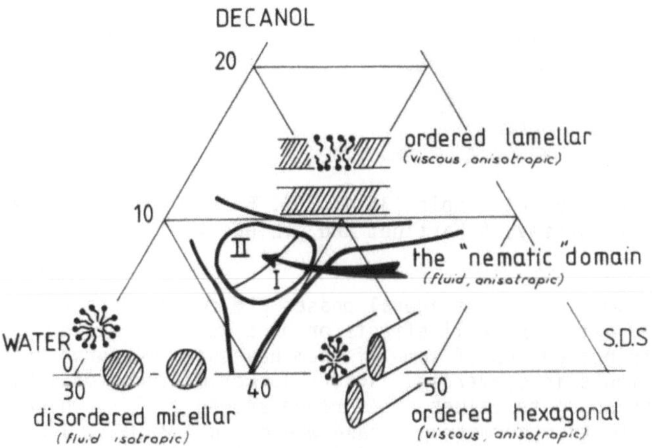

Fig.1 A sketch of the SDS-decanol-water phase diagram in the vicinity of the nematic domain (weight per cent)

Since 1967 these phases have been widely used as orienting media in a magnetic field for NMR determinations of the structures of dissolved molecules with the advantage, when compared to thermotropic phases, that they can dissolve molecules with a range of polarities since they consist of two media, aqueous and paraffinic [3]. Several mixtures have been developed for such one use. This interest for the phases has been of purely utilitarian character and, although they are ten years old, very little efforts have been devoted to the studies of their properties and structures as liquid crystalline phases.

This lecture is a brief review of the few attempts which have been made in these directions. We shall consider the observations which show first the existence of an orientational order, second the absence of translational order, a situation quite similar to that of thermotropic nematics, and finally we shall present the structural data which make the difference in demonstrating that the orientational order does not result from the alignment of isolated molecules but from that of anisotropic aggregates of amphiphilic molecules. We shall focus our attention on SDS-decanol-water samples which have been the objects of most of the studies, suggestions for other mixtures will be found in the references quoted.

2. Orientational Order

2.1 Optical Observations

The samples appear anisotropic between crossed polarizers, "schlieren" textures may be observed, as shown in Fig.2:

Fig.2 Texture of a positive uniaxial sample (36.4% SDS, 52.8% D_2O, 6.6% decanol, 4.2% SO_4Na_2 at 20°C). (Courtesy, A.Martinet and L.Liebert)

(The textures of nematic, lamellar and hexagonal phases are compared in [4]). The samples may be oriented either by wall effects or by a magnetic field (see below), and the birefringences Δn of some of them have been measured by means of conoscopic experiments in converging light [5]. For mixtures of SDS, decanol, water and salt the absolute values of Δn range around 3.10^{-3}, which is very weak when compared to the thermotropic case where $|\Delta n| \simeq 2.10^{-1}$. Here the samples may also be either negatively or positively uniaxial according to composition and temperature. A sample with negative Δn will be henceforth said to be of type I, one with positive Δn of type II [6].

2.2 Alignment of the Director in a Magnetic Field

All samples spontaneously orient in a magnetic field of a few kilogauss. The orientation process is rather slow, its characteristic time may vary from a few seconds to one hour according to the nature of the sample, whereas it is instantaneous in the thermotropic case. The alignment is usually studied by NMR. This technique is of such a wide use in this field [3] [7] that it is worthwhile to describe briefly the essential of its contribution here. Very common NMR probes now are the deuterons either in D_2O or in deuterated methylene groups along the paraffinic chains of the amphiphilic molecules. In an anisotropic environment the deuteron signal is a doublet the splitting of which is proportional to the order parameter $<P_2(\cos\theta)>$, or the degree of anisotropy, and to the orientation factor $P_2(\cos\phi)$, characterizing the orientation of the axis of anisotropy relative to the magnetic field. When a deuterated sample is introduced in the NMR spectrometer, the first spectrum has the usual "powder pattern" appearance indicating a random distribution of ϕ, a certain time later the lines narrow and the powder spectrum evolves into the simple doublet spectrum of a sample with only one orientation ϕ of the anisotropy axis. This evolution is shown in Fig.3.

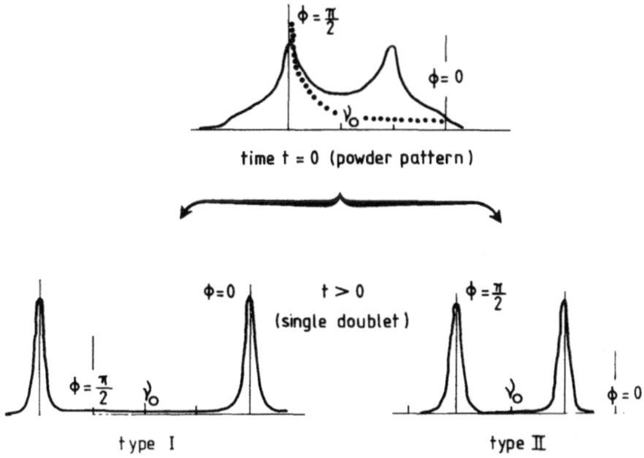

Fig.3 Representations of the evolutions with time of D_2O spectra for the two known types of samples. At time t = 0 the disordered sample is introduced in the magnetic field, all orientations ϕ of the axis of anisotropy are equally probable and the distribution of one component of the doublet within $\phi = 0$ and $\phi = \pi/2$ is indicated by the dotted line, the other is distributed symmetrically about the central frequency ν_0, later on the distribution vanishes and the sample becomes oriented.

The interpretation is that the local axes of anisotropy, or directors, have oriented uniformly in the field. They can orient parallel to the field, $\phi = 0$, as for the type I sample which then has a positive anisotropy of diamagnetic susceptibility $\Delta\chi$, or perpendicular to it, $\phi = \pi/2$, as for the type II sample which then has a negative $\Delta\chi$.

When the nematic director is rotated far from its equilibrium orientation in the field it is known from the study of thermotropic nematics [8] that its relaxation, deduced from the evolution of the NMR spectra, may inform about the rotational viscosity γ_1 of the samples. Similar experiments have been tried with lyotropic nematics. In a first study of SDS samples the relaxation rate depends on the diameter of the sample tube thus revealing the importance of the anchoring at the sample-glass interface [9]. In a more recent study of Cs-perfluorooctanoate samples with positive $\Delta\chi$ no wall effect prevented the measurement of the ratio $\gamma_1/\Delta\chi$, its value varies from 0 in the isotropic phase to $5.10^7 cm^2 s^{-1}$ in the nematic phase before diverging at the approach of a lamellar phase [10]. For this particular sample $\gamma_1/\Delta\chi$ appears to be about one order of magnitude larger than in the thermotropic sample MBBA [11].

2.3 Two Examples of Field Induced Defects and Patterns

A type I sample sheared between glass plates tends to present planar domains of various orientations in the plane of the plates. When a magnetic field of several kilogauss is applied in the plane the domains tend to orient their axes along the field and characteristic lines, shown in Fig.4, appear.

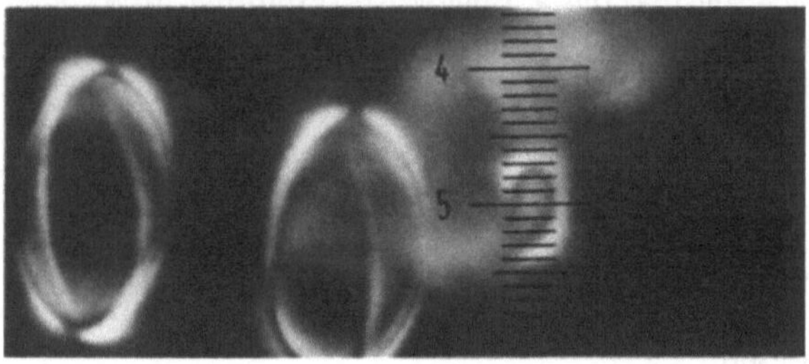

Fig.4 Defects observed during the application of a 20 kgauss field in the plane of the preparation with a type I sample (38.1% SDS, 6.8% decanol, 55.1% D_2O at 20°C). Crossed polarizers, sample thickness 1mm, one large division is 150μ.

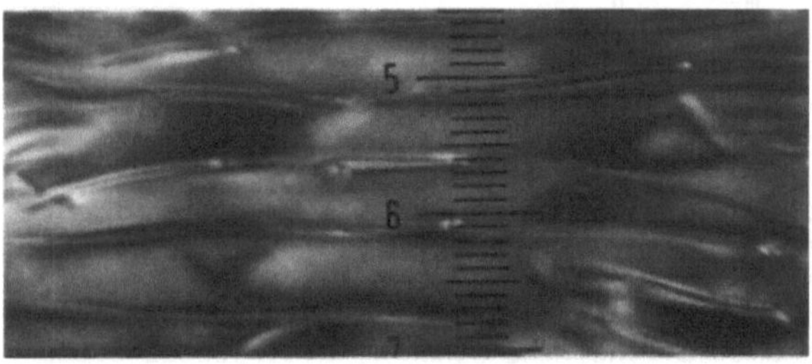

Fig.5 Pattern observed after the application of a 20 kgauss field normal to the planar sample of Fig.4. This organization is transitory under the field.

According to a model developed to explain superficial circular loops in planar thermotropic nematics [12] one might consider that the loop is the border between two regions of the sample where the nematic axes have turned in opposite directions under the action of the field. The directors inside the loop and outside are out of phase by π and the loop pictures their joining. With time the area enclosed by the loops shrinks and the loops disappear.

In a second step, once a homogeneous planar orientation is obtained, the application of the magnetic field normal to the glass plates leads to the transitory appearance of a periodic organization of the isoclines normal to the initial orientation of the nematic axis [13], as shown in Fig.5.

Now the nematic axis rotates in the plane defined by the magnetic field and its original orientation in order to aline along the field. This transient periodic structure is reminiscent of the roll structure observed in the course of the study of the Frederickz transition on a planar thermotropic nematic a.

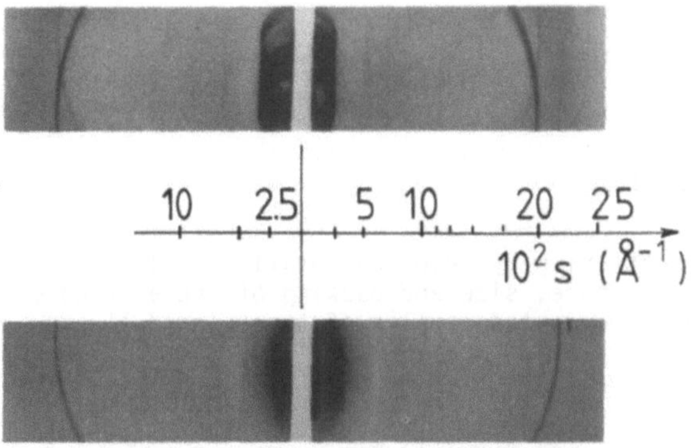

b.

Fig.6 X-ray diagram obtained in a Guinier's camera with non-oriented samples of a lamellar phase of potassium laurate (a), of a nematic phase of type II (b). s = 2sinθ/λ.

film [14]. The model developed in that case is that the decrease of effective rotational viscosity, due to the backflow induced by the rotation of the molecules to aline along the field, makes periodic orientational modes grow faster that the uniform one although they are less stable.

In the case of a type II ($\Delta\chi<0$) homeotropic film the application of a magnetic field normal to the film develops a periodic organization along two dimensions, most likely because the nematic axes reorient away from their original orientation in the homeotropic domains isotropically [13].

3. Translational Disorder

The absence of translational order in the lyotropic nematic phase is readily seen when comparing, in Fig.6, a X-ray powder diagram with that of a lamellar lyotropic phase which is known to be ordered along one dimension.

As it is most often the case with lyotropic liquid crystals the diagrams are to be analyzed in two regions [15]. At large angles, around (4 Å)$^{-1}$, the scattering is due to the short range organization of the paraffinic chains of the amphiphilic molecules within the aggregate, at small angles, about(30 Å)$^{-1}$, it is due to the long range organization of the aggregates. Both diagrams present similar diffuse bands around (4-5 Å)$^{-1}$ and the chains are disordered in the nematic phase as well as in the lamellar one. The diagrams differ in the region around (30 Å)$^{-1}$. This discrepancy is the sign of a deep qualitative difference between the two samples: the narrow lines in (a) reveal the long range ordering of the soap lamellae whereas the diffuse band observed in the nematic case (b) is characteristic of a system without long range order. Nevertheless the very close positions of the band in (b) and of the first order line in (a) show that similar distances prevail for the diffraction in both systems and suggest the presence in the nematic phase of molecular aggregates having one of their dimensions close to the thickness of the lamellar bilayer.

4. Molecular Aggregation

4.1 X-ray Study of the Structures

The study of the small angle scattering by oriented nematic samples has provided some information about the shape, size and packing of the aggregates [13]. The diffraction patterns of oriented samples of types I and II are shown in Fig.7.

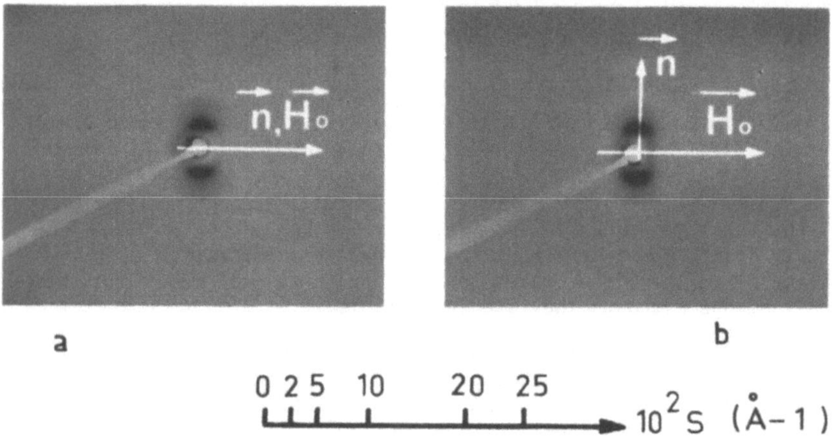

Fig.7 X-ray diagrams obtained in a monochromatic Laue camera with oriented nematic samples of type I (a) and II (b). (38.1% SDS 6.8% decanol 55.1% D_2O for type I, idem + 4% SO_4Na_2 for type II). Temperature 20°C. The X-ray beams are normal to the films. The diagrams are unchanged by rotating the samples around the nematic axes $\vec{n}$. $s = 2\sin\theta/\lambda$.

We shall just summarize here the analysis of the small angle region which is given in [13]. Because of the rotational symmetry of the patterns around $\vec{n}$ the reciprocal structures appear to be a cylinder of axis $\vec{n}$ in the case of type II, a torus of axis $\vec{n}$ in the case of type I. Coming back in real space by Fourier transformation the diffracting objects are therefore flattened cylinders, or discs, of axis $\vec{n}$ for type II, elongated cylinders of axis $\vec{n}$ for type I. In a magnetic field the aggregates aline their largest dimension parallel to it.A schematic representation of the amphiphilic aggregates and their average sizes are given in Table I [16]. Moreover the statistical angular distribution of the axis of each aggregate relative to the nematic axis, the order parameters, must manifest itself on the patterns in fig.7 through the distribution of intensity along the arc of the main reflection. For the type I sample studied here the analysis lead to a rather high estimate of S>0.7 [17]. The analysis is less straightforward in the case of a type II sample because of the limited lateral size of the aggregate which limits the extension of the main reflection, however we tried to estimate the disorientation from the widths of the lateral vertical streaks at distances from the equator and we found the same lower limit.

Finally, from the same experiments [13], the analysis of the distribution of intensity along the large angle diffraction rings of the chains has shown that the chains always orient their elongation axis perpendicular to the magnetic field preferentially.

4.2 NMR Study of the Molecular Behavior

Study of amphiphilic molecules with deuterated methylene groups in lyotropic liquid crystals provides the order parameter of the CD bonds of these groups with respect to the external magnetic field, it informs about the degree of local disorder in the paraffinic medium specifically [15]. However one must not forget in interpreting the experimental data that several motions are liable to contribute to the order parameters: the chain deformations around the local normal to the surfactant-water interface, the reorientation of the chain as a whole when the molecule diffuses along the curved regions of the interface [18], the orientational fluctuation of the overall aggregate. Any attempt to discriminate between those several contributions is doubtful now, because of the insufficient accuracy of the present structural data concerning the curvatures of the interface and the order parameters of the aggregates. The only reliable information at the moment concerns the relative variations of the orientational disorders of the chain links, from the polar head towards the methyl end, which stems from the shape of the curve of DMR splittings (this is an information about intrachain motions only because, considering the disorder of one links with respect to another one, we eliminate the contributions of the overall chain motions). DMR curves for C_{12} chains in lamellar, hexagonal [18] and type II nematic [19] phases are compared in Fig.8.

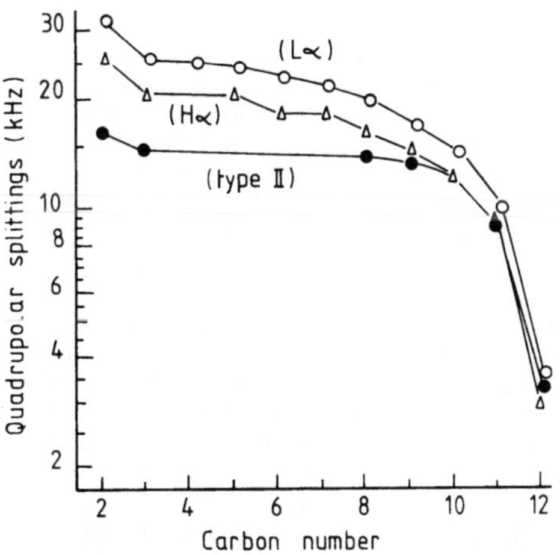

Fig.8 Curves of DMR splittings for deuterated K laurate chains in lamellar Lα, hexagonal Hα lyotropic phases and a nematic type II. The magnetic field is in the plane of the lamella (Lα), along the axis of the cylinder (Hα), in the plane of the disc (II): it stays normal to the long axes of the chains

Whatever the structure is, the disorder increases in very similar ways from the polar heads towards the methyl ends of the chain. The very particular shape of the curves, with a slow decrease from the 3^{d} to the 9th carbon, has been discussed for classical phases [18] [20]: it indicates that the local behavior of the chains is dominated by water-surfactant interactions for the first 3 links then by interchain interactions for the following links. Most likely the local environment of one chain does not change dramatically going from classical to nematic phases. The difference, by a factor 2, between the first splittings for lamellar and nematic phases may reflect differences in the overall motions of the molecules and aggregates [21].

5. Discussion of the Data

The data presented above are summarized in table I.

Sample	Sodium decyl sulfate-decanol-water	
Type	I	II (with sodium sulfate)
Anisotropy of Susceptibility	$\Delta\chi > 0$	$\Delta\chi < 0$
Birefringence	$\Delta n \sim -3.10^{-3}$	$\Delta n \sim +3.10^{-3}$
Aggregates	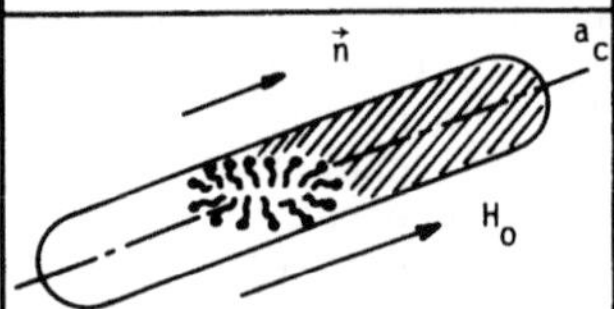 cylinders in water	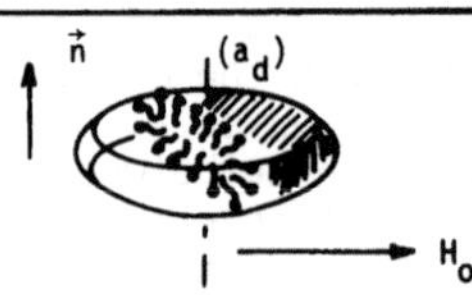 discs in water
Dimensions	length ? average diameter $\sim$ 30 Å	thickness $\sim$ 20 Å average diameter $\sim$ 60 Å
Packing	nearly parallel alignment without translational order	

The numbers quoted in table I refer to a very limited number of studies. These numbers may vary with the composition and temperature of the mixture; also, considering the variety of nematic mixtures, the eventuality of other structures is not to be rejected. However it is interesting to record that the properties associated with the orientational ordering and the structural features may be accounted for by one coherent description which may be then expected to have some general validity.

The lyotropic nematics studied appear to be dispersions of amphiphilic aggregates in water with orientational order but without long range translational order. In our model the aggregates are either prolate (type I) or oblate (type II), their axes (a_c) or (a_d) tend to be parallel and the common directions define the directors $\vec{n}$ of the phases. In the presence of a magnetic field H_o the aggregates aline their axes (a_c) or (a_d) respectively parallel or perpendicular to the field. This is what is to be expected from the magnetic shape anisotropies of elongated or flat objects. In both cases the chains of the molecules tend to keep their stretching axis preferentially perpendicular to the field as it may be expected for isolated saturated hydrocarbon molecules [22]. Therefore the molecular and shape magnetic anisotropies have converging effects to orient the aggregates in the field. On the other hand the birefringence, which changes its sign at almost constant absolute value from one type to another, would correspond mainly to that of saturated hydrocarbons [22] in similar environments but with different orientations of their long axes relative to the director [23].

6. Conclusion

We have described lyotropic phases in which finite and anisotropic aggregates of amphiphilic molecules are dispersed in water keeping some orientational order, but no translational order, just the way thermotropic molecules are organized in a nematic phase. These phases may bring in some new interests owing to the variability introduced in the nematic organization by their lyotropic character. Thus we have seen that it is possible to change prolate aggregates into oblate ones by varying the parameters of the phase diagram, we may also expect to control the interaction potential between the objects through the ionic strength of the aqueous medium. These possibilities make necessary detailed examinations of the phase diagrams in order to describe the transformations of the aggregates and of their organization when going from the ordered phases to the disordered ones through the nematic phases of intermediate order, as suggested in fig.1 [10] [24]. Moreover the ability of these phases to dissolve a wide range of solutes, already quoted in the introduction, has made possible experiments of cholesterization of nematic phases by non mesogen chiral molecules [25] and studies of the coupling between nematic phases and ferrofluids hitherto impossible with thermotropic phases [26].

References and Notes

1. K.D.Lawson and T.J.Flautt, J.Am.Chem.Soc.89,5489 (1967).
2. No "nematic" domain is apparent in the phase diagramms of the systems Na octylsulfate-decanol-water and Na dodecylsulfate-decanol-water shown by P. Ekwall in Liq.Cryst.1,1, Acad. Press (1975). This suggests a critical adjustment of the surfactant and alcohol chain lengths if the nematic domain has not escaped the investigations because of its very small extent.
3. C.L.Khetrapal, A.C.Kunwar, A.S.Tracey and P.Diehl, "Lyotropic liquid crystals", NMR basic principles and progress, Vol.9, P.Diehl, E.Fluck and R.Kosfeld Editors, Springer Verlag Berlin (1975).
4. F.B.Rosevear, J.Soc.Cosmetic Chemists 19, 581 (1968).
5. L.Liébert and A.Martinet, private communication and Colloque RCP "Colloïdes et Interfaces", Paris, Nov.1979.
6. Historically the classification of the nematic lyotropic phases in two types stems from the NMR experiments which discriminated phases with positive (type I) and negative (type II) anisotropies of diamagnetic susceptibility (see below).
7. F.Fujiwara, L.W.Reeves, M.Suzuki, and J.A.Vanin, Proceedings of the National Colloïd Symposium, Knoxville, June 1978, K.Mittal Editor, Plenum Press (1979) and the references contained therein.
8. M.Gasparoux and J.Prost, J.de Phys.32, 953 (1971) and F.M.Leslie, G.R.Luckhurst and H.J.Smith, Chem.Phys.Lett 13, 368 (1972).
9. F.Y.Fujiwara and L.W.Reeves, Can.J.Chem.56, 2178 (1978).
10. N.Boden, K.Mc Mullen and M.C.Holmes, International Symposium of Magnetic Resonance in Colloïd and Interface Sciences, Menton, July 79.
11. W.H.De Jeu, Phys.Lett.69 A, 122 (1978).
12. C.Williams, Thèse de IIIème cycle, Orsay (1973).
13. J.Charvolin, E.Samulski and A.M.Levelut, J.de Phys.Lett.40, L-587 (1979). In this letter the proportions between Figs 3,4 and their scales have not been conserved through the reproduction. The length of the scales shown underneath Fig.3 and 4 are to be multiplied by 2.55 and 1.87 respectively.

14. E.Guyon, R.Meyer and J.Salan, accepted for publication in Mol.Crystals and Liquid Crystals (1980).
15. J.Charvolin, A.Tardieu, "Lyotropic liquid crystals, structure and molecular motions" in Solid State Physics suppl.14 edited by F.Seitz and D.Turnbull, Acad.Press (1978).
16. The type II phase of the same system has also been studied by L.Q.Amaral, C.A.Pimentel, M.R.Tavares and J.A.Vanin, J.Chem.Phys.71, 2942 (1979). These authors also conclude to the existence of lamellar aggregates, but with diameter 10 x larger. Scatterings we have observed are absent from their spectra.
17. We give an inferior limit only because of a possible disorientation induced by the walls of the capillaries containing the samples.
18. B.Mély, J.Charvolin and P.Keller, Chem.Phys.Lip.15, 161, (1973).
19. R.C.Long and J.M.Goldstein, J.Magn.Res.23, 519 (1976).
20. B.Mély, J.Charvolin, in "Physico Chimie des Composés amphiphiles" edited by R.Perron, CNRS Paris (1980).
21. Because of these many reorientational motions, whose effects have not been discriminated yet, the order parameters of the aggregates cannot be determined from NMR studies.
22. Studies of single crystals of analogue amphiphiles yield a negative anisotropy of susceptibility for a single molecule $\Delta\chi_m$ K.Londsale, Proc.Roy. Soc.London, A 171, 541 (1939). We would infer here a reduced value of $\Delta\chi_m$ due to conformational averaging but, nevertheless, because of the residual elongation of the chain a negative $\Delta\chi_m$.
23. Similar values of the birefringences have been measured in type II and lamellar phases showing the limited influence of the size of the aggregate, A.Saupe, Liquid Crystal Conference, Bangalore, déc.1979.
24. Y.Hendrikx and J.Charvolin, these proceedings, p.281
25. K.Radley and A.Saupe, Mol.Phys.35, 1405 (1978).
26. L.Liebert and A.Martinet, J.de Phys.Lett.40, L-363, (1979).

Viscoelasticity and Flow Alignment of Dilute Aqueous Detergent Solutions

S. Hess

Institut für Theoretische Physik, Universität Erlangen-Nürnberg, D-8520 Erlangen, Federal Republic of Germany

1. Introduction

A dynamic Landau theory is presented which allows to treat non-equilibrium phenomena both in the isotropic and nematic phases of a lyotropic liquid crystal. The theory can account for the properties to be stated below which are observed in some dilute detergent solutions, e.g. in aqueous solutions of cetyltrimethylammoniumsalicilate [1]. Firstly, these solutions are viscoelastic if the concentration is above (a rather low) critical value where nonspherical aggregates (micelles or vesicles) are formed [1]. The viscoelasticity which becomes apparent in an oscillatory rather than an overdamped motion in a flow relaxation experiment [1] is surprising because the viscosity of the solution is only slightly larger than that of water. Thus an ordinary Maxwell relaxation model cannot account for this viscoelasticity. Secondly, the viscoelastic solutions exhibit a strong flow birefringence [2] which can persist for some time when the shear is removed [3]. It seems likely that the velocity gradient induces a transition into an orientationally ordered phase which is essentially of nematic type.

2. Physical model

It is assumed that the detergent solution contains nonspherical aggregates of concentration c (not to be confused with the concentration of monomers put into the solution) which, in first approximation, can be considered as rigid. Their orientation, as it becomes apparent in a birefringence experiment, is characterized by the alignment tensor [4,5]

$$\underline{\underline{a}} \sim \langle \overline{\underline{u}\underline{u}} \rangle \equiv \langle \underline{u}\underline{u} - \tfrac{1}{3}\underline{\underline{\delta}} \rangle \qquad (1)$$

where $\underline{u}$ is a unit vector parallel to the figure axis of a particle. The symbol $\overline{\cdot\cdot}$ refers to the symmetric traceless part of a tensor, the bracket $\langle\ldots\rangle$ denotes an average to be evaluated with the orientational one particle distribution function.

There exists a coupling between the alignment and the viscous flow which underlies the flow birefringence. Thus for the study of flow properties, the alignment tensor $\underline{\underline{a}}$ has to be taken into account as a macroscopic variable in addition to the flow velocity $\underline{v}$ and the (symmetric traceless) shear friction pressure tensor $\underline{\underline{p}}$. Equations of change for $\underline{\underline{p}}$ and $\underline{\underline{a}}$ (which are needed in addition to the local conservation equation for the linear momentum) can be derived within the framework of irreversible thermodynamics provided that the dependence of the thermodynamic functions (internal energy, entropy, free energy) on the alignment tensor is known. Here, an ansatz of Landau type is used (dynamic Landau theory).

In particular, it is assumed that the specific internal energy and the specific entropy contain contribution u_a and s_a which depend on the alignment tensor, viz.

$$u_a = -\frac{1}{2}\varepsilon\, \underline{\underline{a}} : \underline{\underline{a}} \quad , \tag{2}$$

$$s_a \sim -\left[\frac{1}{2}\underline{\underline{a}} : \underline{\underline{a}} + ..(\underline{\underline{a}})^3 + ..(\underline{\underline{a}})^4\right] . \tag{3}$$

In (2), ε is a characteristic energy associated with the alignment. The sign has be chosen such that $\varepsilon > 0$ corresponds to a physical situation where $\underline{\underline{a}} \neq 0$ is energetically more favorable than $\underline{\underline{a}} = 0$. In general, ε is a function of the concentration c of the nonspherical aggregates which vanishes for $c \to 0$. More specifically, one has

$$\varepsilon \sim c^{\alpha} \tag{4}$$

with an exponent $\alpha = 1 + \frac{1}{3}\beta$ if the anisotropic part of the long range interaction energy between two particles which leads to (2) has a radial dependence $r^{-\beta}$ (e.g. $\beta = 6$ and $\alpha = 3$ for van der Waals interaction). The ansatz (3) describes the fact that an ordered state with $\underline{\underline{a}} \neq 0$ has a smaller entropy than an orientationally unordered state. Terms (containing scalars) of 3rd and 4-th order in $\underline{\underline{a}}$ have to be taken into account for thermodynamic stability reasons and they are essential for a unified treatment of the isotropic and nematic phases [5,6]. As a consequence of (2) and (3), the specific free energy associated with the alignment is proportional to the Landau potential

$$\Sigma = \frac{1}{2}A\, \underline{\underline{a}} : \underline{\underline{a}} + ..(\underline{\underline{a}})^3 + ..(\underline{\underline{a}})^4 \tag{5}$$

with $A = 1 - \varepsilon (k_B T)^{-1}$ where k_B is the Boltzmann constant and T is the temperature of the liquid. Due to (4), the coefficient A can be written as

$$A = 1 - \left(\frac{c}{c^*}\right)^{\alpha} \tag{6}$$

where c^* is a temperature dependent reference concentration. Notice that $A > 0$ and $A < 0$ for $c < c^*$ and $c > c^*$, respectively. The coefficients of the 3rd and 4-th order terms in (3) and (5) are weakly dependent functions of T and c. The dependence of the excluded volume on the relative orientation of the nonspherical particles can be disregarded for the dilute detergent solutions considered here.

Equations governing $\underline{\underline{a}}$ and $\underline{\underline{p}}$ can be derived by a procedure which has previously been used for thermotropic liquid crystals [5,6]. Here, it suffices to state the resulting equations with the modifications relevant for lyotropic liquid crystals.

3. Equations of change for the friction pressure tensor and the alignment tensor

The local conservation equation for the linear momentum of a fluid with the (total) mass density ρ and the hydrostatic pressure P is

$$\rho \frac{d\underline{v}}{dt} + \underline{\nabla} P + \underline{\nabla} \cdot \underline{\underline{p}} = 0. \tag{7}$$

This equation has to be supplemented by the equations for the friction pressure tensor $\underline{\underline{p}}$ and the alignment tensor $\underline{\underline{a}}$ [5,6,7], viz.

$$-\left(1+\tau_M \frac{d}{dt}\right)\underline{\underline{p}} = 2P_k \tau_p \overline{\underline{\nabla}\underline{v}} + \sqrt{2} P_k \tau_{pa} \left(\frac{d}{dt}\underline{\underline{a}} - 2\overline{\underline{\omega}\times\underline{\underline{a}}}\right), \tag{8}$$

$$-\underline{\underline{\Sigma}}(\underline{\underline{a}}) = \sqrt{2}\tau_{ap} \overline{\underline{\nabla}\underline{v}} + \tau_a \left(\frac{d}{dt}\underline{\underline{a}} - 2\overline{\underline{\omega}\times\underline{\underline{a}}}\right). \tag{9}$$

In (8), $P_k = nk_BT$ is a reference pressure, n is the (total) number density. The phenomenological relaxation time coefficients $\tau_p, \tau_{pa}, ..$ have the properties

$$\tau_{pa} = \tau_{ap} \tag{10}$$

(Onsager symmetry relation) and

$$\tau_p > 0, \quad \tau_a > 0, \quad \tau_{ap}^2 < \tau_a \tau_p \tag{11}$$

(positive definite entropy production); τ_{ap} may have either sign. The (first Newtonian) viscosity η is related to τ_p and to the Maxwell relaxation time τ_M by

$$\eta = P_k \tau_p = G_M \tau_M \tag{12}$$

where G_M is the Maxwell shear modulus. For liquids without network structure one has $G_M \approx P_k$ and consequently $\tau_M \approx \tau_p$. The terms containing

$$\underline{\omega} = \frac{1}{2} \operatorname{rot} \underline{v} \tag{13}$$

in (8) and (9) stem from the rotational motion of the nonspherical particles with an average angular velocity $\underline{\omega}$. It is recalled that $\overline{\cdot\cdot}$ refers to the symmetric traceless part of a tensor, e.g.

$$\left(\overline{\underline{\nabla}\,\underline{v}}\right)_{\mu\nu} = \tfrac{1}{2}\left(\nabla_\mu v_\nu + \nabla_\nu v_\mu\right) - \tfrac{1}{3}\,\underline{\nabla}\cdot\underline{v}\;\delta_{\mu\nu}$$

in cartesian component notation. The quantity $\underline{\underline{\Sigma}}$ in (9) is the derivative of Σ as given by (5) with respect to $\underline{\underline{a}}$, i.e.

$$\underline{\underline{\Sigma}} = \frac{\partial\Sigma}{\partial\underline{\underline{a}}} = A\,\underline{\underline{a}} + ..(\underline{\underline{a}})^2 + ..(\underline{\underline{a}})^3 \,. \tag{14}$$

Eqs. (8,9) are applicable to both the isotropic and nematic phases of a liquid crystal. They contain a number of special cases; e.g. for $\tau_{pa} = 0$ (no coupling between alignment and viscous flow), (8) reduces to the constitutive equation of the Maxwell relaxation model. For $\tau_M = 0$ and $\underline{\underline{\Sigma}}$ replaced by its linear term A $\underline{\underline{a}}$ with $A \sim T - T^*$ instead of (6), (8,9) are essentially the equations proposed by de Gennes [4,8] for the theoretical treatment of the pretransitional behavior of a thermotropic liquid crystal in the isotropic phase. The theory of flow birefringence in molecular gases [9] is based on formally equivalent equation with $\underline{\underline{\Sigma}} = \underline{\underline{a}}$ which have been derived [9,10] from a generalized Boltzmann equation, viz. the Waldmann-Snider equation [11]. The nonlinear terms of $\underline{\underline{\Sigma}}$ are essential for the nematic phase where the alignment has a nonzero value determined by $\underline{\underline{\Sigma}}(\underline{\underline{a}}) = 0$, cf. [5,6]. The Ericksen-Leslie coefficients γ_1 and γ_2 [4] are proportional to τ_a and τ_{ap}, respectively and depend on the Maier-Saupe order parameter as proposed by Helfrich [12]. Eq. (9) can also be derived from a Fokker-Planck equation for the orientational distribution function [13].

4. Viscoelasticity

The viscoelasticity is studied for the isotropic phase where $\underline{\underline{\Sigma}}$ can be approximated by A $\underline{\underline{a}}$ with A given by (6). Elimination of $\underline{\underline{a}}$ from (8,9) then yields,

$$-\left[1 + (\tau_M + \tau)\frac{d}{dt}\right]\underline{\underline{p}} = 2\eta\left[1 + \tau'\frac{d}{dt}\right]\overline{\underline{\nabla}\,\underline{v}} \,. \tag{15}$$

The relaxation times τ and τ' are given by

$$\tau = A^{-1}\tau_a \,, \quad \tau' = \left[1 - \tau_{ap}^2(\tau_a\tau_p)^{-1}\right]\tau < \tau \,. \tag{16}$$

Note that the orientational relaxation time τ can increase dramatically for $c \to c^*$ since the reorientation is slowed down by cooperative effects. The ratio τ'/τ equals η_∞/η where η_∞ is the 2nd Newtonian viscosity which is reached for

$|\underline{\omega}| = \frac{1}{2}|\mathrm{rot}\,\underline{v}| \gg \tau^{-1}$ [14].

Terms containing $\underline{\omega}$ have been disregarded in (15), i.e. $|\underline{\omega}| \lesssim \tau^{-1}$ has been assumed. In rheology, an equation of the type (15) is referred to as Burgers model, a combination of the Maxwell and the Kelvin models.

Next, (8) with (15) are applied to a physical situation encountered in a flow relaxation experiment [1]. For simplicity, a plane Poiseuille flow between two flat parallel plates separated by the distance 2d is considered. The x- and y-directions are chosen parallel to the flow direction and normal to the plates, respectively. It is assumed that the pressure gradient ∇P which drives the liquid is suddenly removed at time t = 0. How does the flow field $v_x(t,\underline{r})$ relax? To study this question, the ansatz

$$v_x = V(t) \cos ky \tag{17}$$

is made where k is chosen such that v_x vanishes at the surface of the plates. One has $k = \pi\ (2d)^{-1}$ for the leading term of a Fourier series expansion. Insertion of (17) into (7,15) yields

$$\dot{V} + (\tau_M + \tau)\ddot{V} + \nu k^2 (V + \tau' \dot{V}) = 0, \tag{18}$$

where $\nu = \eta \rho^{-1}$ is the kinematic viscosity. Thus the amplitude V obeys an equation of the type of a damped harmonic oscillator which can have overdamped and damped oscillatory solutions depending on whether $(\tau_M + \tau)\nu k^2$ is smaller or larger than a critical value (which depends on the ratio τ'/τ). Typically, oscillatory motions occur for

$$\tau_M + \tau \approx \nu^{-1} d^2 \tag{19}$$

where it is recalled that 2d is the distance between the two plates. In other words, the viscoelasticity becomes apparent in a flow relaxation experiment if the (effective) relaxation time $\tau_M + \tau$ is large enough to match the macroscopic relaxation time $\nu^{-1}d^2$. This time is of the order of 1s for a viscosity comparable to that of water and $d \approx 1$ mm. In this case, τ_M is of the order 10^{-11}s. Thus a Maxwell relaxation model corresponding to $\tau = 0$ cannot account for the viscoelasticity observed in the dilute detergent solutions of [1]. The relaxation time τ of the alignment, on the other hand which also occurs in (19), can be large enough. For particles with a linear dimension of 10^{-5} cm (10^{-4}cm) one has $\tau_a \approx 10^{-2}$s (10s), τ is still larger by the factor A^{-1}, cf. (16).

The physical mechanism which underlies the viscoelasticity considered here is the coupling between the alignment and the viscous flow which is also responsible for the flow alignment and its reciprocal phenomenon [15].

5. Flow alignment, shear induced phase transition

A viscous flow, e.g. in a Couette flow experiment, causes an alignment which can be treated theoretically with the help of (9). With $\underline{\underline{\Sigma}}$ replaced by the linear term $A\ \underline{\underline{a}}$, a pretransitional behavior analogous to the case treated in [8] is obtained. This linear approximation, however, breaks down for a strong alignment. The full nonlinear equation has been analysed in [16]. For brevity, it is just mentioned here that a discontinous transition into an orientationally ordered state occurs when the shear rate $\partial v_x/\partial y$ exceeds a certain value which is proportional to τ_a^{-1}. It seems likely that such a shear induced phase transition occurs in the viscoelastic solutions studied in [1-3]. Further experimental investigations which could test this conjecture are desirable.

References

1 S. Gravsholt, J. Colloid Interface Sci 57, 575 (1976)
2 S. Gravsholt, in "Polymer Colloids II", ed.R.Fitch,Plenum, New York 1979
3 S. Gravsholt,"Physico-chemical properties of viscoelastic highly dilute aqueous detergent solutions"; report presented at the conference on Viscoelastic solutions, Dep.of Chem.,Univ.of Bayreuth, Oct.1979
4 P.G. de Gennes, The Physics of Liquid Crystals, Clarendon Press, Oxford 1974
M.J. Stephen and J.P.Strayley, Rev. Mod. Phys. 46, 617 (1974)
5 J. Pardowitz and S. Hess, Physica (1980)
6 S. Hess, Z. Naturforsch. 30a, 728; 1224 (1975)
7 S. Hess, Physica 87A, 273 (1977)
8 P.G. de Gennes, Phys.Lett. 30A, 454 (1969)
9 S. Hess, Phys. Lett.30A, 239 (1969); Springer Tracts Mod. Phys.54, 136(1970) in The Boltzmann Equation-Theory and Applications, E.D.G. Cohen and W. Thirring, eds. Springer, Wien, New York 1973
10 S. Hess and L. Waldmann, Z. Naturforsch. 22a, 1871 (1966)
A.G.St.Pierre, W.E. Köhler and S. Hess,Z.Naturforsch. 27a, 1057 (1972)
11 L. Waldmann, Z. Naturforsch. 12a, 660 (1957);13a, 609 (1958)
12 W. Helfrich, J.Chem.Phys.56, 3187 (1972)
13 S. Hess, Z. Naturforsch. 31a, 1034 (1976)
14 S. Hess, Physica 86A, 383 (1977)
15 S. Hess, Z. Naturforsch. 28a, 1531 (1973)
16 S. Hess, Z. Naturforsch. 31a, 1507 (1976)

Structural Relations Between Lyotropic Phases in the Vicinity of the "Nematic" State

Y. Hendrikx and J. Charvolin

Laboratoire de Physique des Solides, Université Paris-Sud,
F-91405 Orsay, France

Some lyotropic mesophases are anisotropic fluids certain properties of which, orientation in a magnetic field, textures, are similar to those of thermotropic nematics. On the other hand, as for any lyotropic liquid crystals, they must be assemblies of aggregates of amphiphilic molecules but their organizations must differ from those studied so far in classical lyotropic liquid crystals. As a matter of fact in the latter case, anisotropy is only known for viscous well-ordered phases (lamellar or cylindrical) and fluidity is only associated with disordered phases (micellar). A recent X-ray diffraction study of the lyotropic "nematic" phases obtained with the mixture sodium decyl sulfate (SDS)-decanol-water [1] shows that the structural elements of these phases are disc or cylinder shaped aggregates, according to the temperature and the composition. The aggregates are organized with orientational order but without long range translational order.

We are now interested in characterizing the phases in the vicinity of the "nematic" ones in order to define the conditions originating these phases. The "nematic" phases appear located in a small domain of the phase diagram, around 55% H_2O-38% SDS and 7% decanol, at the contact with the lamellar and various cylindrical ordered phases, and with the micellar disordered phase.

The first results suggest well defined structural relations between those phases. When water is added to a cylindrical ordered phase, corresponding to a particular SDS/decanol ratio, transformations are observed first to states without translational order where cylinders, then discs, are orientationally ordered in the "nematic" phases, finally to states without orientational order where the discs are dispersed in the isotropic phase. Otherwise, when decanol is added to a particular binary water-SDS mixture, the cylindrical phases go progressively over to the ordered lamellar phase and in some cases, depending on the water-SDS mixture composition, the transformation goes through the "nematic" phases. These facts suggest that the occurence of the "nematic" phases results from the conjugated actions of water and decanol. Schematically water destroys the stacking of the aggregates and decanol modifies the curvature of their interfaces.

1. J.Charvolin, A.M.Levelut, E.T.Samulski, J.de Phys.Lett., 40, L-587 (1979)

Optical Properties of Lyotropic Nematic Phases

M. Laurent, A. Hochapfel, and R. Viovy
Ecole Normale Supérieure,
Grille d'Honneur - Parc de Saint-Cloud, F-92211 Saint-Cloud, France

1. SOLUTIONS

We have used two kinds of nematic lyotropic solvents giving different optical orientations in the magnetic field, both prepared from sodium decylsulfate. A dye, orange red NO_2 -⟨O⟩- N = N -⟨O⟩- N $(CH_3)_2$, was dissolved as guest molecule in these systems and its optical behaviour due to orientation was examined.

A. The PHASES

The nematic phase of type I known to orient in the magnetic field with its optical axis parallel to the field was the standard phase used in the works of RADLEY and REEVES [1][2][3]: sodium decylsulfate 35.9 %, decanol 7.2 %, deuterium oxide 56.9 %. It will be noted as the ternary system.

The nematic phase of type II known to orient with its optical axis perpendicular to the magnetic field was the standard phase initially proposed by LAWSON [4][5][6]: sodium decylsulfate 36 %, decanol 7 %, sodium sulfate 7 %, deuterium oxide 50 %. It will be noted as the quaternary system.

B. The DYE

Orange red was dissolved in the detergent system at a concentration of 7.10^{-4} M/kg. The absorption maximum of this dye is situated at 476 nm in acetone and at 493 nm in the detergent solution. Spectral data indicated that the dye dissolved in the system in a monomeric from and that no chemical alteration took place.

C. EXPERIMENTAL

The components were weighed into well stoppered erlenmeyer flasks, homogenized by stirring at 25-35°C and allowed to stand at room temperature for one week.

The dye was dissolved in the detergent system in constricted sealed test tubes by alternative heating at 35°C and mixing by centrifugation of the returned tube.

The solutions were filled into spectrophotometric cells of path 1 mm for the magnetic field orientation. When aligning by surface treatment, the solutions were introduced by capillarity between glass plates separated by spacers of 150 μ. The filling was done in a glove box in water saturated atmosphere at 28°C.

2. MICROSCOPIC OBSERVATIONS

A sample drop between slide and cover was observed in polarized light on a Leitz Dialux Pol heating stage microscope.

The ternary system melted at 18°C to a fluid almost homeotropic preparation which was maintained up to 27°C. Then various textures appeared depending on the sample and they remained unchanged up to about 45°C where a new texture grew on the stage. According to Charvolin [7] the system is between 18 and 27°C in the nematic type II state of disc-shaped aggregates and above in the nematic type I state of cylindric aggregates. The texture of the type I state was changed to homogenous alignment when the slides were treated by rubbing or SiO coating.

The quaternary system melted at 25°C to an almost homeotropic preparation containing some ribbon-texture. No change appeared until isotropic melting at about 80°C. The texture was not altered by surface treatment.

ROSEVEAR [8] has shown textures of the "middle phase" similar to those of the type I state and of the "neat phase" more like those of the type II state.

3. OPTICAL PROPERTIES

A. DICHROISM

Results concerning the ternary phase only because no dichroism was observed in the quaternary system. The measurments were made on the absorption maximum of the dissolved dye.

1°- Transmission has been plotted against the angle ϕ between polarized light and the orienting field $\vec{B}$. Fig.1

In all cases, we obtained a straight line, in good agreement with the theoritical law:

$$\mathcal{T}(\phi) = [\mathcal{T}(0) - \mathcal{T}(90°)] \cos^2 \phi + \mathcal{T}(90°)$$

This means that:
- medium is uniaxial
- mean direction of orientation is that of the magnetic field

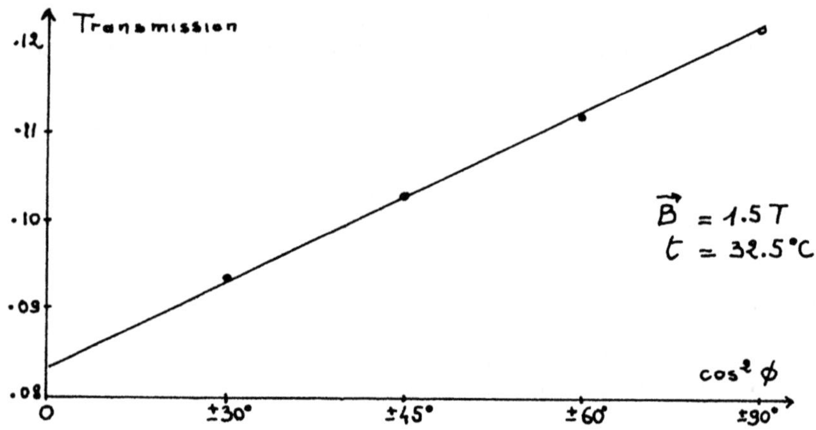

Fig. 1

2°- Kinetics of orientation in magnetic field (or disorientation after removing the field) Fig.2

We can see:

a- The slope at the origin is strongly dependent on field strength (about proportional to B^2). The most important part of orientation takes place within one hour.

b- After this period, the orientation continues to increase slowly (during several hours). This could be due to anchorage effect.

c- For each curve, the saturation value is a function of the field Fig. 3. A field of 1,5 Tesla leads nearly to the highest possible value of orientation.

3°- Comparing different orientation techniques.

Surface treatment is a common way of orienting thermotropic nematics. Our experiments show that the nematic aggregates of the detergent systems are also oriented by these techniques. The dye molecules which are supposed to have a fixed position within each aggregate, become oriented because of the orientation of the aggregates. Rubbing leads to orientation of the dye molecules perpendicular to the director in the same way as the magnetic field. SiO coating leads to parallel orientation.

4°- Dichroism against temperature

The dichroism is not changed at the phase transition at 27 - 28°C Fig.4.

However, due to change in diffusion, this transition modifies strongly the value of transmitted light (with or without dye) Fig.5.

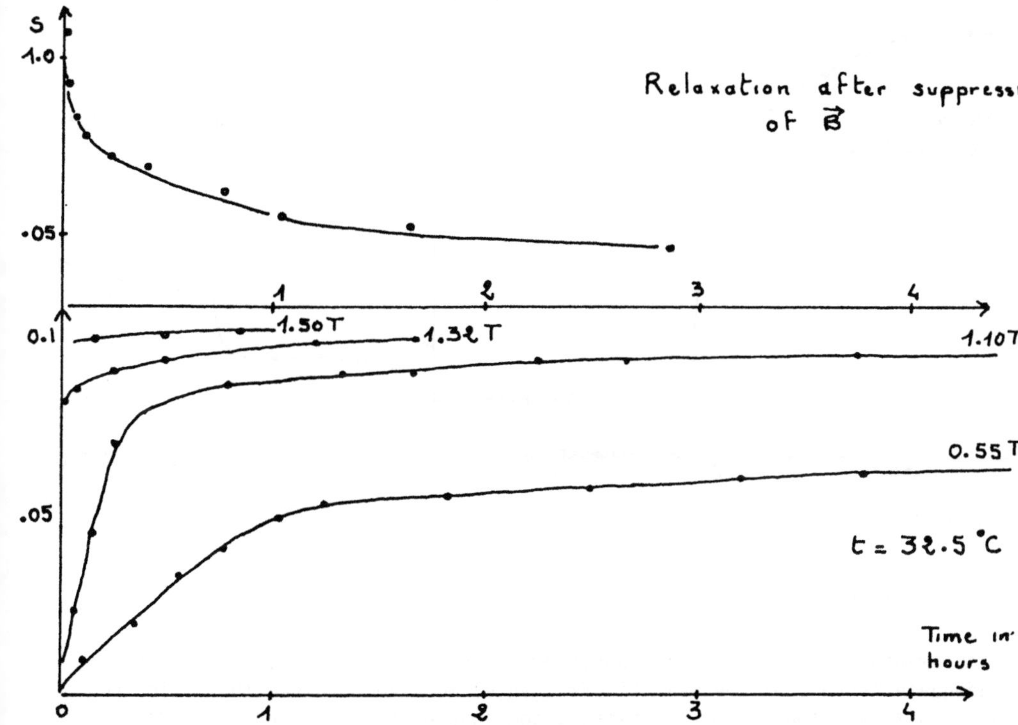

Fig. 2

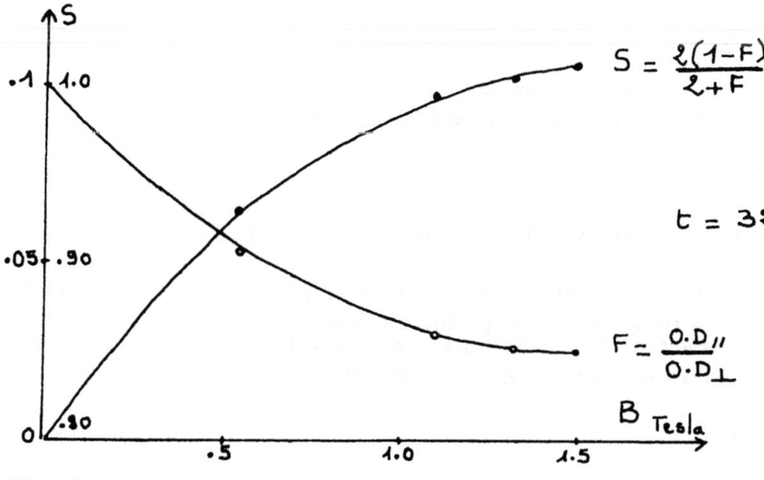

Fig. 3

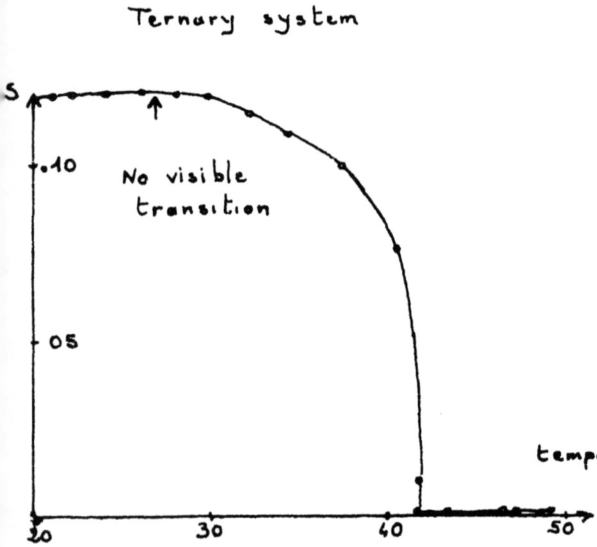

Fig. 4

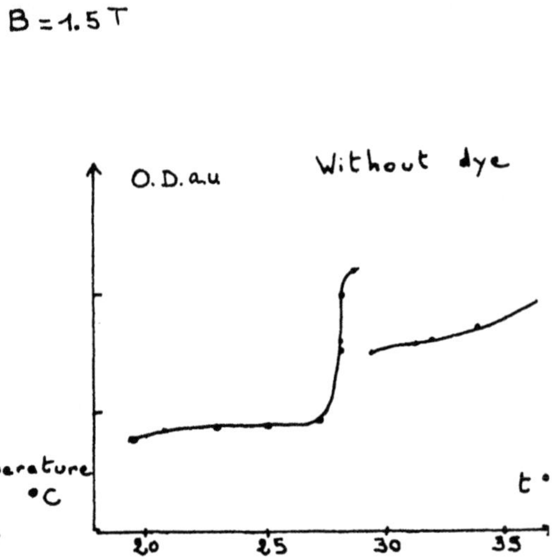

Fig. 5

B. BIREFRINGENCE

Measurements with a small prism angle or the channeled spectrum method are inadequate because of the weak birefringence. We have obtained the following values with a method using a Babinet's compensator (Cell thickness: 2 mm).

For the ternary system:

$$\Delta n = n_{/\!/} - n_{\perp} = -1{,}97.10^{-3} \text{ at } 32°C$$

$$\Delta n = \qquad -1{,}83.10^{-3} \text{ at } 22°C$$

(mean value of the measurements at 454, 494, 551, 595 nm).

4. INTERPRETATION

We adopt, for the ternary system, a model of cylinders [3][7].

For the determination of order parameter in relation to dichroism, we have used the classical result relative to the dye included in nematics[9]. In this model the moment of transition makes an angle β with the molecule. The molecule makes an angle Θ with the director Fig.6a.

$$O.D. = K \langle (E.\mu)^2 \rangle \qquad\qquad O.D = \text{optical density}$$

$$F = \frac{O.D_{\parallel}}{O.D_{\perp}} = \frac{2\ (1-S) + 6\ S \cos^2 \beta}{S + 2 - 3\ S \cos^2 \beta}$$

F = dichroic ratio
S = order parameter

where $S = \frac{1}{2} \langle 3 \cos^2\Theta - 1 \rangle$

Here, we have made the following assumptions:

- Cylinders are statistically distributed around the director, by an angle Θ Fig.6 b
- All the dye molecules make an angle of 90° with the axis of the cylinders Fig.6 c
- Within the cylinder section, the probability for finding a dye molecule is the same in all directions. That means circular section of the cylinder.

The formulas have the same form but:

- molecules are replaced by cylinders
- transition moment by dye molecule so $\beta = \frac{\pi}{2}$, $S = \frac{2\ (1-F)}{2 + F}$

S here becomes the order parameter of the aligned cylinders

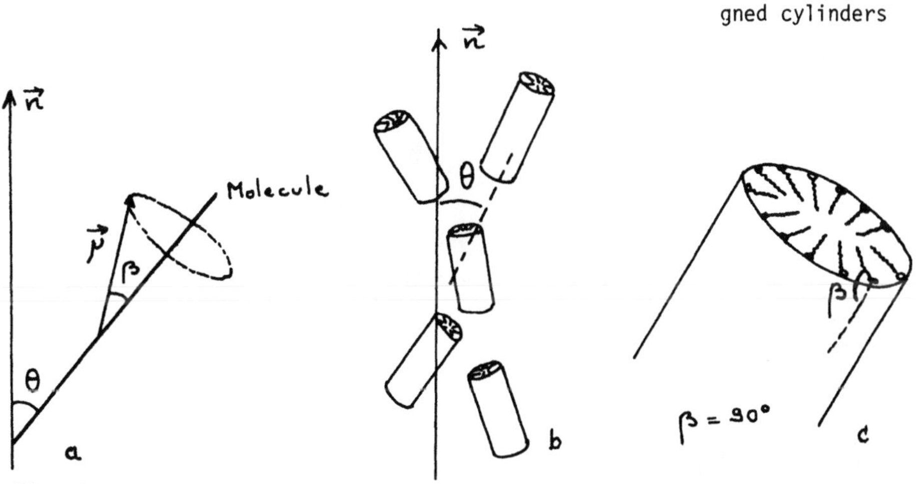

Fig. 6

1. K. RADLEY and L. REEVES Can.J. Chem. 53 (1975) 2998
2. K. RADLEY, L. REEVES and A. TRACEY J. Phys. Chem. 80 (1976) 174
3. L. LIEBERT Colloque CNRS n° 938 Physicochimie des Composés Amphiphiles(1978)
4. J. LINDON and B. DAILEY Mol. Phys. 20 (1971) 937
5. D. CHEN, F. FUJIWARA and L. REEVES Can. J.Chem. 55 (1977) 2396 and 2404
6. K. LAWSON and T. FLAUTT J. Am. Chem. Soc. 89 (1967) 5489
7. J. CHARVOLIN, A. LEVELUT and E. SAMULSKI J. de Phys. Lett (1979) preprint
8. F. ROSEVEAR J. Am. Oil; Chem. Soc. 31 (1954) 628
9. R. JOURNEAUX and R. VIOVY Photochem. and Photobiol. 28 (1978) 243

ANNEX

Microphotographs of the observed textures are shown (magnification X 100).

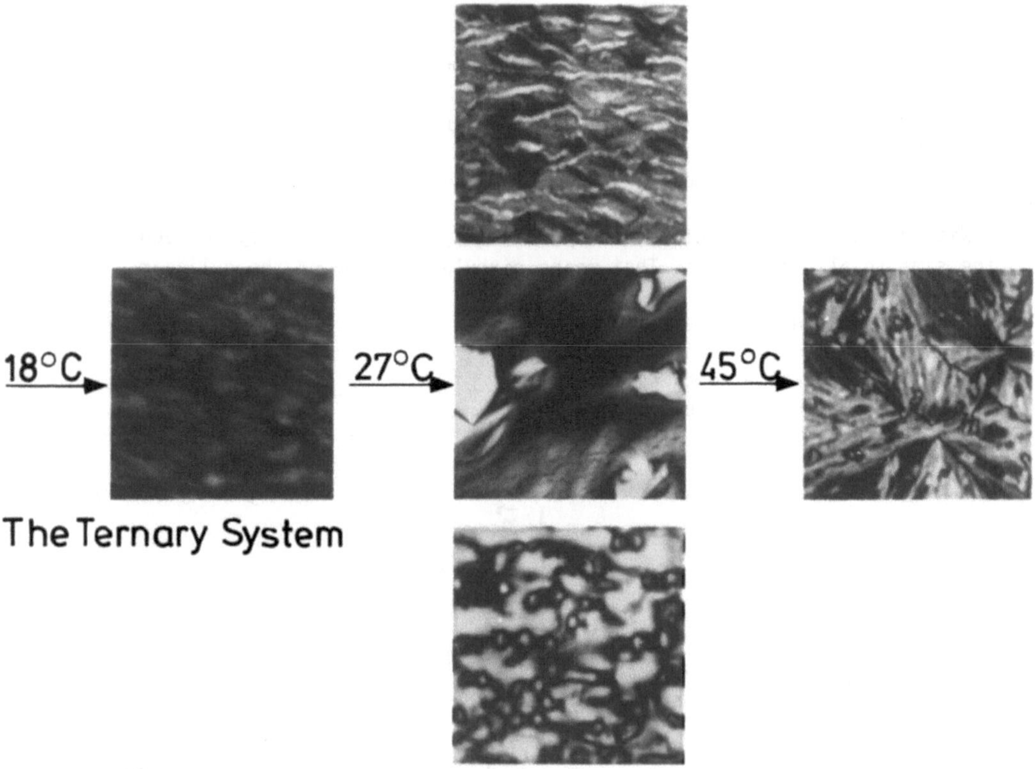

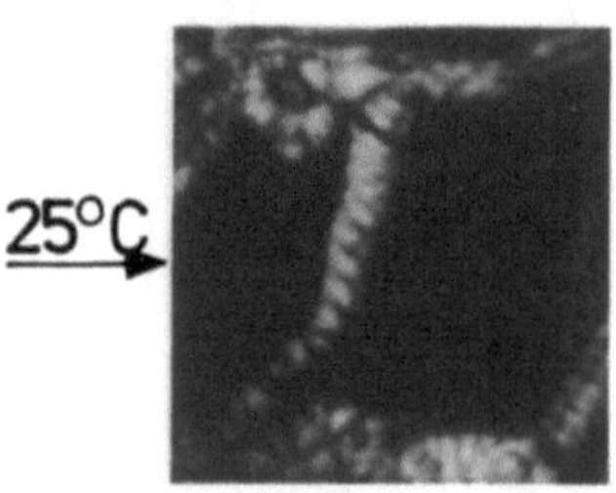

The Quaternary System

Phase Transitions in a Solution of Rod-Like Particles with Different Lengths *

H.N.W. Lekkerkerker and R. Deblieck

Fakulteit Wetenschappen, Vrije Universiteit Brussel,
B-1050 Brussels, Belgium

The Onsager theory for the isotropic-nematic phase transition in a solution of rodlike particles is extended to the case of mixtures of such rods with different lengths. For the case of a mixture of particles with two different length to diameter ratios (L/D) it is found that there is a significantly higher molefraction of long rods in the anisotropic phase than in the isotropic phase. This is clear from the Π-x diagram (Π : osmotic pressure) of the coexisting phases presented in Fig. 1. Further the orientational order of the long rods in the anisotropic phase is considerably enhanced compared to the monodisperse system of such rods. This is illustrated in Fig.2. The relevance of these results for the explanation of the isotropic-hexagonal transition in soap solutions is indicated.

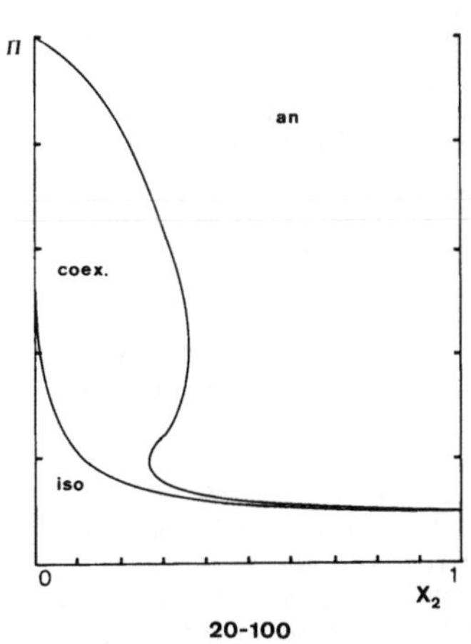

Fig.1 : Π-x diagram for the coexisting phase $(L/D)_1 = 20$, $(L/D)_2 = 100$

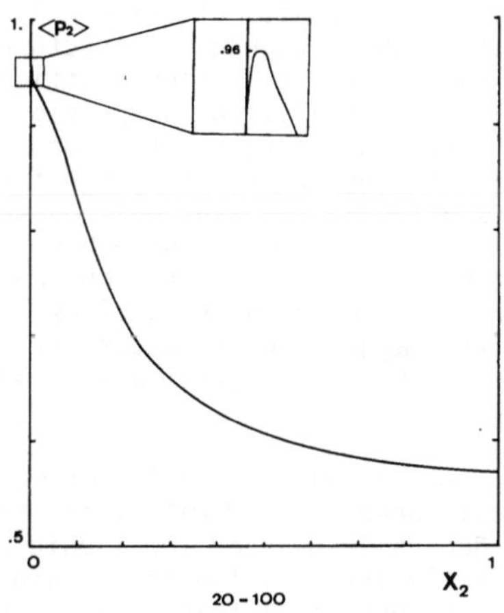

Fig.2 : order of the long rods in the anisotropic phase

★ Submitted for publication to Journal de Physique.

Aggregate Structure and Ion Binding in Amphiphilic Systems Studied by NMR Diffusion Method

P.-O. Eriksson and G. Lindblom

Department of Physical Chemistry 2, Chemical Centre,
P.O.B. 740, S-220 07 Lund, Sweden

1. Abstract

The pulsed magnetic field gradient NMR method has been used to study the diffusion of all species in the isotropic phase of the ternary system lithium octanoate-butanol-water. It was found that an increase in the butanol concentration eventually leads to a disruption of the normal micelles and reversed micelles were not observed at any butanol content. The Li^+ diffusion coefficients in the lamellar phase of the lithium octanoate-decanol-water system were shown to be independent of the composition of the sample. This was interpreted to be due to a constant fraction of bound ions in accordance with the ion condensation model previously used to describe the ion binding in lyotropic liquid crystals.

2. Introduction

The NMR diffusion method has proved to be useful for studies of systems composed of aggregates of colloidal dimensions [1-10]. We have shown that information about aggregate structure can be obtained from a quantitative comparison between diffusion coefficients measured for various liquid crystalline phases [3-7]. In our previous studies we have investigated amphiphile diffusion, [3-7] and to some extent water diffusion [4,8] and in this work we also include counterion diffusion. Hence the diffusion coefficients of all species in the isotropic phase of the ternary system lithium octanoate-butanol-heavy water have been determined. This phase extends continuously from the water to the butanol corner in the phase diagram. The main objectives in these investigations have been to get information of the amphiphilic aggregates structure at various butanol concentrations and some experience of the potential use of NMR lithium ion diffusion in amphiphilic systems.

The translational diffusion of counterions in micellar solutions has previously been studied with trace techniques [11,12]. However this method cannot generally be used for liquid crystalline samples and we have therefore briefly investigated the possibility of using NMR diffusion techniques for measurements of lithium ion diffusion in a lamellar liquid crystalline phase.

3. Materials

Lithium octanoate was prepared from octanoic acid (specially pure, BDH). A solution of octanoic acid in ethanol and water was neutralized with an aqueous solution of lithium hydroxide. The ethanol was evaporated in a vacuum desiccator and the remaining water was removed by freeze drying.

In order to reduce the influence of water diffusion on the spin-echo attenuation of the amphiphile and the butanol it was necessary to exchange the hydroxylic protons of butanol with deuterons. This was achieved by treating the alcohol twice with excess of heavy water followed by drying with anhydrous K_2CO_3 (Kebo Grave puriss). Deuteriumoxide was purchased from Ciba Geigy (>99.7% atomic purity). Butanol (p.a.) and decanol (p.a.) were purchased from Merck AG.

The samples of the lithium octanoate-butanol-heavy water system were prepared by weighing appropriate amounts of the components into 7 mm NMR tubes which were then sealed. The lamellar phases of the lithium octanoate-decanol-water system were prepared as previously described [13].

4. Methods

The diffusion coefficients were measured with the conventional pulsed magnetic field gradient technique developed by STEJSKAL and TANNER [2]. These studies were performed with a Bruker 322-s pulsed NMR spectrometer operating at 13.8 MHz for ^2H at 34.9 MHz for ^7Li and at 60.8 MHz for ^1H. The spectrometer was equipped with a homebuilt pulsed magnetic field gradient unit. Fig. 1 shows an illustration of the pulse sequences in the NMR experiment for diffusion studies.

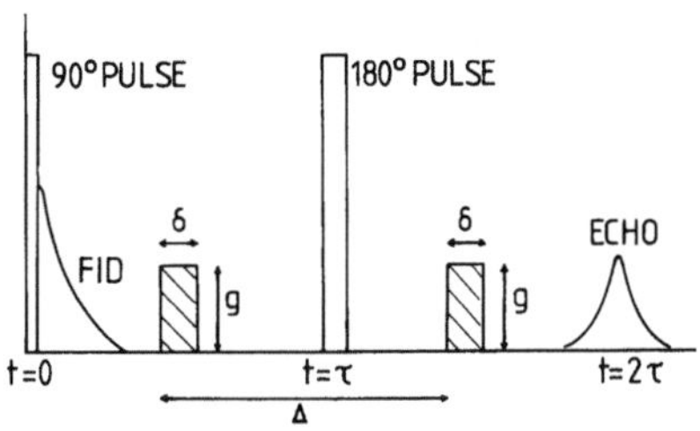

Fig. 1. Schematic representation of the pulse sequences in the NMR experiment for diffusion studies.

Two magnetic field gradient pulses are applied at each side of the 180° pulse. Due to diffusion the spin echo amplitude at time 2τ will be attenuated according to the equation

$$\ln(E_g/E_o) = -(\gamma\delta g)^2(\Delta-\delta/3)\cdot D \qquad (1)$$

where E_o and E_g stand for the echo magnitudes in absence and in presence of the field gradient pulses respectively. γ is the magnetogyric ratio. A plot of $\ln E_g$ versus $(\gamma\delta g)$ $(\Delta-\delta/3)$ gives the diffusion coefficient, D, from the slope of the straight line. The diffusion coefficients were determined by varying g in discrete steps between approximately 0.3 and 3 T m^{-1} while keeping the pulse width, Δ, and the time between pulses, δ, constant at 26 ms and 0.5 ms respectively for protons, at 65 ms and 1.0 ms for deuterons and at 130 ms and 1 ms for lithium. The waiting time between pulse sequences was sufficiently long to let the spin system relax back to its equilibrium magnetization. Since the magnitude, g, of the gradient pulses is difficult to measure precisely absolute values of D were determined relative to standard samples with known diffusion coefficients. As standard samples 2 M aqueous solution of LiCl($D_{Li,20^oC}$ = 8.4 · 10^{-10} m^2s^{-1}) [14], pure ordinary water (D_{19^oC} = 1.98, 10^{-9} m^2 s^{-1}) [15] and heavy water (D_{33^oC} = 2.28 · 10^{-9} m^2 s^{-1}) [15] was used.

The diffusion of octanoate and of butanol both contribute to the attenuation of the proton spin echo. The echo amplitude then conforms to a biexponential function of the form

$$\ln E_g \sim \ln \{n_A \cdot \exp(-2\tau/T_{2A})\exp(-(\gamma\delta g)^2(\Delta-\delta/3)\cdot D_A) + + n_B \cdot \exp(-2\tau/T_{2B})\exp(-(\gamma\delta g)^2(\Delta-\delta/3)\cdot D_B)\} \tag{2}$$

where n_A and n_B are the relative numbers of nuclei from alcohol and amphiphile respectively. T_{2A} and T_{2B} are the mean transversal relaxation times of the two components respectively. Assuming that T_2 for butanol and for octanoate have the same value[1], estimates of the butanol and the octanoate diffusion coefficients were made from a least square fit of the experimental points to a biexponential plot.

For the lithium ion diffusion studies of the lamellar phases, a signal averaging technique was employed using a Varian C-1024 computer to accumulate 25 transients.

We estimate the error in the water and lithium ion diffusion coefficient to be about 10% while for the butanol and for the octanoate the error is larger.

5. Results and Discussion

5.1 Diffusion in the Isotropic Phase of the Lithium Octanote-Butanol-Water System

The translational diffusion coefficients at 33±2°C for all the components in the ternary system lithium octanoate-butanol-heavy water were measured as a function of butanol concentration. The weight ratio between lithium octanoate and heavy water was kept constant at 4:1. The phase diagram of the system studied is not yet known but the corresponding phase equilibria of the system sodium octanoate-butanol-water has been reported by Ekwall and coworkers [16] (see fig. 2). The phase diagram of our system is assumed to be similar and the composition of the samples studied here are indicated in fig. 2.

Figure 3 summarizes all the diffusion coefficients measured. From this figure it can be inferred that the diffusion coefficient of both water and lithium ions decreases with increasing butanol content while the diffusion coefficient of butanol approaches monotonically the diffusion coefficient of

[1] This crude assumption does not affect our qualitative use of these diffusion coefficients in our interpretation of the data in terms of aggregate structure. The absolute values of the diffusion coefficients may not be correct however.

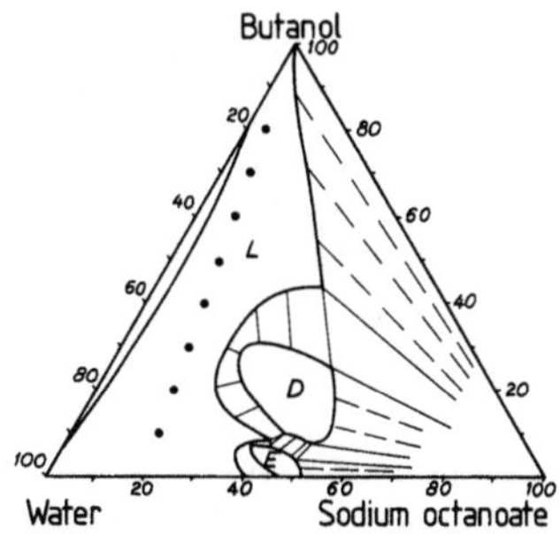

Fig. 2. The phase diagram of the ternary system sodium octanoate-butanol-water. Corresponding lithium octanoate samples are indicated in the figure.

Fig. 3. The translational diffusion coefficients obtained as a function of butanol concentration: ● heavy water, ○ lithium ions, ▲ butanol and ■ octanoate.

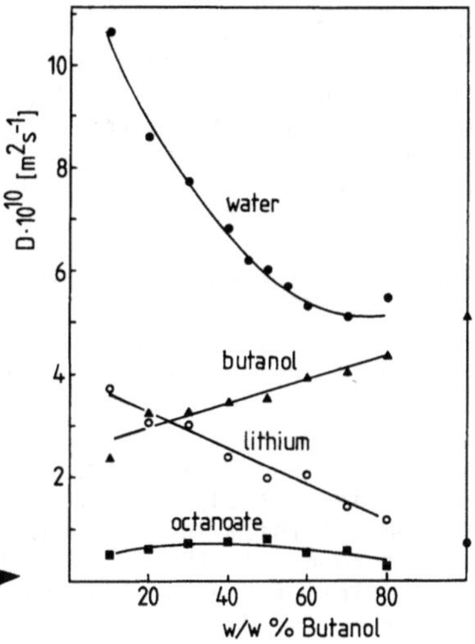

pure butanol. Likewise the diffusion coefficient of lithium ions approaches the diffusion coefficient of lithium ions dissolved in pure butanol.

The lithium ion diffusion at high alcohol content in the reversed micellar phase of the ternary system lithium octanoate-decanol-water has also been determined and it was found to be about one order of magnitude smaller ($1 \cdot 10^{11}$ m^2s^{-1}) than the diffusion coefficient of lithium ions in the butanol system at high butanol content. Kamenka et al. [17] reported a water diffusion coefficient in reversed micelles in the decanol system of $6 \cdot 10^{-11}$ m^2s^{-1} which is one order of magnitude smaller than what we found for water in the butanol system at high alcohol content.

For all the samples the effective diffusion time (Δ) of lithium ions and water was varied and it was found that no restricted diffusion occurred on the time scale studied (10-100 ms). It should however be noted that for very small aggregates like e.g. reversed micelles restricted diffusion of counterions and water cannot be detected with our method due to technical problems at very small Δ-values.

The addition of butanol to an octanoate micellar solution will lead to a part solubilization of butanol in the micellar aggregates. Since butanol is soluble in water up to 20% it will also be present in the aqueious intermicellar solution and thus water and lithium ions will interact with both micelles and butanol. The lithium ions probably become partly solvated by butanol which explains the decrease in the water and counterion diffusion coefficients with increasing butanol content. The high butanol concentration in the intermicellar region makes it also "more hydrophobic" leading to a destabilization of the micellar aggregates. Thereby the amphiphile monomer concentration will increase with increasing butanol concentration.

Thus addition of butanol to an octanoate micellar solution will eventually disrupt all the micellar aggregates in the solution ending up with just an isotropic mixture of amphiphile, water and alcohol. The conclusion that no reversed micellar aggregates forms on addition of butanol is strongly supported by the fact that the water and lithium ion diffusion coefficients at high butanol concentration are one order of magnitude larger than in the reversed micellar phase of the decanol system.

5.2 Lithium Ion Diffusion in a Lamellar Phase

The appearance of static effects like quadrupole splittings of many ions in anisotropic liquid crystals and of very short T_2-relaxation times often complicate a direct measurement of the ion diffusion coefficient of lamellar mesophases using the ordinary spin echo method. However, Li NMR spectra usually show a rather narrow central peak also for nonoriented lamellar phases, making it possible to study lithium ion diffusion. Recently, Tiddy [9] studied, using pulsed NMR, lithium ion diffusion of a lamellar phase containing a perfluoronated detergent. In this work we have studied lithium ion diffusion of nonoriented samples of the lamellar phase composed of lithiumoctanoate, decanol and water. The data obtained are summarized in figure 4, also showing part of a tentative phase diagram extrapolated from the corresponding sodium octanoate system [16]. It can be inferred from this figure that the measured ion diffusion coefficient is independent of the composition for all samples studied in the lamellar phase.

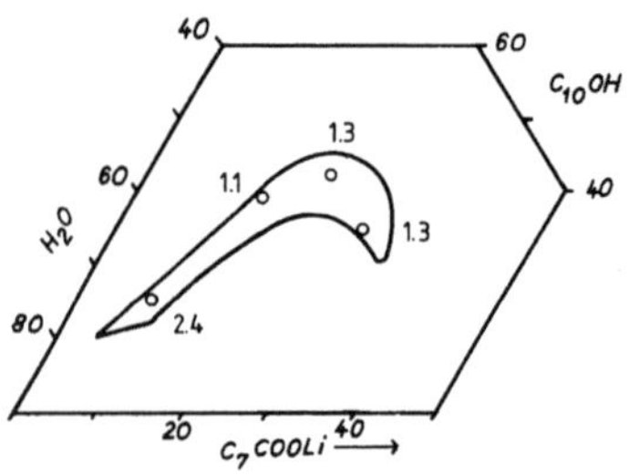

Fig. 4. Part of a tentative phase diagram of the system lithium octanoate-decanol-water. Measured Li^+-diffusion coefficients are inserted in units of 10^{-10} m^2s^{-1}.

The observed diffusion coefficient is given by the expression

$$D_{obs} = p\, D_b + (1-p)D_f \qquad (3)$$

where p is the fraction of bound ions. D_b and D_f are the diffusion coefficients of bound and free ions, respectively. Then, the experimental finding of an almost constant value of D_{obs} leads to the conclusion that the fraction of bound ions does not vary appreciably with composition in the lamellar phase. This is in agreement with recent NMR studies of Li quadrupole splittings [13,18] obtained for this system. The observation of a constant fraction of bound ions (independent of composition at high water contents) in lamellar mesophases was recently explained by an ion condensation model [19,20].

An estimation of the local diffusion coefficient for the ion diffusion along the lamellae can also be made. The echo amplitude E_g for a nonoriented sample depends on the angle, θ, between the normal to the lamellae and the applied magnetic field [8].

If it can be assumed that lamellae oriented at the so called magic angle contributes most to the echo attenuation upon the application of the magnetic field gradients, then the lithium ion diffusion along the lamellae, D_L, is equal to 3/2 D_{obs}. In a forthcoming paper we will report on whether the assumption made is valid or not by studying macroscopically aligned samples.

6. References

1. Stejskal, E.O., Adv.Mol. Relaxation Processes 3, 27 (1972).

2. Stejskal, E.O. and Tanner, E.J., J.Chem.Phys. 49, 288 (1965).

3. Lindblom, G. and Wennerström, H., Biophys.Chem. 6, 167 (1977).

4. Lindblom, G., Larsson, K., Johansson, L.B.-Å., Fontell, K. and Forsén, S., J.Amer.Chem.Soc. 101, 5465 (1979).

5. Arvidson, G., Fontell, K., Johansson, L. B.-Å., Lindblom, G., Ulmius, J. and Wennerström, H., Ber.Bunsenges.Phys.Chem., 82, 977 (1978).

6. Wieslander, Å., Rilfors, L., Johansson, L. B.-Å., Lindblom, G. Biochemistry, submitted for publication.

7. Söderman, O., Johansson, L. B.-Å., Lindblom, G. and Fontell, K., Mol.Cryst. Liquid Cryst. in press and these proceedings, p.297

8. Lindblom, G., Wennerström, H. and Arvidson, G., Int.J. Quantum Chem. Suppl. 2, 12, 153 (1977).

9. Tiddy, G.J.T., J.Chem.Soc. Faraday I, 73, 1731 (1977).

10. Lindman, B., Kamenka, N., Kathopoulis, T.-M., Brun, B. and Nilsson, P.-G., to be published.

11. Lindman, B., and Brun, B., J.Coll.Interf.Sci. 42, 388 (1973).

12. Wennerström, H. and Lindman, B., Phys. Report, 52, 1 (1979).

13. Persson, N.-O. and Lindblom, G., Chemica Scripta

14. Bakulin, E.A. and Zavodnaya, G.E., Zh.Fiz.Khim. 36, 2261 (1962).

15. Mills, R., J.Phys.Chem., 77, 685, (1973).

16. Ekwall, P. in "Advances in Liquid Crystals (ed. G.H. Brown), vol. 1, p. 1, Acad. Press, N.Y., 1975.

17. Kamenka, N., Fabre, H. and Lindman, B., C.R. Acad.Sc. Paris, 281, 1045 (1975).

18. Lindblom, G., Lindman, B. and Tiddy, G.J.T., J.Amer.Chem.Soc. 100, 1199, (1978).

19. Wennerström, H., Lindman, B., Lindblom, G. and Tiddy, G.J.T., J.Chem.Soc. Faraday I, 75, 663 (1979).

20. Engström, S. and Wennerström, H., J.Phys.Chem. 82, 1711, (1978).

Ion Binding in Liquid Crystals as Studied by Chemical Shift Anisotropies and Quadrupole Splittings

O. Söderman, A. Khan, N.-O. Persson and G. Lindblom

Division of Physical Chemistry 2, Chemical Centre,
P.O.B. 740, S-220 07 Lund, Sweden

1. Abstract

NMR quadrupole splittings and/or chemical shift anisotropies have been studied for several counterions (^{19}F, ^{133}Cs and ^{7}Li) in various liquid crystalline systems. It was observed that these parameters were dependent on concentration of the components in the system and on the phase structure. For the decylammonium fluoride - heavy water system, chemical shift anisotropy changed sign between hexagonal and lamellar liquid crystalline phases and the values of the shift anisotropy in the lamellar phase were two times greater than in the hexagonal phase. The phase diagram following this principle has been determined for the system [1].

The critical micellar concentration of 0.25 molal for octylammonium fluoride - heavy water system has been determined from the measurement of isotropic chemical shift of ^{19}F using aqueous solution of NaF at infinite dilution as reference.

The ratio between the values of the quadrupole splitting and chemical shift anisotropy of ^{133}Cs for lamellar liquid crystals has been found to remain constant [2]. This finding has been interpreted in terms of changes in the fraction of ions bound to the amphiphile. A preferential binding of Cs^+ ions to the amphiphile over K^+ ions has also been observed from the measurement of ^{133}Cs quadrupole splittings and chemical shift anisotropies in an experiment where Cs^+ ions were partially replaced by K^+ ions.

^{7}Li quadrupole splittings for lamellar liquid crystals were independent of surfactant concentrations at high constant water contents but dependent on concentrations at low constant water contents [3]. This feature of the splittings can be explained by an ion condensation model [4].

Reference

[1]. O. Söderman, A. Khan & G. Lindblom, J.Magn. Reson., 36, 141 (1979).
[2]. N.-O. Persson & G. Lindblom, J.Phys.Chem. In Press.
[3]. N.-O. Persson & G. Lindblom, Chemica Scripta In Press.
[4]. S. Engström & H. Wennerström, J. Phys. Chem., 82, 2711 (1978).

The Structure of a Lyotropic Liquid Crystalline Phase that Spontaneously Orients in a Magnetic Field

O. Söderman, L.B.-Å. Johansson, G. Lindblom, and K. Fontell

Department of Physical Chemistry 2, Chemical Centre,
P.O.B. 740, S-220 07 Lund, Sweden

1. Abstract

About ten years ago LAWSON and FLAUTT [1] found that some lyotropic liquid crystalline phases could be macroscopically aligned by an applied magnetic field. Two types of these phases exist, either having positive diamagnetic anisotropy (denoted type I) or negative diamagnetic anisotropy (denoted type II).

The structure or the shape of the aggregates building up these phases is still not completely unraveled. We have used deuteron NMR, polarized absorption and diffusion studies in attempts do elucidate the structure for samples of this kind. The type I sample contained potassium laurate, potassium-chloride and heavy water (34.0/2.3/63.7 wt %). Type II contained potassium laurate, potassium-chloride, decanol and heavy water (30.0/4.0/6.0/60.0 wt %).

The alignment of samples in a magnetic field can be followed by observing the change in the ^2H_2O quadrupole splitting. The splitting provides information about both the microscopical and macroscopical order in the sample. To obtain more information about the symmetry of the liquid crystal, samples were macroscopically aligned between glassplates [2] and the ^2H NMR spectra are recorded at different angles between the plates and the magnetic field. These measurements imply that the directors for the type I sample are oriented as a powder in the plane of the glassplates, whereas the director for the type II samples is perpendicular to the glassplates.

Using retinal as probe the polarized absorption was studied for type I samples. The order parameter was calculated from the experimental data and was compared with the order parameter for retinal solubilized in a lamellar phase. The results show that the order parameter of retinal in the type I phase is about - ½ of that for the lamellar phase. This indicates that the aggregates in the type I phase have cylindrical symmetry.

Finally, the diffusion coefficient for the amphiphiles along the aggregates was determined with pulsed NMR with pulsed fieldgradients. From this experiment it can be concluded that the lengths of aggregates are over 1000 nm. Taken together all our findings strongly indicate that type I phase consists of long rodlike aggregates. The results obtained for type II are compatible with a disc-like structure.

References

1. Lawson, K.D. and Flautt, T.J., J.Amer.Chem.Soc. 89, 5489 (1976); Black, P.J., Lawson, K.D. and Flautt, T.J., Mol.Cryst.Liquid Cryst. 7, 201 (1969).

2. Söderman, O., Lindblom, G., Johansson, L.B.-Å. and Fontell, K., Mol. Cryst. Liquid Cryst. In press.

A Lyotropic Phase from Tetracarboxylated Copperphathalocyanines

S. Gaspard, A. Hochapfel and R. Viovy

Laboratoire de Physico-Chimie des Pigments Végétaux et Substances Modèles de l'Ecole Normale Supérieure de Saint-Cloud, Grille d'Honneur - Parc de Saint-Cloud, F-92211 Saint-Cloud, France

We have shown the existence of a new lyotropic phase from phthalocyanine compounds having a plane tetraazaporphine macrocycle with hydrophilic lateral substitutions. Basic solutions of these gave birefringent textures which were oriented by rubbing technics [1]. The textures were similar to those in the neat phases [2].

Molecules having configurations similar to our compounds are known to give thermotropic mesophases [3].

The optical spectra of oriented samples from these solutions show a weak dichroic effect from which the order parameter has been determined.

Microphotographs of the various textures are presented. The photographs are available on request.

Systematic investigations of the following parameters have been made
a- Influence of the cation of the basic solution.
We have tested the following hydroxydes

Li^+, Na^+, K^+, Ca^{2+}, NH_4^+ and tetrabutylammonium. Only the two first members gave the birefringent textures.

b- Influence of the central metal
We have tested the following phthalocyanines

Cu Pc, Co Pc, Ni Pc, Pd Pc and Pt Pc. The copperphthalocyanine was the only one which led to lyotropic organization.

c- Influence of the lateral substitution
When the carboxyl group was replaced by a sulfonic group, the birefringent textures disappeared.

CONCLUSION

The sodium or lithium hydroxyde solutions of tetracarboxylated copperphthalocyanine were the only ones to give the lyomesophase. A column model of molecular stacking is proposed.

1. S. GASPARD, C.R. Acad. Sci. Série C 289 387 (1979)
2. F. ROSEVEAR, J. AMER. OIL CHEM. SOC. 31 628 (1954)
3. S. CHANDRASEKHAR, PRAMANA 9 471 (1977)

Reorientation of the Director of an Amphiphilic Nematic Mesophase in a Static Magnetic Field

N. Boden and K.J. McMullen

Department of Physical Chemistry, The University, Leeds, United Kingdom

M.C. Holmes

Division of Physics, The Polytechnic, Preston, United Kingdom

G.J.T. Tiddy

Unilever Research, Port Sunlight, United Kingdom

The spontaneous alignment of the director in a strong static magnetic field is a distinctive property of a nematic mesophase. The motion of the director with respect to the mesophase is, as given by the continuum theory [1],

$$\lambda_1(\dot{\phi} - \Omega) - \tfrac{1}{2}\chi_a \frac{B_o^2}{\mu_o\mu_r} \sin 2\phi - \rho_1\ddot{\phi} + T(r,\phi) = 0 \qquad (1)$$

where ϕ is the angle between the director and the direction of the magnetic field $\underset{\sim}{B} = \underset{\sim}{k}B_o$, λ_1 the twist viscosity coefficient, Ω the angular velocity of the mesophase, ρ_1 the moment of inertia per unit volume and $T(r,\phi)$ the elastic torque acting on the director. The mesophase is assumed to have positive diamagnetic anisotropy $\chi_a = \chi_\parallel - \chi_\perp > 0$ and $\mu_r \approx 1$. For thermotropic nematogens, the third term is negligibly small as is the fourth provided $B_o >> B_c$, the critical magnetic field strength for the Freedericksz effect [1]. The same assumptions have been invoked in recent studies of the magnetohydrodynamic properties of nematic lyomesophases [2,3]. BODEN et al. [2] have determined λ_1/χ_a in an unusually fluid mesophase by measuring the equilibrium orientation ϕ when the tube containing the sample is rotated with angular velocity Ω about an axis perpendicular to $\underset{\sim}{B}$: (1) gives

$$\sin 2\phi = \Omega/\Omega_c \quad \text{with} \quad \Omega_c = -\chi_a B_o^2/2\mu_o\lambda_1 .$$

Both BODEN et al. [2] and FUJIWARA and REEVES [3] have reported corresponding measurements in more viscous mesophases of the motion of the director following an instantaneous rotation of the sample tube to some start angle $\phi(0)$: (1) predicts

$$\tan\phi(t) = \tan\phi(0) \exp(-t/\tau_D) \qquad (2)$$

where

$$\tau_D = -\mu_o\lambda_1/\chi_a B_o^2 .$$

In these studies it was observed that for small $\phi(0)$ the alignment of the director remained homogeneous throughout the relaxation process, whilst for large $\phi(0)$ a dispersion in the director orientation was found. The mechanism of this dispersion process is still unclear, though FUJIWARA and REEVES [3] have suggested it has its origin in wall effects.

The nematic mesophase formed by mixtures of caesium perfluoro-octanoate and water is particularly suited to studies of director relaxation because the phase diagram is known [4] and λ_1/χ_a may be varied with temperature over eight orders of magnitude [2]. The mesophase consists of disk shaped micelles which orient with their minor axes parallel to the field. The magnetic relaxation of the director in this mesophase is the subject of this paper. The orientation and distribution of directors are monitored by recording the deuterium NMR spectrum of ^2H_2O. In a homogeneously aligned mesophase the spectrum is a simple doublet with separation

$$\Delta\nu(\phi) = \Delta\nu(0^o)(\frac{3}{2}\cos^2\theta - \tfrac{1}{2}) .$$

Sample tube rotation was controlled by a stepping motor and was effected in times less than 0.5 s; other experimental details are as reported elsewhere [4].

Three categories of director relaxation are observed depending upon the values of λ_1/χ_a and $\phi(0)$ as follows.

(i) Small $\phi(0)$ and λ_1/χ_a: the director relaxes homogeneously, i.e. without dispersion, according to (2) [see Fig.1].

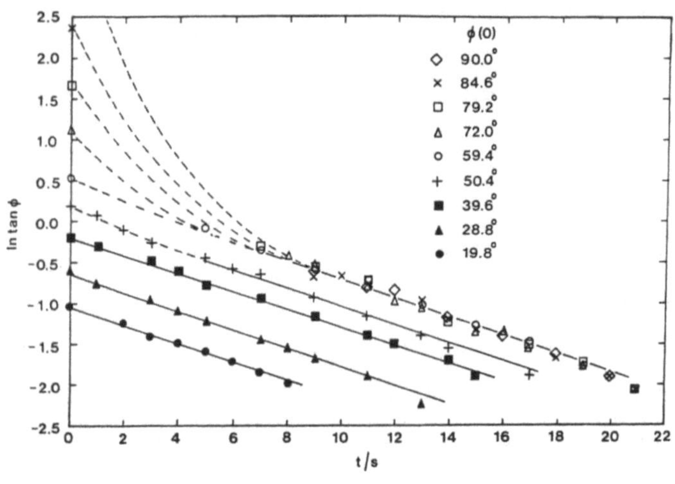

Fig.1 ln tanϕ versus time plots for various start angles $\phi(0)$ in nematic CsPFO/^2H_2O (60.7% by weight ^2H_2O) at 303.95 K: $\lambda_1/\chi_a = 3.15 \times 10^7$ kg m^{-1} s^{-1}. The data points correspond to well defined doublets and the broken curves denote the interval of director dispersion.

(ii) Large $\phi(0)$ and λ_1/χ_a: the director relaxes with a distribution of rates leading to rapid dispersion [Fig.2].

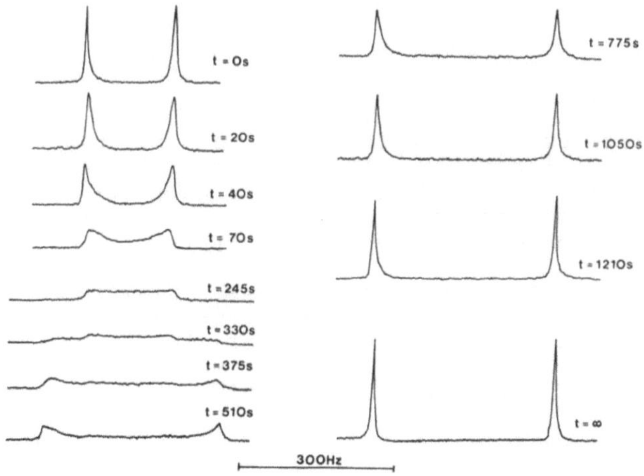

Fig.2 ^2H spectra for sample as in Fig.1 at 292.65 K with $\phi(0)$ = 90°.

(iii) Intermediate $\phi(0)$: the director initially disperses, but is refocussed at long times where it behaves according to (2) [Fig.3]. Note how the extrapolated value of $\phi(0)$ [Fig.1] is independent of $\phi(0) > \phi_c(0)$. This initial defocussing followed

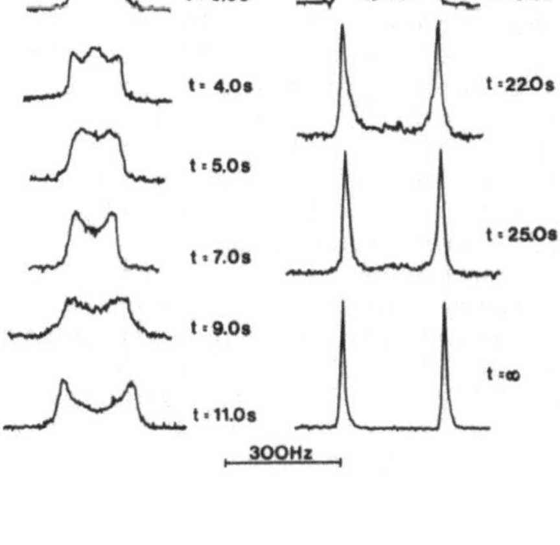

Fig.3 ^2H spectra for a sample containing 64.8% by weight ^2H_2O at 295.70 K with $\phi(0)$ = 79.2^o: λ_1/χ_a = 2.12×10^7 kg m^{-1} s^{-1}.

by refocussing of the director is explained by (1). Suppose some process generates a distribution of director orientations with a half width Δ; then, from (1), we obtain

$$\dot{\Delta} = \dot{\phi}\Big|_{\phi+\frac{1}{2}\Delta} - \dot{\phi}\Big|_{\phi-\frac{1}{2}\Delta} = \frac{\chi_a B_o^2}{2\mu_o \lambda_1}\Delta\cos 2\phi .$$

This expression predicts defocussing ($\dot{\Delta}<0$) will occur for $\phi > 45^o$ and refocussing ($\dot{\Delta}>0$) for $\phi < 45^o$ (λ_1 is by definition negative). Any dispersion of the director at large ϕ will, therefore, tend to be refocussed when $\phi(t) < 45^o$. The extent of refocussing will be dependent on both Δ and λ_1/χ_a.

Defocussing of the director requires a process involving a distribution of relaxation times τ_D. It is necessary to examine the third and fourth terms of (1) as possible origins for this process. PIERANSKI et al. [5] have solved (1) including the elastic term $T(r,\phi)$: its only effect is to modify the expression for τ_D as follows

$$\tau_D = -\mu_o\lambda_1/\chi_a(B_o^2 - B_c^2)$$

where B_c is a critical magnetic field. Thus, although B_c is expected to approach B_o as the nematic-lamellar transition is approached causing a divergence of τ_D, it cannot generate a dispersion of the director at large ϕ. On the other hand, the term representing the inertia of the director could lead to a distribution of relaxation times because ρ_1 is a function of the radius of gyration of the reorienting "units". A distribution in their sizes could give rise to the observed behaviour. We have, therefore, solved (1) numerically and were able to reconcile the corresponding $\phi(0)$ and linear parts of the ln tanϕ versus t plots. However, the values obtained for ρ_1/χ_a were found to depend on $\phi(0)$ which is physically unrealistic; moreover, the values estimated for the mean radius of gyration are similar in magnitude to the size of the bulk sample.

There must be two distinct mechanisms for the relaxation of the mesophase director. Hydrodynamic relaxation which is described by (1) and a second process characterised by a distribution of relaxation times as nicely shown by the spectra reproduced in Fig.2. The spectral intensity corresponding to the director oriented perpendicular to the field ($\phi = 90^o$) decays with a time constant of roughly 100 s which is much smaller than the value 1124 s obtained for τ_D from measurements with $\phi(0) < \phi_c(0)$. At long times, t > 750 s, the intensity at $\phi = 0^o$ increases with a time constant corresponding to τ_D indicating that the second process has become insignificant. The rate of this process must therefore decrease with decrease in ϕ, i.e. it is a function of the orientational magnetic energy $U(\phi) = \frac{1}{4}\chi_a B_o^2(1 - 2\cos^2\phi)$. The actual mechanism of the process is not clear. FUJIWARA and REEVES [3] have observed in other lyonematic mesophases that for $25^o < \phi(0) < 45^o$ initial hydrodynamic relaxation of the director was followed at longer times by a dispersion of the directors; this behaviour they attributed to surface produced inversion walls migrating into the bulk mesophase [6]. Their observation seems to differ from those reported herein for the caesium perfluoro-octanoate system. Clearly, further studies of this subject are required.

References

1. See for example S. Chandrasekhar, "Liquid Crystals", Chapter 3 (1977), Cambridge, University Press.
2. N. Boden, K. McMullen, M.C. Holmes, Proc. "Magnetic Resonance in Colloid and Interface Science", Menton, 1979.
3. F.Y. Fujiwara, L.W. Reeves, Can. J. Chem., (1978) 56, 2178.
4. N. Boden, P.H. Jackson, K. McMullen, M.C. Holmes, Chem. Phys. Lett. (1979) 65, 476.
5. P. Pieranski, F. Brochard, E. Guyan, J. Phys. (1973) 34, 35.
6. P.G. de Gennes, J. Physique, 32, 789 (1971).

NMR Measurements of W/O Type Microemulsions Formed by Large Bolaform Ions

H. Spiesecke

Joint Research Center, Ispra, Italy

Recently functionalisized micelle forming surfactants have attracted interest as membrane models and in energy conversion systems. Whilst the overall conformation of the hydrocarbon chain of single headed detergent molecules is obvious, surfactants with two headgroups can be visualized in a number of different arrangements in solution [1]. Menger and Wrenn have shown that the long chain bolaform ion Me_6C_{12} folds at an air water interface [2]. More recently Zana et al. investigated the association behavior of this compound in dilute aqueous solution by chemical relaxation, conductivity, and emf measurements. They could not find any evidence for micelle formation, however [3].

Considering that micelles are not formed by detergent molecules with a chain length shorter than eight these results are not surprising. We therefore synthesized bis-trimethylammonium-docosane dibromide from docosadiene to obtain at least eight to ten straight chain segments assuming the molecule would fold in aqueous solution. We investigated its carbon-13 nuclear magnetic resonance spectrum and its solubility behaviour in ternary systems. Of the twelve theoretical lines eight can be resolved. They were assigned with the aid of the homologuous series Me_6C_4 - Me_6C_6 and comparison with other hydrocarbons [4]. Read from low to high magnetic fieldthey correspond to C-1, the N-CH_3 group, C_{7-11}, C_5, C_6, C_4, C_2, and C_3 The distinction between C-5 and C-6 is ambiguous.

Qualitatively the nuclear relaxation of the individual carbon atoms along the chain should give an indication of the motional behavior of the molecule [5]. The relaxation rate decreases from the head group which is "anchored" in the Stern layer monotonously along the chain as it can move more freely towards the interior of the micelle. In Table 1 the results of our T_1 measurements for Me_6C_{22} are compared with literature values for octyl-trimethylammonium bromide and hexadecyl-trimethylammonium bromide [5].

As one can see the bolaform ion does not follow this line but the relaxation rate increases again towards the middle of the chain. This stiffening of the segmental motions one would expect if the molecule would form a loop inside the micelle.

Dissolving the free radicakdi-t-butyl nitroxide in the aqueous solution resulted in a broadening of the C-7,11 line indicating a preferential solubilisation towards the middle of the hydrocarbon core of the aggregates.

Table 1 Relaxation rates for bis-trimethylammoniumdocosane dibromide, octyl-trimethylammonium- and hexadecyl-trimethylammonium bromide

	Me_3N	C_1	C_2	C_3	C_4	C_5	C_6	C_{7-11}	C_8	C_{14}	C_{16}
conc.											
.1m	1.60	2.12	2.14	2.09	1.72	1.82	2.03	2.55			
.25m	1.23	1.67	1.88	1.73	1.69	1.95	2.07	2.40			
.2 m	1.58	2.13	2.13	-	2.13	1.28			0.77		
.4 m	.55	1.85	1.47							.83	.119

Besides the nuclear magnetic resonance measurements we studied the solubilisation of a .25 molar aqueous solution of Me_6C_{22} in toluene using Triton N-101 as surfactant. This non ionic detergent forms reversed micelles with water in toluene. Since bis-trimethylammoniumdocosane dibromide is practically insoluble in non polar organic solvents the formation of a homogeneous solution in the pseudo ternary system toluene,Triton N-101, aqueous solution of Me_6C_{22} can only occur if one assumes that both head groups of the molecule are dissolved in the water droplets inside the micelles requiring a folding of the hydrocarbon chain. A stable system was obtained using the following quantities: Toluene 1.842g (51%), Triton N-101 1.219g (33.8%),H_2O .471g (13%), Me_6C_{22} .078g (2.1%). Although it was possible to identify the lines due to $N-CH_3$ and $N-CH_2$ the other carbons could not be clearly identified being largely covered by Triton N-101 lines. This prevented the use of Mn ions as relaxing agent inside the water pools.

In conclusion one may say that the experimental evidence given above favours a looped chain conformation for bolaform ions having a chain length of at least twenty carbon atoms.

1 Y.Okahata,T.Kunitake, J.Am.Chem.Soc.101,5231,1979
2 F.M.Menger,S.Wrenn, J.Phys.Chem.78,1387,1974
3 S.Yiv,K.M.Kale,J.Lang,R.Zana, J.Phys.Chem.80,2651,1976
4 P.A.Couperus,A.D.H.Clague,J.P.C.M.van Dongen,Org.Mag.Res.11, 590, 1978
5 D.Doddrell,A.Allerhand, J.Am.Chem.Soc. 93,1558,1971

Part VIII

Interfaces, Bi- and Monolayers and Biological Applications

Defect Structure and Texture of Isolated Bilayers of Phospholipids and Phospholipid Mixtures

E. Sackmann, D. Rüppel and C. Gebhardt

Department of Biophysics, University of Ulm,
D-7900 Ulm, Federal Republic of Germany

1. Introduction

In order to deal with physical properties of lipid layers one has to distinguish between two extreme cases: (1) multilamellar systems of coupled monolayers containing variable amounts of water and (2) isolated bilayers (vesicles) in bulk water. The elastic properties of the first are determined by a large contribution associated with a change in the interlayer spacing and a much weaker curvature energy [1]. The behaviour of isolated bilayers is mainly determined by the curvature elastic energy while the system is free to escape in the third dimension perpendicular to the membrane surface. In crystalline bilayers one has to take into account also the elastic contribution associated with a shear in the direction of the membrane normal.

In our review we are dealing with defects of the texture of isolated lipid layers (vesicles) as revealed by freeze fracture electron microscopy. This texture is due to a variation in the local membrane curvature. In analogy to the situation in liquid crystal physics one can obtain detailed structural information from the symmetry of the defect pattern of the texture. As a second point the stabilization of the defect structure by solutes is discussed. The reconstruction of the 3-dimensional microscopic surface profile from freeze fracture electron micrographs can provide additional structural information.

2. On Bilayer-Phases and the Possibility of Ferroelectricity

As the multilamellar systems (liposomes at excess water), isolated bilayers of lecithins with saturated hydrocarbon chains exhibits three distinct phases: one fluid smectic A like phase (called L_α) and two crystalline modifications (called $L_{\beta'}$ and $P_{\beta'}$). The same holds for other lipids which exhibit a tilt of the chains with respect to the membrane normal in the crystalline state such as phosphatidyl glycerol [2]. The pretransition $P_{\beta'}$ $L_{\beta'}$ is of first order with a small heat of transition: $\Delta H \simeq 8$ kJ/mole. From x-ray studies it is well known that the $L_{\beta'}$- phase is planar while $P_{\beta'}$ exhibits a wave like superstructure [2,3]. The structural change at the pretransition is not understood yet. There is some experimental evidence that the $L_{\beta'} \rightarrow P_{\beta'}$ transition is associated with a partial decoupling of the two monolayers which are tightly interlocked in the $L_{\beta'}$-phase.

Thus, hydrophobic molecules (such as pyrene) incorporated in the membrane interior exhibit a considerable increase in mobility at the $L_{\beta'} \rightarrow P_{\beta'}$ transition [5]. However, the pretransition also strongly affects the state of the polar head groups. This is clearly demonstrated in Deuterium-NMR experiments. Fig. 1b shows the quadrupole splitting (or the order parameter) of the choline head group of DPPC as a function of temperature. Starting at low temperature one observes a sharp decrease in splitting by about 16% at the pretransition (T_p = 33°C) followed by a further decrease of about the same amount at the main transition.

The change in order at the pretransition is only sharp at increasing temperature while it becomes rather smeared out if one starts from the fluid phase. As shown below this may be explained in terms of a high defect density existing in freshly formed

a

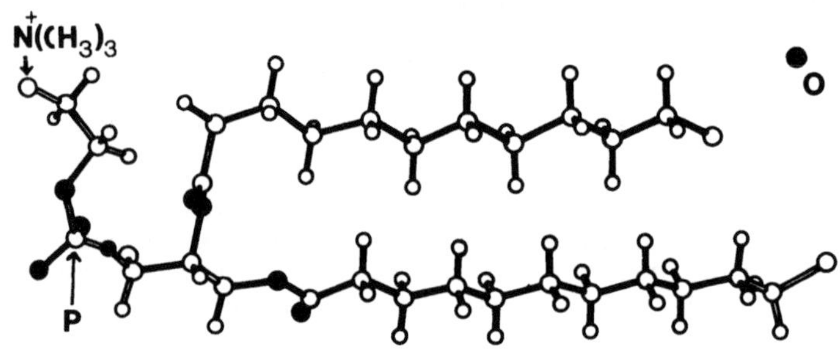

b

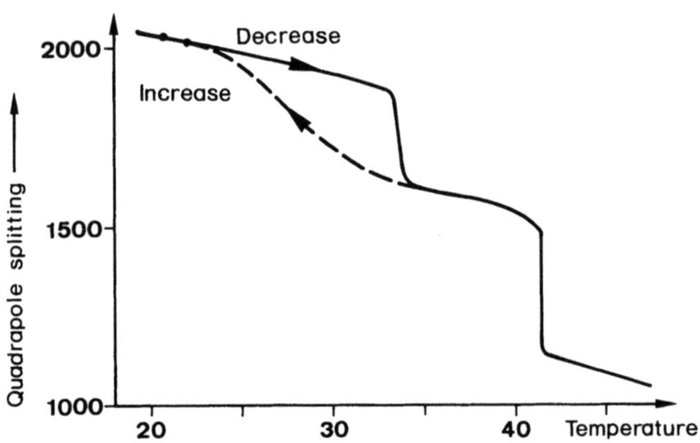

Fig. 1 (a) Structure of lecithin in bilayer according to x-ray [4] and neutron small angle scattering [6]. Note that the direction of the hydrocarbon chain and the polar head group are coupled.
(b) Quadrupole splitting of $N^+(CD)_3$-group of dipalmitoyl phosphatidyl choline (DPPC). Measurement performed with macroscopically oriented multilayers. (Courtesy of H.H. FÜLDNER, Thesis Ulm 1980)

$L_{\beta'}$-phase. The large change in order parameter seems to contradict the results of BÜLDT et al. [6] who came to the conclusion that the phosphocholine head is oriented parallel to the membrane both in the $L_{\beta'}$ and $P_{\beta'}$-phase. Now, the NMR order parameter is a dynamic quantity which depends on the time scale of the molecular motion. A decrease in order parameter could thus originate in a faster molecular motion.

As suggested some time ago by CHAPMAN et al. [7], the $P_{\beta'}$- to -$L_{\beta'}$-transition could be due to a strong ferroelectric (or antiferroelectric) coupling of the choline dipoles which are disordered in the $P_{\beta'}$-phase. Strong evidence for a collective ferroelectric polarization comes from dielectric relaxation measurements by KAATZE [8] . This author observes a collective motion of clusters of cooperatively polarized dipoles containing between 30 and 80 molecules for DPPC below the pretransition. The motional correlation time is of the order of 10^6 Hz.

It should be noted that a contribution to a collective ferroelectric polarization could come both from the electric dipoles of the choline head group as well as of the C = O groups of the glycerol backbone. In a tilted phase, a spontaneous polarization is expected even if the molecules could rotate fast about their long axes (cf reference [9]).

3. Freeze Fracture Electron Microscopy and Texture of Lipid Phases (a Technique of Picture Analysis):

The freeze fracture EM allows to detect structural features of lipid bilayers which lead to a variation of local curvature. The application of the method for structural studies is hampered by two main difficulties:

(1) Changes in structure may occur during the cooling process and, more seriously, ice crystals may form during cooling if this is not achieved fast enough.

(2) One obtains a two-dimensional projection of the three-dimensional vesicle (or cell).

The first problem may be overcome by application of a sandwich technique. The sample is brought between two thin metal plates which are rapidly cooled by dipping into fluid Freon. Another equally fast method is the spraying technique [10]. Cooling rates of the order of 10^4C/sec may be achieved.

In order to overcome the 2nd problem, a method was developed which allows to reconstruct the three-dimensional surface profile of vesicles [11]. It is based on the analysis of the platinum layer thickness. The principle of the method is summarized in Fig. 2. The sample is prepared in a two-step procedure. Fracturing of the frozen vesicle preparation exposes the interior of the bilayer. The subsequent evaporation of water lays free the outer membrane surface (Fig. 2, Step A). Then a platinum layer of an average thickness of 30 Å is deposited under an oblique angle $\partial_o = 45^o$. This layer is then stabilized by deposition of a 300 Å thick carbon layer in a direction $\partial_o = 90^o$.

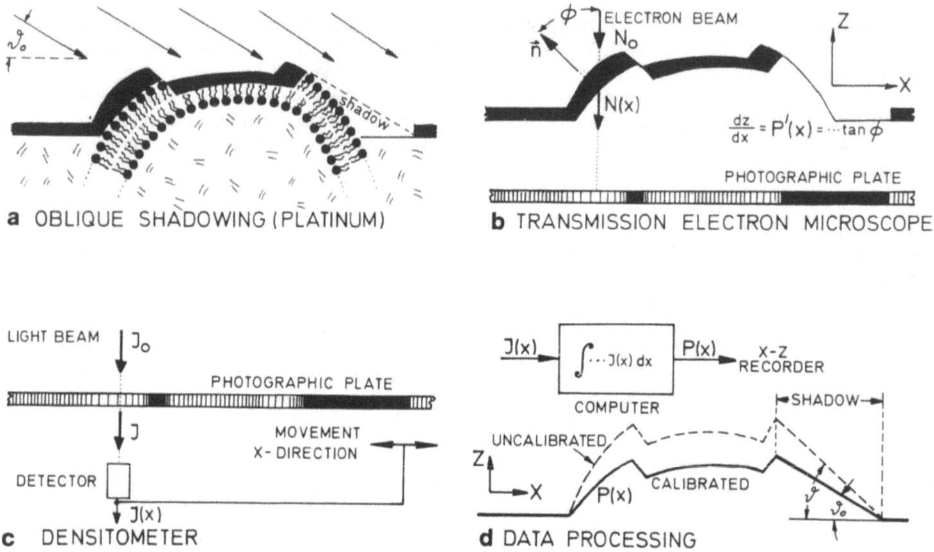

Fig.2 Reconstruction of three-dimensional surface profiles of vesicles. Schematic representation of the method.

The lipid and water is removed by solvent. The thickness of the Pt-layer is a direct function of the orientation of the local tangent to the membrane surface with respect to the shadowing direction and contains the information concerning the three-dimensional surface profile. The Pt-layer thickness is determined by measuring the electron beam absorption (Step B of Fig.2). This is done in an indirect way by densitometric evaluation of the negative films (Step C of Fig.2). In order to measure the third dimension (in the z-direction) a calibration procedure is necessary. This can be achieved by adjusting the height of the reconstructed surface in such a way that the slope of any region of complete shadow corresponds to the shadowing angle (Step D in Fig.2).

4. Texture and Microscopic Surface Profiles of Pure Lipids

a) L_α-phase and nucleation of $P_{\beta'}$: The L_α-phase exhibits a smooth surface if the sample is cooled rapidly from a temperature above the main transition. At slower cooling rates one may observe the nucleation of the P_β -phase as shown in Fig.3a. It exhibits smooth areas and worm-like clusters with triangular surface profiles (cf Fig.3b). The latter are attributed to nuclei of crystallized $P_{\beta'}$-lipid. The escape in the third dimension of the lipid in the $P_{\beta'}$-phase minimizes the elastic energy associated with the splay elastic deformation at the boundary between fluid and tilted crystalline lipid. The wavelength exhibits large fluctuation whereas a well defined ripple distance is observed well below the transition temperature (Fig.3b).

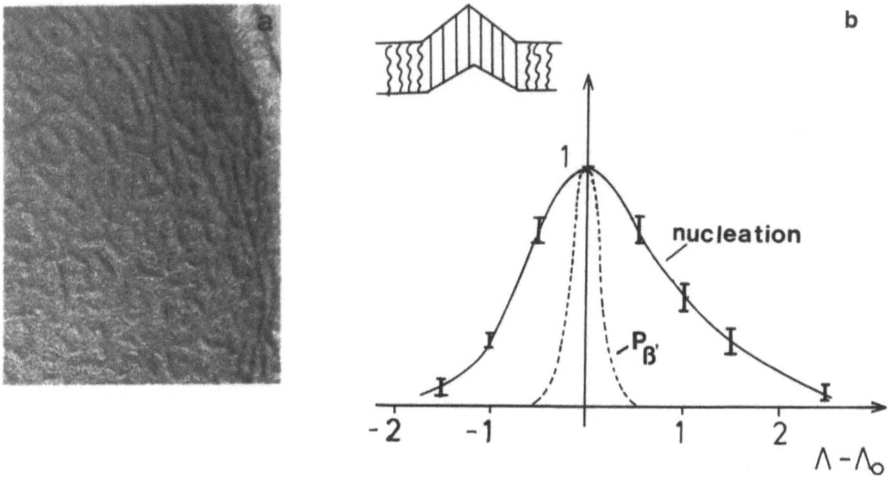

Fig.3 (a) Nucleation of $P_{\beta'}$-phase after cooling from L_α-phase just above the transition.
(b) Probability distribution of distance between ripples.

b) Two types of $P_{\beta'}$-ripple-phase: The well known ripple phase is observed when the lecithin vesicles are cooled from a temperature between the main- and the pretransition. As shown in Fig.4 there are two types of ripple structures which will be called the Λ -phase and the Λ /2-phase in the following. They are clearly dinstinguished by (1) the distance between the ripples (wavelength Λ or Λ /2), (2) the form and symmetry of the microscopic surface profiles and (3) the defect structure of the surface texture. The two structures are sometimes coexistent at the same large vesicle (cf Fig.4a). In this case the parallel ripple lines of the two phases form closed loops and do not cross. A characteristic feature of the Λ-phase clearly visible in the micrograph (Fig.4a) is the appearance of disclinations in the ripple pattern of strength $s = +1/2$ and $s = -1/2$. In contrast, the Λ/2-phase is characterized by the formation of wall-defects. This is due to the asymmetry of the ripple profile. Two regions of ripples of opposite orientation meet at the wall. Fig.4b shows the surface profiles of the two structures in a direction perpendicular to the ripple line. The profile of the hydrocarbon interface of the outer monolayer is shown. The ripples of the Λ-phase have grooves along the maxima and exhibit mirror symmetry with respect to planes parallel to the ripples and going through the grooves or the minima, respectively. The Λ/2-phase is characterized by an asymmetric sawtooth-like profile with sides a and b (cf Fig.11). The ratio of the projections of the sides on the plane of the membrane is on the average $r = 3:1$. Thus the profile of the Λ/2-phase agrees with the electron density calculations of TARDIEU et al. [3]. These authors also propose a ratio, r, of about 3:1.

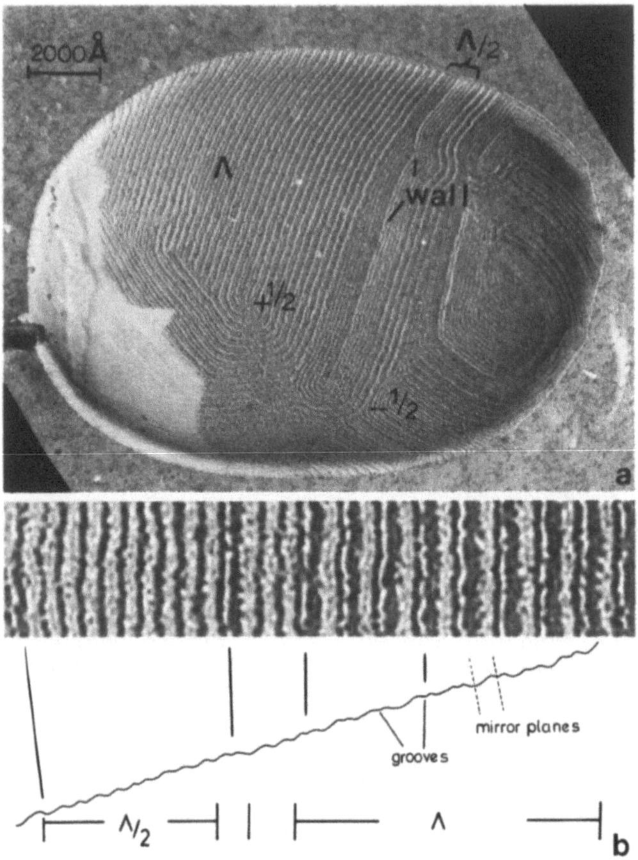

Fig. 4 (a) Texture of large vesicle cooled from $P_{\beta'}$-phase. Two types of textures are observed: The " Λ-phase" with a ripple distance of Λ and the "Λ /2-phase" with a ripple distance of Λ /2. Note the s = + 1/2 and s = - 1/2 defects of the Λ -phase and the wall defect of the Λ /2-phase.
(b) Microscopic surface profile obtained by reconstruction of the freeze etch EM. A cross-section in a direction perpendicular to ripples is shown for a vesicle with coexistent Λ and Λ /2 phase.

For the Λ/2-phase the vesicle surface is nearly planar along broad bands parallel to the ripples. Planar bands of different orientation intersect along the walls in such a way that the latter form the crest of a roof-like average surface profile. Values of the wavelengths are summarized in Table I. Judged from observation of a large number of preparations, the Λ-phase seems to be more stable for small vesicles (∿1 μm diameter) while the Λ /2-phase is mainly observed with large vesicles.

c) Defect structure of biaxial $L_{\beta'}$-phase: The $L_{\beta'}$-texture depends on the time for which the vesicle was kept at a temperature below

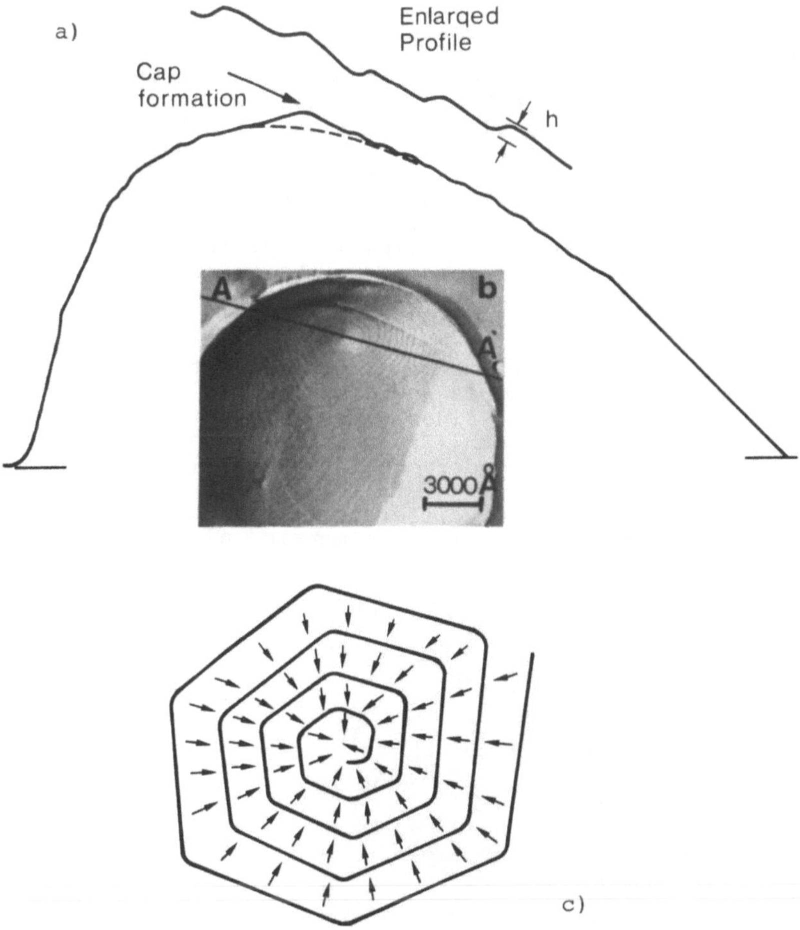

Fig.5 Texture and surface profile of L_{β}'-phase of lecithins.
(a) Microscopic surface profile along cross-section A-A' through the center of the defects and enlarged profile in direction perpendicular to defect lines. Amplitude of steps h $\simeq$15 Å. The broken line indicates the spherical approximation of the vesicle contour in the region about the defect centre.
(b) Electron micrograph of typical defect pattern of vesicle in L_{β}'-phase observed shortly after cooling from P_{β}'-phase.
(c) Form of defect line and orientation of tilted lipid. The splay deformation at the centre is relaxed by escape in a direction perpendicular to the membrane plane. The smooth regions between the defect lines are planar in the direction of the tilt (arrows).

the pretransition and is thus largely determined by kinetic effects. The pattern of Fig.5b is observed when the sample has been kept for less than 1 - 2 hours at $T < T_p$. Spiral-shaped defect lines originating at different sites on the vesicle surface are observed while the lines exhibit hexagonal symmetry. The reconstructed surface profile along a section going through the spiral centre section A-A' of Fig.5a exhibits two general features: (1) The microscopic profile in a direction perpendicular to the lines exhibits a step like shape with an amplitude h = 10 Å. (2) At the center of the spiral, the average surface deviates from a spherical shape (broken line in Fig. 5b) and forms a conical cap. This indicates an escape in the third dimension. At those areas where two different spirals meet, the surface profile becomes scaly.

When the sample is kept slightly below T_p for an extended period of time (≃ 10 hours) the spiral is wound up that is the widths of the smooth areas between two lines widens. In some cases the surface becomes completely smooth. In general one observes extended smooth regions encircled by defect lines with a triangular micro profile (cf Fig.6.) This suggests that they are due to Néel walls separating regions of different tilt directions, that is the membrane escapes into the third dimension along lines.

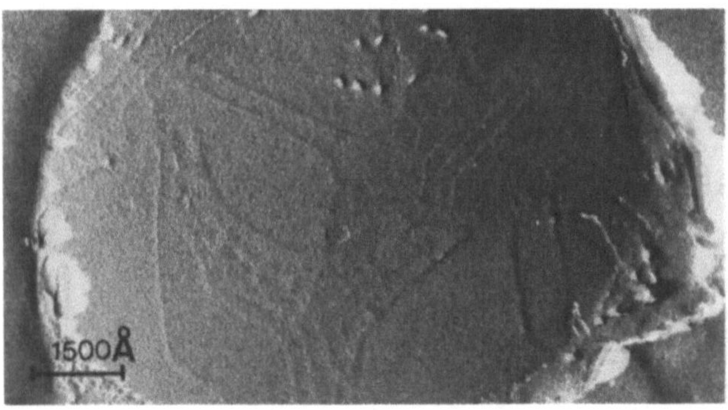

Fig.6 Electron micrograph of DMPC $L_{\beta'}$-phase after equilibration time of about 5 hours. Some defect lines exhibit a triangular profile indicating Neel walls.

The surface profile shows sharp steps at the defect line (cf. Fig.5a, enlarged profile). The height of the step is about equal to the monolayer thickness. The sharpness of the steps indicates that the bilayers have been mutually displaced in a direction parallel to the local director. This shows that the $L_{\beta'}$-phase is a crystal structure of high stiffness, which tends to be completely planar. Now, the formation of the $L_{\beta'}$-phase is a fast process. Therefore in a first step the spherical shape of the vesicle will be approximated by a steplike arrangement of the flat bands. Obviously the spiral-type pattern has the fastest growing rate. In a second step the $L_{\beta'}$-phase adjusts to the vesicle curvature. This could be achieved by a uniform distribution of edge dislocations with the dislocation lines parallel to the membrane plane. Such dislocations could be formed by expulsion of molecules from the inner monolayer. As a second possibility, bilayer curvature may be introduced if lipid molecules along lines perpendicular to the tilt direction are rotated by 90° about an axis normal to the membrane. If this is only done in one monolayer one may also obtain a spontaneous bilayer curvature.

5. Stabilization of Defects by (Small Amounts) of Solute

Drastic changes in the defect pattern are observed if small amounts of amphiphatic impurities are added. In analogy to metal physics one expects that such solutes may stabilize defects. Examples are shown in Fig.7 and 8. Fig.7 shows the ripple structure of DMPC containing 1 mole % cholesterol. Clearly, the texture is identical to the $\Lambda/2$-structure. As a general result we found that very small amounts of cholesterol ($\sim$ 0.5 mole %) stabilize the $\Lambda/2$-structure. The same was observed for other solutes such as cholate. The same sawtooth-like profile of the ripples as in pure DMPC is observed. Values of the wavelength and the ratio, r, of the projected widths of the sawtooth profile are given in table I. These values do not change appreciably up to 6 mole % of cholesterol. This indicates that up to this limit, the cholesterol can be incorporated into the defect regions of the pure lipid. This is consistent with the finding that the pretransition becomes absent at this concentration. A rather regular defect pattern is observed (Fig.7b) which is stable for many hours. It consists of pairs of $s = -1/2$ and $s = +1/2$ defects, respectively, which are connected by walls. The defect pattern may also be interpreted as $s = +1$ and $s = -1$ disclinations which are dissociated into $(+1/2; +1/2)$- and $(-1/2; -1/2)$- pairs, respectively. It should be noted that only disclination lines of integer strength (e.g. $s = +1, -1$) are topologically allowed.

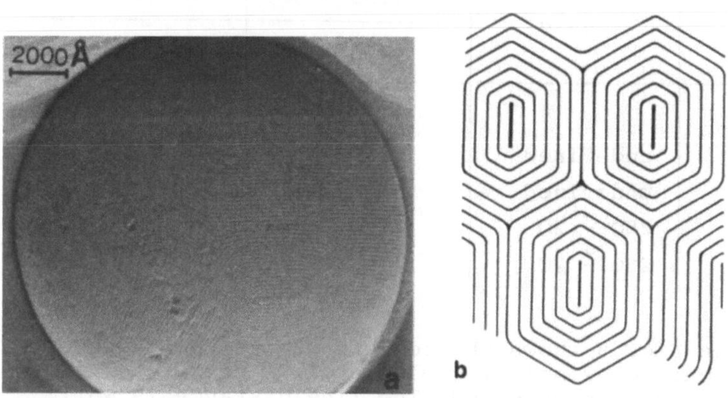

Fig.7 (a) Ripple structure of DMPC at $19^{o}C$ ($P_{\beta'}$-phase) containing one mole % of cholesterol.
(b) Typical stabilized defect pattern of pairs of $s = -1/2$ and $s = +1/2$.

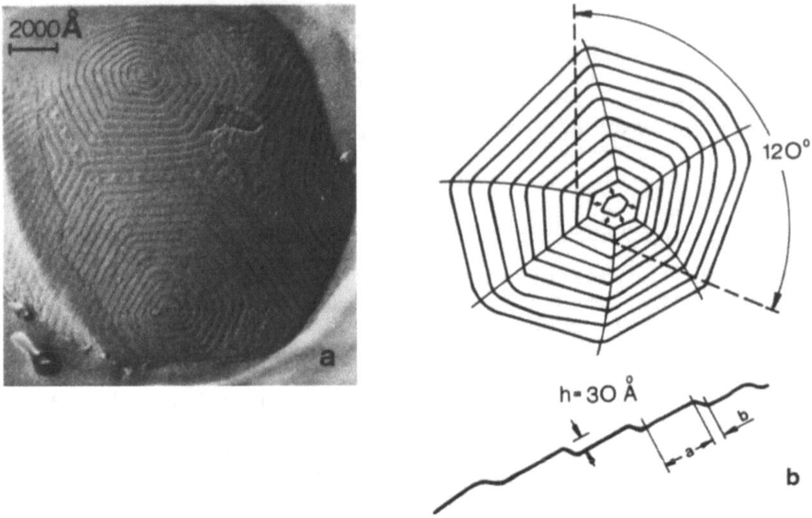

Fig.8 (a) Stabilized defect structure of $L_{\beta'}$-phase of DMPC in presence of two mole % cholesterol.
(b) Contour of closed defect line and enlarged profile in direction perpendicular to these lines.

Table I Wavelength of ripple phase in Å for DMPC in absence and presence of cholesterol.

System	wavelength of phase $L_{\beta'}$ [Å]	wavelength of phase $P_{\beta'}$ [Å]	ratio $r = {}^{a}/b$ cf Fig.11a
DMPC pure	irregular	$\Lambda/2$:130 ± 20	3
		Λ :220 ± 20	-
DMPC + Cholesterol			
0.5 mole %	undefined	120	3
1 mole %	undefined	120	3
4 mole %	250 ± 20	120	3
6 mole %	250 ± 20	150	3

The micrograph in Fig. 8 shows the texture of the $L_{\beta'}$-phase of DMPC in the presence of 2 mole % cholesterol. A pattern of closed hexagonal defect lines is observed. The line distance is well defined between 2 and 6 mole % of cholesterol (cf table I). In contrast to the situation for pure DMPC, the defect pattern is stabilized by the solute for many hours. At very low cholesterol content ($<$0.5 mole %) the lines are also closed although the line distance becomes irregular and the defects heal out after an extended period of time. The same step-like microscopic profile as for pure DMPC is found. Above 6 mole % the textures above and below the pretransition become identical. This is expected since the latter vanishes.

The stabilization of the periodic defect pattern of the $L_{\beta'}$- and the $P_{\beta'}$-phase by small amounts of impurities is an intriguing problem. It is understandable that the defect pattern is stabilized by the solute molecules. However, we cannot give an explanation yet for the strict periodicity of the pattern.

6. Texture of Lipid Mixtures Undergoing Phase Separation (Spinodal Decomposition)

The freeze fracture EM is a powerful technique to establish phase diagrams of lipid mixtures [13,14]. It gives simultaneously information on the lateral organization of components. Rapid cooling is a prerequisite for such experiments. When the solute concentration is increased, it can no longer be incorporated into the defect region of the ripple phase and lateral phase separation is observed. While a well defined wavelength, Λ , of the $P_{\beta'}$- phase is observed at low solute concentrations, it exhibits a sharp increase in fluctuations if a phase line is crossed.

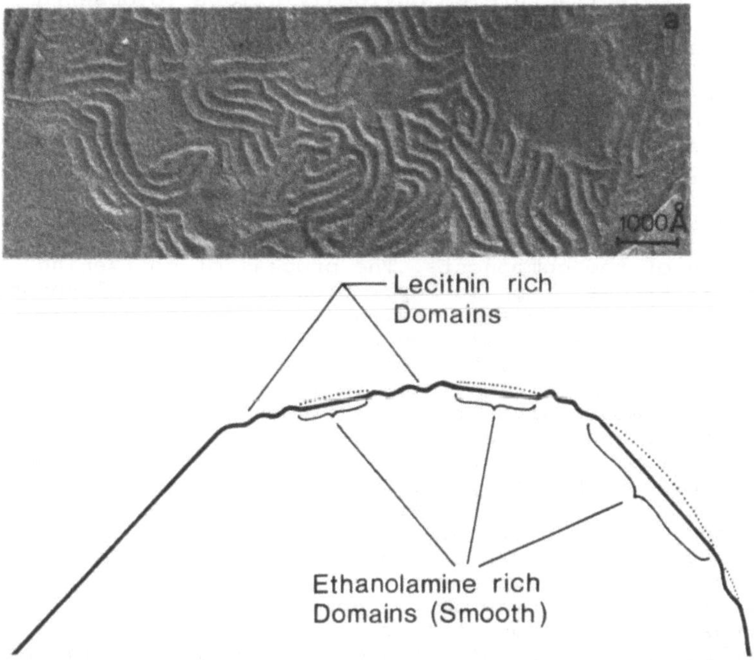

Fig.9(a) Texture of vesicle from 1:1 mixture of DMPC and dimyristoyl ethanolamine (DMPE). The vesicle was quenched from a temperature where a fluid phase enriched in DMPC (80 %) and a crystalline phase enriched in DMPE (80 %) are coexistent.
(b) Surface profile of the above mixture. The smooth domains of DMPE enriched crystalline phase are indicated.

An example is shown in Fig. 9a. The mixture of DMPC and dimyristoyl phosphatidyl ethanolamin (DMPE) has a "cigar"-like phase diagram in which a fluid phase enriched in DMPC coexists with a crystalline phase enriched in DMPE. The 1:1 mixture of Fig. 9a was cooled from a temperature (37°C) where a fluid, DMPC enriched phase (80 % DMPC) coexists with a crystalline phase containing 80 % DMPE. The preparation was kept for 10 hours at this temperature in order to establish thermal equilibrium. A domain like lateral organization of smooth and wavy patches is observed. The first are due to the crystalline DMPE-enriched phase. The undulation of the latter is caused by the additional process of phase separation when the sample is rapidly cooled for the preparation. One then runs through a part of the coexistence region until the solidus line is reached. The undulation may be explained as follows: In the crystalline DMPE-enriched phase the chains are oriented perpendicular to the membrane surface while the nearly pure DMPC which is formed exhibits a tilt. In order to minimize the splay energy at the phase boundaries the latter component escapes in the third dimension.

One can learn two important things from this experiment:

1) The process of lateral phase separation can be explained in terms of the theory of spinodal decomposition in two-dimensions [15,16,17]. The freeze fracture EM obviously allows to observe the spatial variation in composition in the plane of the membrane through the variation in local curvature. The technique thus provides a method to measure distribution functions of the wavelength of concentration fluctuations as a function of cooling rate. Lipid mixtures are therefore interesting model systems to study the physics of spinodal decomposition (in two dimensions) which is a subject of great present interest [16].

2) An outstanding property of the two-dimensional lipid alloys undergoing phase separation is the formation of a stable domain like organization of the components. The process of coarsening after the initial stages of phase separation which is well known from three-dimensional systems is obviously prevented in isolated lipid layers. This has been explained previously in terms of the difference in spontaneous curvature of the lipid components: a local variation in composition leads to a corresponding change in the local spontaneous curvature. This causes a splay elastic force field at the transition between the domains which provides the restoring force preventing the coarsening process [17]. The system may also become unstable. One then observes the detachment of small vesicles from the bilayers reminiscent of the process of phagocytosis. An example of the different spontaneous curvatures of segregated phases is provided by the DMPC/DMPE mixture. The reconstructed surface of the 1:1 mixtures is shown in Fig. 9b. It is seen that the smooth DMPE-enriched crystalline domains are nearly planar (curvature $c_o = 0$) while the fluid undulated patches show the spontaneous curvature of the vesicles ($c_o = 10^{-4}\ cm^{-1}$). Fig. 9b thus provides experimental evidence for the previous model of domain-structure stabilization [17].

7. Two Models of Ripple-Structure

It was first pointed out by LARSSON [18] that the ripple structure is the consequence of two conflicting requirements. The first is due to the difference in the areas occupied by the chains (C) and the head group (H) of the lipid molecules. In order to balance this difference the chains may be tilted by a tilt angle $\cos\Theta = C/H$ for $H > C$. The second is the optimal fitting of the chains in the all-trans configuration of the crystalline phase (cf Fig. 10a). This requirement tries to adjust the tilt angle to a fixed value of $\tan\Theta_h' = 2n\,\Delta/a$, where a is the interchain distance, Δ is the distance between two adjacent CH_2-groups and where n is an integer. According to LARSSON $\Theta<\Theta'$ which leads to coexistence of tilted and non-tilted phase in the ripple structure. A third requirement is that the chains must be parallel at the interface of the coexistence region. Two models of the ripple phase have been proposed subsequently.

In a model proposed by GEBHARDT et al. [17] the undulated structure of well defined wavelength is attributed to a balance between a spontaneous curvature energy of each monolayer and the compression energy of the bilayer. The model starts from the idea that in the $P_{\beta'}$-phase the monolayers are decoupled and as a consequence they may curve spontaneously in order to adjust to the above mentioned structural requirement. This leads to a periodic variation of compression and dilatation as indicated in Fig. 10b. In this model the tilt angle varies periodically between $\Theta = \Theta_o$ and $\Theta = 0$. However, the average tilt of the chains is expected to be zero. This seems to be in contrast to the x-ray results [3,4]. The structure predicted by this model is, however, in agreement with the defect structure of the Λ-phase. The existence of s = + 1/2 and s = - 1/2 disclinations and the absence of wall-defects lines is only consistent with a uniaxial phase. The model is also in agreement with the surface profile of the Λ-phase [11]. It should be emphasized that this model ignores the interlocking of the zigzag-chains and assumes that the molecules may be rather freely shifted in a direction perpendicular to the plane of the membrane. The model would therefore also allow for a relatively large value of the vesicle curvature. This could therefore explain the finding that the Λ-phase prevails in small vesicles with radii below 1 µm. In this model the pretransition may be explained in terms of a coupling-decoupling mechanisms of the monolayer, together with a change in order in the polar head group region.

The DONIACH model [19] starts from the idea of a tightly packed crystal structure of the bilayer with strong interlocking of the monolayers below the main phase transition. The basic assumptions are: (1) The paraffin chains are in a rigid extended closed packed configuration and the zigzag chains are interlocked in such a way that they fit nicely into one another as shown in Fig. 10a. (2) The bilayer as a whole exhibits a spontaneous curvature. This is explained in terms of a polarization of the lipid heads by electrostatic coupling to the water dipole.

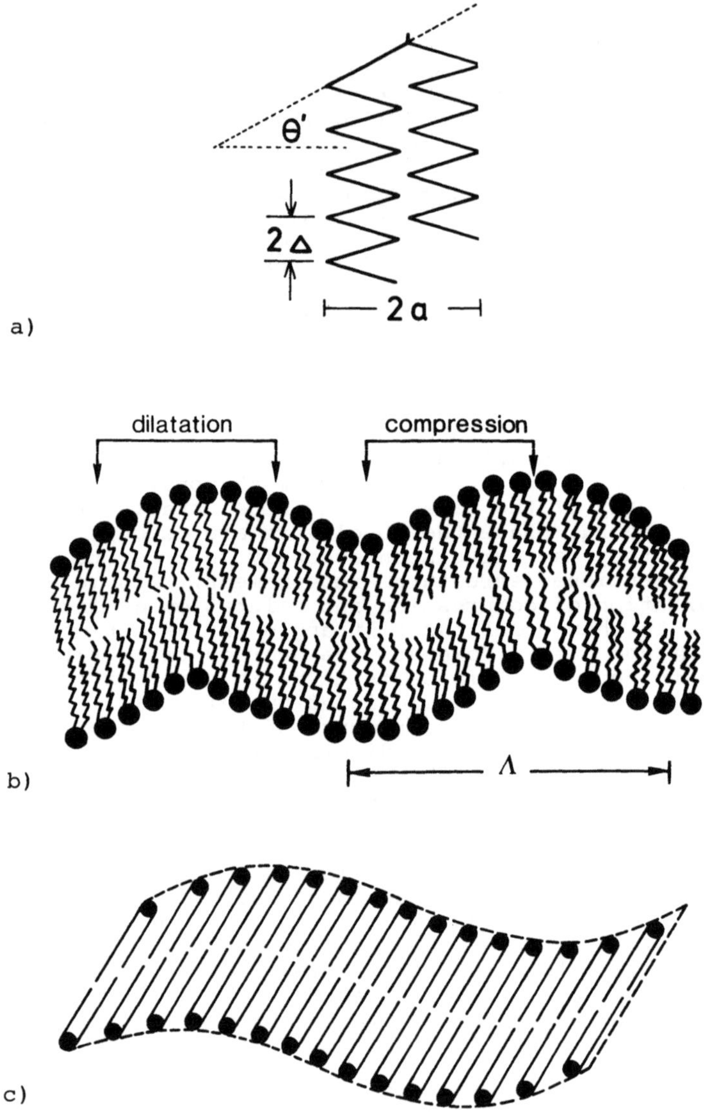

Fig.10 Molecular structure of ripple phase
(a) Tilt angle induced by fitting of fully extended paraffin chains.
(b) Model of Gebhardt et al [17]. Spontaneous curvature of monolayer.
(c) Model of spontaneous bilayer curvature by Doniach.

The interlocking tendency leads to a free energy

$$U(\Theta) = \gamma(\Theta - \Theta_o)^2$$

with a minimum for a tilt angle $\Theta = \Theta_o$. The spontaneous local curvature leads to an additional energy since the tilt angle varies by an increment $\Delta\Theta$ between two adjacent paraffin chains. The energy associated with the curvature is given by

$$K = \sum_i \lambda [(\Delta \Theta_i)^2 - (\Theta_o')^2]^2$$

where the sum extends over all lipid molecules i. λ and Θ_o' are adjustable parameters.

Since the bilayer as a whole must remain planar, the spontaneous local curvature must oscillate between positive and negative values as a function of distance along the bilayer. The structure of minimum energy is determined by a dimension less spontaneous curvature parameter

$$\eta = (\Theta_o'/\Theta_o)^2$$

For $\eta < 1$, $\Delta\Theta_i$ oscillates between two extreme values ($\pm\Theta_o\sqrt{1+\eta}$ and $\pm\Theta_o\sqrt{1-\eta}$) while one obtains an average tilt $\langle\Theta_o\rangle$. This structure is shown in Fig. 10c. For $\eta > 1$, the average tilt angle is zero. There is, however, a local tilt which varies between positive and negative values. The structure looks rather similar to that of the model of spontaneous monolayer curvature.

While the monolayer-curvature model predicts an uniaxial phase (with local biaxiality), and would only be consistent with $s = \pm 1/2$ defects of the texture, the TARDIEU-DONIACH model allows for biaxiality and could thus lead to wall defects. However, both models predict essentially symmetric surface profiles and thus do not explain the sawtooth-like profile of the $\Lambda/2$-structure. Both the Λ and the $\Lambda/2$-phase could be understood on the basis of the original Larsson-criteria of the close-packing of the chains and head groups. These criteria require (1) that the zig-zag chains must fit nicely into one another and (2) that the areas of the fully extended chains (C) and the polar head group (H) are identical. Difficulties arise if C and H are different. There are several possibilities to adjust to such a mismatch. (1) Free volume may be created in the hydrocarbon region by the formation of chain defects (e.q. kinks). (2) The monolayers or the bilayer may curve spontaneously. (3) The chains may be tilted by an angle Θ with respect to the membrane normal. (4) The area occupied by the head groups could be adjusted by a change in orientation and/or the amount of bound water. Assumptions (2) and (3) are the basis of the above mentioned models. In principle both models could be modified in order to account for an asymmetric profile. For crystalline phases with fully extended chains problems arise with the spontaneous curvature models. Any curvature would be accompanied by a substantial free volume which may lead to a prohibitively high energy. The same is expected to hold for the 1st mechanism. It appears that the most natural way to adjust to the mismatch of C and H in the crystalline phase is the combination of (3) and (4), that is structural change in the head group region together with the chain tilting. LARSSON [17] proposed a combination of tilted and non-tilted chains. This is in contrast to monolayer measurements. Comparison of pressure/area diagrams of lecithins and phosphatidic acids leads to the conclusion that $C \simeq 40$ Å^2 and $H \simeq 49$ Å^2. This would lead to a tilt angle of $\theta = 35^{\circ}$.

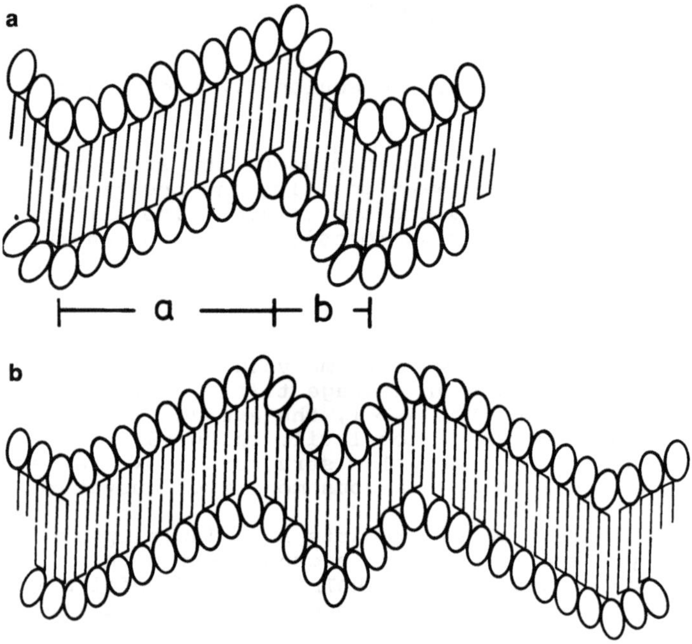

Fig.11 (a) Possible molecular structure of $\Lambda/2$-phase. Combination of subphases of tilt angles $\theta_3' = 28^{\circ}$ and $\theta_5' = 46^{\circ}$. The tilt angle with respect to the normal of the average membrane plane is 10°. (b) Proposed structure of Λ-phase. It consists of two oppositely oriented $\Lambda/2$-sawteeth which are slightly inclined so that the chains at the boundary of the sawteeth are parallel.

According to

$$\tan \theta_n' = \frac{2n\Delta}{a} \quad \text{with } \Delta = 2{,}5 \text{ Å} \quad a = 5 \text{ Å}$$

for the tilt angle as a result of the shift of n zig-zag units (CH_2-groups). It should be noted that two adjacent chains of the same lipid molecule are fixed at $n = 1$. Shifting two adjoining molecules by $n = 0, 1, 2, 3, 4$, zig-zag units gives values of $\theta_1' - \theta_2' = 0^{\circ}, 15^{\circ}, 28^{\circ}, 38^{\circ}, 46^{\circ}$. One thus expects that the P_{β}'-phase consists of a combination of two types of tilted subphases (one type at one ripple side) in appropriate amounts.

This would indeed explain the asymmetric profile of the $\Lambda/2$-phase (r = 1:3). The combination of $\Theta_3'=28^o$ and $\Theta_5'=46^o$ gives the best fitting to our experimental findings. We thus propose the structure presented in Fig.11a.

In this picture the Λ-phase results from the $\Lambda/2$-phase by rotating every second saw-tooth by 180°. In order to fit the orientation of the chains at the interfaces the saw-teeth have to be slightly inclined.

Judged from the microscopic surface profile of the $\Lambda/2$-phase and the above considerations concerning the area adjustment a combination of non-tilted and 46°-tilted sub-phases would also be possible. However, the surface profile of the Λ-phase rules out this possibility. The mirror symmetry of the profile and the absence of walls requires that the chains are oriented parallel to the mirror plane and perpendicular to the average plane of the membrane. The non-tilted sub-phase would have to be parallel to the average membrane plane in contrast to our findings.

An open question is the formation of a well defined ripple wavelength in the above picture. It should be noted that the chain and head areas are only matched on the average whereas there is still a residual mismatch on a local scale. In the sub-phase with $\Theta_3' = 28^o$ H >C while H <C for $\Theta_5' = 46^o$. In a closely packed crystalline phase there exists a high stiffness against changes in the inter-chain distance. On the other side, the head groups may yield more easily to a lateral pressure by a change in their average inclination with respect to the membrane plane. The appearance of a well defined wavelength could thus be explained in terms of a periodic variation of compression and dilatation in the head group region. This enables the head groups of the sub-phase Θ_3' to expand partially into the Θ_5'-phase. This would lead to a partial reduction in the strain energy. This situation is indicated in Fig. 11a where the orientation of the head groups is varied periodically.

Literature

[1] Kléman, M., Williams, C.E., Costello, M.J. and Galik-Krzywicki, T.
Phil. Mag. 35, 33 (1977)

[2] Marsh, D.
Biochim. Biophys. Acta 510, 63 (1978)

[3] Tardieu, A., Luzatti, V. and Reiman, F.C.
J. Molec. Biol. 75, 711 (1972)

[4] Janiak, M.J., Small, D.M. and Shipley, G.G.
Biochem. 15, 4575 (1976)

[5] Sackmann, E.
Ber. Bunsenges. Phys. Chem. 82, 891 (1978)

[6] Büldt, G., Gally, H.U., Seelig, A. and Seelig, J.
Nature 271, 182 (1978)

[7] Chapman, D., Williams R.M and Ladbrooke, B.D.
Chem. Phys. Lipids 1, 445 (1967)

[8] Kaatze, U.
Progr. Colloid and Polymer Sci, 29. Hauptversammlung 1979

[9] Sackmann, E.
Light-Induced Charge Separation in Biology and Chemistry, eds. H. Gerischer and J.J. Katz, pp. 259-285
Berlin: Dahlem Konferenzen 1979

[10] Bachmann, L. and Schmitt, W.W.
Proc. Natl. Acad. Sci. USA, 68, 2149 (1971)

[11] Krbecek, R., Gebhardt, C., Gruler, H. and Sackmann, E.
Biochim. Biophys. Acta 554, 1-22 (1979)

[12] Vervegaert, P.H.J., Verkleij, A.J., Elbers, P.F. and van Deenen, L.L.M.
Biochim. Biophys. Acta 311, 320 (1973)

[13] Luna, E., McConnell, H.
Biochim. Biophys. Acta 509, 462-473 (1978)

[14] Stewart, T.P., Hui, S.W., Portis, A.R.Jr., Papahadjopoulos, D.
Biochim. Biophys. Acta 556, 1 (1979)

[15] Cahn, J.W.
Trans. Metallurgical Soc. AIME 242, 166 (1968)

[16] Binder, K., Stauffer, D.
Advances in Physics 25 V.4, 343 (1976)

[17] Gebhardt, C., Gruler H. and Sackmann, E.
Z. Naturf. 32c, 581 (1977)

[18] Larsson, K.
Chem. Phys. Lipids 20, 225 (1977)

[19] Doniach, S.
Chem. Phys. 70, No. 10, 4587 (1979)

[20] Petrov, A.G., Seleznev, S.A. and Derzhanski, A.
Acta Physica Polonica A 55, 385 (1979)

[21] Helfrich, W.
Z. Naturf. 28c, 693 (1973)

The Crenation of Lipid Bilayers and of the Membrane of the Human Red Blood Cell

F.R.N. Nabarro[1], A.T. Quintanilha and K. Hanson[2]

Lawrence Berkeley Laboratory, University of California

1. Introduction

Phospholipid bilayers in an aqueous medium consist of two sheets of long phospholipid molecules, the polar heads in each sheet, usually negatively charged, being in contact with the aqueous medium, and the ends of the fatty-acid chains in one monolayer lying close to the corresponding ends in the other layer (e.g.[1]). The two layers may be identical, yielding a symmetrical bilayer, or different, yielding an asymmetric bilayer. Biological membranes are usually asymmetric, and contain much non-phospholipid material such as cholesterol (which is closely associated with the phospholipid molecules) and proteins which may be disposed in various ways with respect to the two phospholipid layers. In membranes with a high metabolic activity, these proteins are abundant, and probably contribute largely to the structural properties of the membrane. The red blood cell has a very low metabolic level (MARTIN and FUHRMAN [2] show an extremely low oxygen consumption, but the red cell also performs anaerobic glycolysis), and there is only one protein molecule for every 90 molecules of phospholipid or cholesterol (though the masses of protein and lipid are roughly equal) [3(p.27), 4]. The total number of phospholipid and cholesterol molecules in one membrane is about 5×10^8 (e.g.[5]). The protein spectrin, which is not found in other cells, forms some kind of network on the cytoplasmic side of the membrane [6,7]. It probably interacts with actin which is present in the membrane, and this spectrin layer appears to be almost entirely responsible for the resistance of the membrane to shear in its plane [8] and largely responsible for its resistance to change of area [9,10]. However, the bending modulus of the membrane, considered as a mechanical shell, seems to be largely controlled by the lipid bilayer [11-13].

Red cells may be haemolysed by two different routes. The first route, which we shall not discuss in detail here, is described by PONDER [14] as shape changes accompanied by volume changes. It occurs when the cell is placed in a hypotonic medium. The cell increases in volume while keeping its surface area almost constant, and lyses when it has distended to a spherical form. The second route is that of "shape changes unaccompanied by volume changes". This route has two branches. On either branch, the membrane becomes grossly distorted, the volume of the cell remaining almost constant. The effective area of the membrane decreases, and the cell becomes approximately spherical, ultimately lysing when the membrane shrinks tightly on to the contents. On the first branch, the outer of the lipid

---

[1] Permanent address University of the Witwatersrand,Johannesburg,South Africa
[2] Now at ETEC Corp., 3392 Investment Blvd.,Hayward CA 94545

layers seems to expand more than the inner. Spicules are formed, and the cell crenates to form an echinocyte. As the process develops, the crenations become finer. The process seems to be reversible or nearly so unless the surface area of the cell is permanently reduced by the breaking-off of spicules [15,16]. At a certain stage, the appearance of the cell in the optical microscope changes abruptly from a glistening sphere to a dusky "prolytic sphere". This change is irreversible. On the second branch, the inner of the lipid layers seems to expand more than the outer. The cell develops a cup-shaped stomatocyte form. Normally, only one invagination is formed during the reversible stages. The process becomes irreversible when parts of the membrane become incorporated within the cell as the boundaries of vesicles [16,17]. Our aim is to provide a rough quantitative theory of the deformation along the branches of this second route. In seeking this, we shall first develop a formal theory of crenation and then estimate the relevant parameters for four models of the red blood cell. Finally, we comment briefly on a situation in which invaginations may be formed.

DEUTICKE [18] showed that crenation is usually induced by anionic or non-ionized compounds, while cationic substances usually induce the formation of cups. SHEETZ and SINGER [19] rationalized these observations, and explained some time-dependent effects, by assuming that the distorting molecules were absorbed on the negatively charged cytoplasmic side if they were cationic and could diffuse (in the neutral form) sufficiently rapidly across the membrane; anionic molecules and all impermeant molecules would absorb in the outer layer, repelled by the negative charge or merely unable to penetrate. We take this as evidence that the primary action of these active molecules is to intercalate into one or other of the lipid layers.

The observations of RAHMAN et al.[20,21] are of particular interest. The red cell is heavily distorted during its passage through the spleen (e.g. [22]). The spleen contains phospholipase A, which removes one lipid tail from a phospholipid molecule, leaving a wedge-shaped molecule with a large hydrophilic head and a long narrow hydrophobic tail. After a certain number of passages through the spleen, a concentration of molecules is attained which causes the cell to crenate. The duration of its passage through the spleen is then increased, the attack of the phospholipase is enhanced, and after a few more passages the cell is unable to penetrate the spleen and is removed from the circulation. If this interpretation of the observations is correct, crenation is not merely an interesting artefact, but forms an essential part of a mechanism of planned obsolescence whereby red cells, which have no protein repair mechanisms, are removed from the circulation after a predetermined life span.

2. The Elastic Interactions

Our most general model of the membrane consists of two lipid layers, each of thickness h, lined with a layer of spectrin of thickness γh (Fig.1). We model the two lipid layers and the spectrin layer by sheets of elastic continuum. We neglect other forms of energy, such as electrostatic interactions. While these may not be small, their effects will usually be included in the measured "elastic" parameters. The lipid layers have no resistance to shear. We consider the two lipid layers to be either connected or unconnected in the sense of EVANS [23]. We assume that each layer is flat when unstressed, and that the inner layer and the spectrin (if present) adhere at all times without sliding. The inner lipid layer and the spectrin have the same areas when unstressed, and their actual area exceeds their area when unstressed by

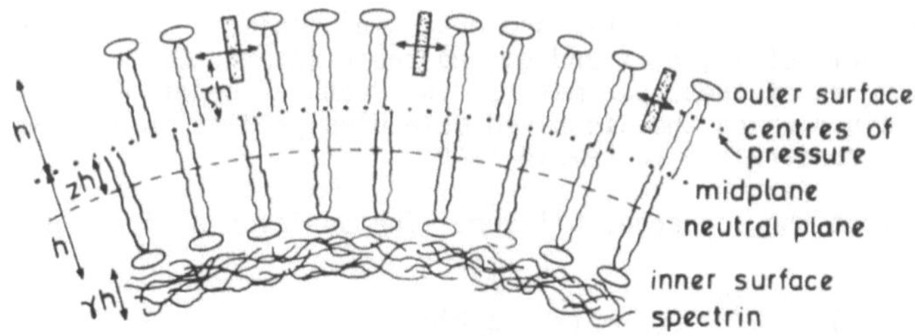

Fig.1 Two lipid layers, each of thickness h, with a layer of spectrin of thickness γh. Absorbed molecules exert their centres of pressure at a height ζh above the midplane of the bilayer. The neutral surface for bending the bilayer is a distance zh below the midplane.

a fraction Δ_i. The area of the outer layer exceeds that which it has when unstressed by a fraction Δ_o; we do not assume that the inner and outer layers are unstressed at the same value of Δ. The layers may be unconnected, so that Δ_o and Δ_i are independent, or connected, so that one layer cannot slide over the other.

If the membrane is bent, each layer has the same curvature (sum of principal curvatures) K. When each layer bends under the action of couples but no resultant tangential tractions, it has a neutral surface in which there is no change of dimensions. The height of this neutral surface above the midplane of the lipid bilayer is $z_o h$ for the outer lipid layer, $z_i h$ for the inner lipid layer coupled to the spectrin, and $z_c h$ for a connected membrane in which one lipid layer cannot slide over the other. We define Δ_o and Δ_i to be the dilatations in the corresponding neutral surfaces. For a connected membrane,

$$d(\Delta_o - \Delta_i)/dK = (z_o - z_i)h. \qquad (1)$$

We assume that when unit thickness of the material of a lipid layer is stretched in its own plane to an areal strain of δ (its thickness being free to reach the configuration of lowest energy) it acquires an energy per unit area $\frac{1}{2}m_L\delta^2$. Then in this state it has an isotropic tension $m_L\delta$ per unit thickness. The elastic modulus for stretching the spectrin-free lipid bilayer in its plane is

$$A_L = 2m_L h, \qquad (2)$$

while the bending modulus of the connected spectrin-free bilayer is

$$k = \tfrac{2}{3}m_L h^3 . \qquad (3)$$

The elastic modulus for stretching the spectrin layer in its plane is similarly

$$A_S = \gamma\, m_S h. \qquad (4)$$

If we write $A_S/A_L = \lambda$ (5)

we find $m_S = (2\lambda/\gamma)m_L$. (6)

We can now express the elastic properties of connected and unconnected membranes in terms of m_L, h, λ and γ, and their interaction with absorbed molecules in terms of these parameters, the depth of penetration ζ, and the areal strain Δ which the dissolved molecules produce in a lipid monolayer which is constrained to remain flat.

Consider first the connected bilayer. The position z of the neutral surface is given by

$$\int_{-h}^{h} m_L(x - z_c h)dx + \int_{-(\gamma+1)h}^{-h} m_S(x - z_c h)dx = 0, \tag{7}$$

which with the aid of (6) gives

$$z_c = -\lambda(2 + \gamma)/2(1 + \lambda). \tag{8}$$

The bending modulus k_c may be obtained by considering the energy stored in bending to unit radius. We find

$$k_c = \int_{-h}^{h} m_L(x - z_c h)^2dx + \int_{-(\gamma+1)h}^{-h} m_S(x - z_c h)^2dx. \tag{9}$$

Using (3), (6) and (8), this becomes

$$k_c/k = E/4(1 + \lambda), \tag{10}$$

where $E = 4 + 16\lambda + 12\,\lambda\gamma + 4\lambda\gamma^2 + \lambda^2\gamma^2$. (11)

Now consider the unconnected bilayer. For the single lipid layer

we find $$z_o = \tfrac{1}{2} \tag{12}$$

and $$k_o/k = \tfrac{1}{8}. \tag{13}$$

For the inner layer connected to the spectrin we find that the position z_i of the neutral surface above the midplane of the lipid bilayer is given by

$$\int_{-h}^{0} m_L(x - z_i h)dx + \int_{-(\gamma+1)h}^{-h} m_S(x - z_i h)dx = 0, \tag{14}$$

which leads to

$$z_i = -(1 + 4\lambda + 2\lambda\gamma)/2(1 + 2\lambda). \tag{15}$$

The bending modulus of the inner lipid layer connected to the spectrin is given by

$$k_i = \int_{-h}^{0} m_L(x - z_i h)^2dx + \int_{-(\gamma+1)h}^{-h} m_S(x - z_i h)^2dx. \tag{16}$$

We evaluate this expression with the aid of (3), (6) and (15), and obtain

$$k_i/k = F/8(1 + 2\lambda)^2, \tag{17}$$

where $F = 1 + 10\lambda + 12\lambda\gamma + 8\lambda\gamma^2 + 16\lambda^2 + 24\lambda^2\gamma + 20\lambda^2\gamma^2 + 8\lambda^3\gamma^2$. (18)

We shall be concerned with a membrane of area $4\pi R^2$, of which a fraction ρ has a curvature K_ρ and a fraction $1 - \rho$ a curvature $\bar{K}_\rho$. The areal extensions Δ_o and Δ_i of the outer and inner layers are each uniform throughout the membrane. The mean curvature is $\rho K_\rho + (1 - \rho)\bar{K}_\rho$, and in accordance with

(1), (12) and (15) we may write

$$\Delta_o - \Delta_i = [\rho K_\rho + (1 - \rho)\bar{K}_\rho - K_r]\, jh/(1 + 2\lambda), \tag{19}$$

where $\quad j = 1 + 3\lambda + \lambda\gamma.$ (20)

Here the term K_r represents the existence of a spontaneous curvature; Δ_o and Δ_i do not vanish if the membrane is kept connected, and flattened. The curvature K_0 of the relaxed membrane minimizes the sum of stretching and bending energies and is less than K_r.

Since we believe that there is no hydrostatic pressure within the cell, the total tension in the outer and inner layers must vanish. Thus

$$\Delta_o + (1 + 2\lambda)\Delta_i = 0. \tag{21}$$

Solving (19) and (21) we find

$$\Delta_o = -(1 + 2\lambda)\Delta_i = [\rho K_\rho + (1 - \rho)\, \bar{K}_\rho - K_r]\, jh/2(1 + \lambda). \tag{22}$$

The energy associated with the stretching of the two layers is

$$E_s = 4\pi R^2\, \tfrac{1}{2} m_L h\, [\Delta_o{}^2 + (1 + 2\lambda)\Delta_i{}^2]$$

$$= \pi m_L h^3 [\rho\kappa_\rho + (1 - \rho)\bar{\kappa}_\rho - \kappa_r]^2\, j^2/(1 + \lambda)\,(1 + 2\lambda), \tag{23}$$

where $\kappa_\rho = K_\rho R$, etc. (24)

The bending energy is

$$E_b = 4\pi R^2 \tfrac{1}{2} [\rho K_\rho{}^2 + (1 - \rho)\bar{K}_\rho{}^2](k_o + k_i)$$

$$= \pi m_L h^3 [\rho\kappa_\rho{}^2 + (1 - \rho)\bar{\kappa}_\rho{}^2] G/(1 + \lambda)(1 + 2\lambda), \tag{25}$$

where $G = \frac{1}{3}(1 + 5\lambda + 6\lambda\gamma + 4\lambda\gamma^2 + 2\lambda^2\gamma^2)(1 + \lambda).$ (26)

The total energy is $E_s + E_b$.

We now provisionally determine κ_r by requiring that for a uniform membrane with $\rho = 1$ the total energy is least when κ_ρ is equal to κ_0, the spontaneous curvature. This yields

$$\kappa_r/\kappa_0 = H/3j^2, \tag{27}$$

where $H = (1 + 2\lambda)\,(4 + 16\lambda + 12\lambda\gamma + 4\lambda\gamma^2 + \lambda^2\gamma^2).$ (28)

It is easily verified that $\kappa_r > \kappa_0$ when $\lambda \geq 0$ and $\gamma \geq 0$.

We now consider N foreign molecules absorbed into the outer lipid layer. Our model of such a molecule is an incompressible cylinder which extends from the outer surface of the membrane to a depth $2(1 - \zeta)h$ where $\frac{1}{2} \leq \zeta \leq 1$. We specify the cross-sectional area of the molecule by saying that when c molecules are absorbed in unit area of a free lipid monolayer, the area increases by $c\Delta$. Each molecule has an elastic energy of absorption; there may be an energy of elastic interaction between neighbouring absorbed molecules; there is an elastic interaction between the strain field of each molecule and the stress field existing in the lipid layer. We shall not be concerned with the energy of a molecule in an unstressed lipid layer, which plays a part in determining the partitioning of molecules between the membrane and the surrounding medium. There may well be an elastic interaction between absorbed molecules [24]. We do not know how to calculate it, it will vanish when the concentration of absorbed molecules is small, and we neglect it. To estimate the interaction energy between an absorbed molecule and the stress field in the membrane we note that this stress field is a linear function of distance from the midplane. If the hydrostatic pressure at a height ζh above the midplane is $p(\zeta)$, the interaction energy e_i of a molecule with the stress field is simply $p(\zeta)$ times the volume of the molecule, or

$$e_i = 2(1 - \zeta)\, h\Delta p(\zeta). \tag{29}$$

The stress at height ζ arises both from the stretching of the outer lipid layer Δ_0 and, from its bending, $Kh(\zeta - z_0)$. Using (12), (22) and (29), we find that the interaction energy of a single molecule in a region of curvature K in the membrane already considered (where $K = K_\rho$ or $K = \bar{K}_\rho$) is

$$e_i = -2m_L h^2 (1 - \zeta)\Delta\{[\rho K_\rho + (1 - \rho)\bar{K}_\rho - K_r]j/2(1 + \lambda) + K(\zeta - \tfrac{1}{2})\}. \tag{30}$$

Now suppose that the concentration of absorbed molecules in the fraction ρ of the membrane is c_ρ, while the concentration in the fraction $1 - \rho$ is $\bar{c}_\rho$. Then the total interaction energy E_i of the molecules with the stress field of the lipid layers is

$$E_i = -[4\pi(1 - \zeta)jm_L h^2\Delta R/(1 + \lambda)] \times$$

$$\times \{\rho c_\rho[(\rho+\beta)\kappa_\rho + (1-\rho)\bar{\kappa}_\rho - \kappa_r] + (1-\rho)\bar{c}_\rho[\rho\kappa_\rho + (1-\rho+\beta)\bar{\kappa}_\rho - \kappa_r]\}, \tag{31}$$

where $\beta = 2(\zeta - \tfrac{1}{2})(1 + \lambda)/j$. (32)

We shall consider two extreme cases. The first is that of molecules such as detergents, with a chain length roughly equal to that of a lipid molecule. Here $\zeta = \tfrac{1}{2}$, $\beta = 0$. The second is that of small molecules such as anaesthetics, for which $\zeta \simeq 1$, and we take $(1 - \zeta)\Delta$ to have a finite limit D, while $\beta = (1 + \lambda)/j$. When $\beta = 0$, E_i depends only on the mean concentration and the mean curvature. There is no coupling between the spatial distribution of concentration and the spatial distribution of curvature.

3. The Model and Its Qualitative Behavior

We first consider the case of crenation, in which the outer lipid layer is expanded with respect to the inner. We specify that the area of the membrane, after it has absorbed the crenating molecules, is $4\pi R^2$. We approximate the initial form of the cell by a sphere of that radius. HELFRICH and DEULING [25] showed that bending energy alone would make the biconcave shape of a red blood cell the form of minimum energy of the unperturbed cell only if the membrane had a natural curvature K_0 which was opposed to the actual mean curvature of the cell. Suppose the number of crenating molecules is N. We consider the formation of a group of ν spicules, which are modelled by hemispheres of radius r ($r \ll R$) (Fig.2). The total area of the membrane is assumed to remain constant during this process.

We write $r = \rho_\nu R$. (33)

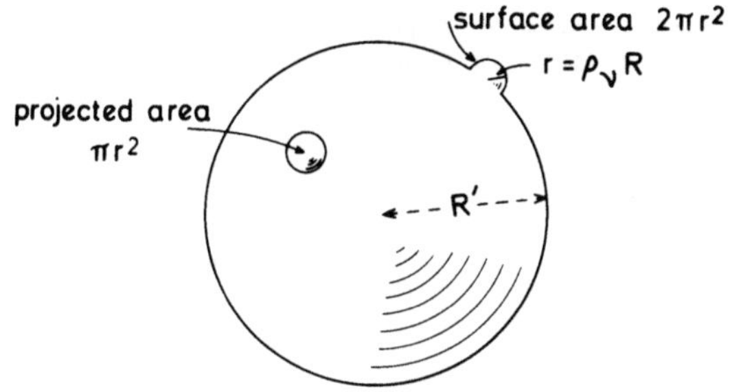

Fig.2 A cell with two spicules, each a hemisphere of radius r. The radius R' of the cell is now less than its initial radius R.

If the distribution of molecules was controlled predominantly by their elastic energy, and the molecules are not so long that $\beta = 0$, their concentration in the spicules could be much higher than that in the bulk of the membrane, so that the induced curvature of the spicules would be close to their actual curvature, $2/\rho_\nu R$. However, the bending modulus k of the membrane is very small. HELFRICH and DEULING calculate the total bending energy of the red cell in its equilibrium form to be $8\pi k(1 - \frac{1}{2}\kappa_0)^2$; they estimate that $\kappa_0 = -3$, and we may take for k the value 2.3×10^{-12} erg obtained [13] for egg lecithin. We find the bending energy to be 3.6×10^{-10} erg, or 8400 times $k_B T$, the thermal energy at T = 310 K. Here k_B is Boltzmann's constant, 1.38×10^{-16} erg K^{-1}. This energy can displace 8400 absorbed molecules into a region of appreciably higher concentration. It is difficult to estimate how many molecules are absorbed during the process of crenation. Experimentally, PONDER [14(p.35)] estimated about 10^7 molecules of a long-chain detergent, while SEEMAN [26] gives 5×10^{13} anaesthetic molecules cm^{-2} as the limit of "saturation", which with a membrane of surface area 150 μm^2 gives 7.5×10^7 anaesthetic molecules. These estimates are roughly compatible, but one cannot be certain that all of the molecules are fully intercalated into the membrane. The recent measurement by MOHANDAS and FEO[27] that 10.8×10^8 molecules of a phenothiazine derivative per cell are necessary to induce the prolytic state makes it clear that not all of the molecules associated with the cell in their experiments can be intercalated into the outer layer of the membrane, since this outer layer contains only about 2.5×10^8 lipid-like molecules. If we assume that the total number of absorbed molecules involved in the crenation of a red blood cell is 10^6 or 10^7, we see that the elastic energy could displace only about one molecule in 120 or 1200 if the concentration in the spicules was several times greater than that in the rest of the membrane.

Consider the case in which the concentration of absorbed molecules in the spicules is ℓ times that in the initial uniformly loaded sphere. The total number of absorbed molecules in the spicules is $\nu\ell N\rho_\nu^2/2$, and the number remaining in the spherical surface is $N(1 - \nu\ell\rho_\nu^2/2)$. The concentration in the spherical surface is $N(1 - \nu\ell\rho_\nu^2/2)/4\pi R^2(1 - \nu\rho_\nu^2/2$, and the radius of the spherical surface is R', where it can be seen from Fig. 2 that

$$R'^2 = R^2(1 - \nu\rho_\nu^2/4), \tag{34}$$

provided that $\rho_\nu \ll 1$.

We may now insert the parameters of this model into the expressions for the elastic energy developed in §2. We have

$$\rho = \tfrac{1}{2}\nu\rho_\nu{}^2, \tag{35}$$

$$\kappa_\rho = (2\nu/\rho)^{\frac{1}{2}}, \tag{36}$$

$$c_\rho = \ell N/4\pi R^2, \tag{37}$$

$$\bar{\kappa}_\rho = 2/(1 - \rho/2)^{\frac{1}{2}}, \tag{38}$$

and

$$\bar{c}_\rho = N(1 - \ell\rho)/4\pi R^2(1 - \rho). \tag{39}$$

We write

$$x = (2\nu\rho)^{\frac{1}{2}} \tag{40}$$

and

$$\eta = \pi m_L h^3/(1 + \lambda)(1 + 2\lambda). \tag{41}$$

Then

$$E_s/\eta = j^2[x + 2(1 - \rho)/(1 - \tfrac{1}{2}\rho)^{\frac{1}{2}} - \kappa_r]^2, \tag{42}$$

$$E_b/\eta = G[x^2/\rho + 4\ (1 - \rho)/(1 - \tfrac{1}{2}\rho)], \tag{43}$$

and

$$E_i/\eta = -n\{(1 + \beta\ell)\ x + 2[1 - \rho + \beta - \beta\ell\rho]/(1 - \tfrac{1}{2}\rho)^{\frac{1}{2}} - \kappa_r\}, \tag{44}$$

where

$$n = j(1 + 2\lambda)\ (1 - \zeta)\Delta N/\pi Rh. \tag{45}$$

In deriving this result, we have neglected the energy of the junctions between the spicules and the main sphere. These junctions are not abrupt, but each consists of a transition region of area less than that of the hemispherical cap, and of smaller curvature, so that its energy is a modest fraction of that of the hemisphere.

The formation of a spicule requires large shear strains. The elastic behaviour under large shears is non-linear. From the data of EVANS [28], we see that if the elastic constant for small shear strains of the membrane is μ, a hemispherical bulge of radius r can be formed by suction into a pipette of the same radius by a negative pressure of $0.75\ \mu/r$. We estimate the work done to be about $E' = (2\pi r^3/9)\ (0.75\ \mu/r) \simeq \tfrac{1}{2}\mu r^2$, and add

$$E'/\eta = m\rho, \tag{46}$$

where

$$m = (1 + \lambda)(1 + 2\lambda)\mu R^2/\pi m_L h^3. \tag{47}$$

In addition, the crenated sphere has $N\ell\rho$ molecules at a concentration $\ell N/4\pi R^2$ and $(1 - \ell\rho)N$ molecules at a concentration $N(1 - \ell\rho)/4\pi R^2(1 - \rho)$, while the uncrenated sphere had N molecules at a concentration $N/4\pi R^2$. This non-uniform distribution gives an entropic contribution to the free energy of

$$E_e = Nk_BT\{\ell\rho\ \ln\ell + (1 - \ell\rho)\ln[(1 - \ell\rho)/(1 - \rho)]\}. \tag{48}$$

We write

$$E_e/\eta = 2\alpha n\ \{\ell\rho\ \ln\ell + (1 - \ell\rho)\ \ln\ [(1 - \ell\rho)/(1 - \rho)]\}, \tag{49}$$

where

$$\alpha = (1 + \lambda)\ Rk_BT/2\ j\ m_L\ h^2(1 - \zeta)\Delta. \tag{50}$$

The total free energy is

$$F(\rho, x, \ell) = E_s + E_b + E_i + E' + E_e. \tag{51}$$

The increase in free energy on forming the crenated sphere from the uncrenated sphere is given by

$$F_c/\eta = F(\rho, x, \ell)/\eta - F(0, 0, 1)/\eta \tag{52}$$

$$= j^2\{[x + 2(1 - \rho)/(1 - \tfrac{1}{2}\rho)^{\frac{1}{2}} - \kappa_r]^2 - [2 - \kappa_r]^2\}$$

$$+ G[x^2/\rho - 4\rho/(2 - \rho)]$$

$$-n\{(1 + \beta\ell)x + 2[1 - \rho + \beta - \beta\ell\rho]/(1 - \tfrac{1}{2}\rho)^{\frac{1}{2}} - 2(1 + \beta)\}$$

$$+ m\rho + 2\alpha n\{\ell\rho \ln \ell + (1 - \ell\rho)\ln[(1 - \ell\rho)/(1 - \rho)]\}. \tag{53}$$

The equilibrium state of the cell is obtained by minimizing F_c with respect to variations of ρ, x and ℓ. We first minimize with respect to x, obtaining

$$(2\nu\rho)^{\frac{1}{2}} = x = \tfrac{1}{2}I(\rho)\rho\{n(1 + \beta\ell) - 2j^2[2(1 - \rho)/(1 - \tfrac{1}{2}\rho)^{\frac{1}{2}} - \kappa_r]\}, \tag{54}$$

and

$$F_c/\eta = j^2\{GI(\rho)[2(1 - \rho)/(1 - \tfrac{1}{2}\rho)^{\frac{1}{2}} - \kappa_r]^2 - (2 - \kappa_r)^2\}$$

$$- 4G\rho/(2 - \rho)$$

$$+ j^2(1 + \beta\ell)I(\rho)[2(1 - \rho)/(1 - \tfrac{1}{2}\rho)^{\frac{1}{2}} - \kappa_r]\rho n$$

$$- 2[(1 - \rho + \beta - \beta\ell\rho)/(1 - \tfrac{1}{2}\rho)^{\frac{1}{2}} - (1 + \beta)]n$$

$$- \tfrac{1}{4}(1 + \beta\ell)^2 I(\rho)\rho n^2 + m\rho$$

$$+ 2\alpha n\{\ell\rho \ln \ell + (1 - \ell\rho)\ln[(1 - \ell\rho)/(1 - \rho)]\}, \tag{55}$$

where

$$I(\rho) = 1/(G + j^2\rho). \tag{56}$$

When $\beta > 0$ and the concentration of absorbed molecules is high, our model predicts that the energy can be reduced indefinitely if the absorbed molecules are concentrated into a large number of very small spicules, which cover only a small part of the total surface of the cell. We let $\ell \to \infty$, $\ell\rho \to 1$. Then all the terms in (55) are bounded, except for the term in n^2, which $\sim -\ell n^2\beta^2/4G$, and the entropy term, which $\sim 2\alpha n \ln \ell$. The former dominates. Also $x \sim n\beta/2G$, $\nu \sim \ell n^2\beta^2/8G^2$ and $\rho_\nu \sim 4G/\ell n\beta$. These values are all compatible with the assumptions of our model. A physical limit is set by the requirement that the radius of a spicule, $\rho_\nu R$, should be appreciably greater than the thickness of a lipid layer, h. This requires

$$\rho_\nu > h/R \simeq 10^{-3} \tag{57}$$

and

$$\ell < 4GR/n\beta h. \tag{58}$$

Even with this restriction on ℓ, the numerical value of the term in n^2 in (55) is $-n\beta R/h$. With $R/h \simeq 10^3$, this term usually dominates unless $n\beta$ is very small. In general, the stable state (which may not necessarily be readily achieved) is one in which the foreign molecules are concentrated into small spicules, which in practice bud off as microvesicles. The experimental evidence (on cells crenated by addition of calcium ions or depletion of ATP) (e.g. [29-31]) shows that such microvesicles are indeed formed, but that they are free of spectrin. This possibility is not included in our theory. In principle, quantitative experiments could be analyzed to estimate the cohesive energy between spectrin and the lipid bilayer.

We now seek a realistic upper limit to n. It follows from (45) that the total cross-sectional area of the absorbed molecules is given by

$$\Delta N = \pi Rhn/j(1 + 2\lambda)(1 - \zeta). \tag{59}$$

With the typical values $\lambda = 2$, $\gamma = 0$, $\zeta = \frac{3}{4}$ this becomes

$$\Delta N = 4\pi Rhn/35. \tag{60}$$

When n = 1, this cross-sectional area is about 1/35 of the area of lipid exposed in a central section of the cell. The curvature induced in a connected membrane is $\Delta N/4\pi R^2(\zeta - z_c)h$. With the parameters just selected, this becomes n/50R. The model breaks down unless ΔN is substantially less than the surface area of the cell, $4\pi R^2$. Thus

$$n < 35\ R/h \simeq 3.5 \times 10^4. \tag{61}$$

This condition is always satisfied in our calculations. However, if a spicule is enriched in absorbed molecules by a factor ℓ, we must also satisfy (58). With the parameters just selected, this becomes

$$\ell n < 2 \times 10^5. \tag{62}$$

Numerical calculations with plausible values of the parameters show that this region of negative F_c with ℓ large and $\ell\rho \simeq 1$ does not appear for $\ell < 101$ unless n is rather large. When $\beta > 0$, a minimum of F_c appears in the range $0 < \rho < 1$, $1 < \ell < 2$, the value of ρ at the minimum increasing with increasing n, until the minimum leaves the region $0 \leq \rho \leq 1$, $\ell \geq 1$ at the point (1,1). The energy contours around this minimum, or the contours remaining when the minimum has disappeared, are joined to the contours representing a decrease of F_c as $\ell \to \infty$ and $\ell\rho \to 1$ by a saddle point. As n increases, this saddle point moves in the direction of decreasing ℓ. The model thus predicts that, as n increases, the cell develops crenations in a reversible manner. When n reaches a value which is large, an irreversible transition occurs to a region with ℓ large, though our formulae are valid only if (62) is satisfied, and (62) is violated by a factor 3-4 in the latter region.

The occurrence of reversible crenation, followed by an abrupt and irreversible transition as n increases, corresponds to the abrupt appearance of the prolytic sphere as described by PONDER [14 (pp. 28-29, 35-37)]:- "After half a minute or so, some of the cells begin to crenate ... the crenations become smaller and more numerous until they cover the whole surface and the discoid shape is lost; as the cell turns over in the fluid, it will appear no longer as a crenated disc, but as a crenated sphere. This stage does not last long, for the crenations become finer and finer and soon are not resolvable as crenations at all; the cell then has the appearance of a smooth glistening sphere. Quite quickly, this stage is succeeded by that of the prolytic sphere; the glistening property of the surface is replaced by a uniform duskiness." "... to bring about perfect sphering ... . Up to this point, the shape change is reversible ... the sphere then passes into the prolytic sphere ... . These last stages of the process are irreversible ... . Forms such as the crenated disk, the crenated sphere, and the smooth spherical form ... change to forms with either more, or less, development of surface for the same volume ... according to the conditions established in their environment ... once the prolytic sphere has been formed, however, the effects seem to be irreversible, and attempts to dislodge the sphering agent from the cells result in hemolysis".

4. Estimation of the Parameters

For the thickness h of a single lipid layer we take 3.7 nm, and for the radius R of the sphere which has the same area as the cell 3.6 μm [32] (JAY's measurement of the surface area [33] leads to R = 3.3 μm). HELFRICH and DEULING [25] estimate the intrinsic curvature K_0 of the bilayer, measured in units of 1/R, to be -3.

The bending modulus of the bilayer k has been defined as that of the connected spectrin-free bilayer. The value 2.3×10^{-12} erg was obtained [13] for egg lecithin by observing ripple-like Brownian motions, and may represent the unconnected modulus with matter flowing freely between the crests and the troughs of the ripples. We would then have $k = 4 \times 2.3 \times 10^{-12} = 9.2 \times 10^{-12}$ erg for the connected spectrin-free bilayer. In experiments in which they destroyed the "structural matrix", which we identify with the layer of spectrin, EVANS et al. [9] found for the compressive modulus of the bilayer $A_L = 95$ dyne cm^{-1}. From (2) and (3) with h = 3.7 nm we obtain $k = 4.3 \times 10^{-12}$ erg. However, the optical flicker observations of BROCHARD and LENNON [12] on whole red blood cells led them to lower values of k even in the presence of the spectrin layer. From the correlation length of the fluctuations they estimated a bending modulus for the whole membrane (presumably unconnected) of $k_o + k_i = 2.3 \times 10^{-13}$ erg. This value is surprisingly low, since [34] the experimental value for egg lecithin is of the order expected theoretically. Perhaps the natural membrane, being less ordered, is more flexible. On the other hand, BROCHARD and LENNON's calculated mean amplitude of motion in the middle of the cell, with an effective bending modulus of 5×10^{-13} erg [12,Eq.(3.2)], is 0.3 μm [12,Eq.(3.7)]. The observed value [12,Eq.(2.5)] is 0.08 μm, corresponding to an effective modulus of $5 \times 10^{-13} \times (0.3/0.08)^2 = 7 \times 10^{-12}$ erg. We would expect the effective modulus for this motion to be about $k_o + k_i$. On the assumption that the spectrin layer is strong but thin, with $\lambda = 2$ and $\gamma = 0$, (13), (17) and (18) lead to $k = 6.85(k_o + k_i)$, so that the two interpretations of the observations of BROCHARD and LENNON would give $k = 1.6 \times 10^{-12}$ erg and 4.8×10^{-11} erg. For a strong and thick spectrin layer with $\lambda = 2$ and $\gamma = 3$, we find $k = 0.498(k_o + k_i)$, leading to $k = 2.5 \times 10^{-13}$ erg and 3.5×10^{-12} erg. The extreme estimates are 2.5×10^{-13} erg and 4.8×10^{-11} erg, with the other estimates lying between 1.6×10^{-12} erg and 9.2×10^{-12} erg. The most acceptable estimate seems to be $k = 4 \times 10^{-12}$erg, within a factor of 2; the calculations of §5 assume $k = 4.336 \times 10^{-12}$erg. A possible complication is that there might be sufficient "slack" in the spectrin network for it to make little contribution to the energy of Brownian distortions, while contributing appreciably to the energy of the larger distortions involved in crenation.

Electron micrographs of replicas of the inner face of the membrane [7] suggest that the spectrin forms a thin network with $\gamma \approx 0$, while the micrograph by MARCHESI et al. [6] of sectioned ghosts suggests that the spectrin layer is not thin, but extends to a depth of about 3h into the cytoplasm, corresponding to $\gamma = 3$. For the ratio λ of the areal modulus A_S of the spectrin to that of a double layer of lipid A_L we use the results [9] $A_S = 193$ dyn cm^{-1}, $A_L = 95$ dyn cm^{-1}, leading to $\lambda = 2$. For the shearing modulus μ of the spectrin layer we take the value [8] of 7×10^{-3} dyn cm^{-1}. The ratio ζ of the distance of the centre of pressure of an absorbed molecule from the midplane of the bilayer to the thickness h of a single lipid layer may range from almost 1 for a small anaesthetic molecule to almost $\frac{1}{2}$ for a molecule of lysolecithin.

We do not attempt to calculate the effective surface area Δ or the effective volume of an absorbed molecule from first principles, since, as is emphasized by SEEMAN [26] for biological membranes, the expansion of a lipid layer by an absorbed molecule is about 10 times that which would be estimated from van der Waals volume of the molecule if the layer expanded isotropically. HENDRICKSON [35] has shown that typical local anaesthetic molecules have partial molecular surface areas Δ in a phosphatidylcholine monolayer of between 0.66 and 2.15×10^{-14} cm^2. Multiplying by the monolayer thickness of 3.7×10^{-7} cm, assumed constant, we find large effective volumes of about 4×10^{-21} cm^3. For example, SEEMAN estimates the van der Waals volume of the molecule of procaine to be $2.4 \times 10^{-22} cm^3$, whereas HENDRICKSON's measurement corresponds to about $2.15 \times 10^{-14} \times 3.7 \times 10^{-7} = 8 \times 10^{-21}$ cm^3, or 34 times the van der Waals estimate. TRUDELL [36] has produced evidence that increases in the area of the membrane are accompanied by decreases in the thickness, so that the increase in volume is comparable to that predicted from the van der Waals volume of the dissolved molecules. The difficulty in estimating the increase in area a priori remains. We may compare HENDRICKSON's values of Δ with those derived from SEEMAN's studies of whole blood cells. For a membrane "saturated" with anaesthetic 5×10^{13} absorbed molecules cm^{-2} produce an areal strain of about 6 per cent. If we assume that the molecules are absorbed only in the outer layer, the areal expansion of a free lipid layer would be 12 per cent for 5×10^{13} absorbed molecules cm^{-2}. Hence $\Delta = 0.12/5 \times 10^{13} = 0.24 \times 10^{-14}$ cm^2, a value about $\frac{1}{4}$ that estimated for a lipid monolayer. As in the experiments of MOHANDAS and FEO [27] the foreign molecules associated with the membrane may at these high concentrations not all be fully absorbed into the membrane, and we shall take $\Delta = 10^{-14}$ cm^2. In estimating the effective volume of a small molecule with $\zeta \simeq 1$, we assume that the surrounding lipid molecules are forced apart for a quarter of their length, leading to an effective volume of $\frac{1}{4}h\Delta = 9 \times 10^{-22}$ cm^3.

Finally, the ratio of the actual number of absorbed molecules N to the normalized number n is given by (45) as $N/n = \pi Rh/j(1 + 2\lambda)(1 - \zeta)\Delta$, and when $\zeta \simeq 1$ we take $\zeta = \frac{7}{8}$. Numerically,

$$\frac{N}{n} = \frac{4.2 \times 10^4}{(1+3\lambda+\lambda\gamma)(1+2\lambda)(1-\zeta)}. \tag{63}$$

5. Numerical Calculations

We take two models containing spectrin, with the membrane and absorbed molecule properties estimated in §4, with $\lambda = 2$ in both models. In one model the spectrin layer is very thin, and $\gamma = 0$, while for the other model the spectrin layer is thick, and $\gamma = 3$. The absorbed molecules are assumed to be short, with $\zeta = \frac{7}{8}$ and an effective volume $\frac{1}{4}h\Delta = 9 \times 10^{-22}$ cm^3. The parameters in the two models are then

$$\alpha = 150,\ \kappa_r = -3.67,\ m = 666,\ \beta = 0.321,\ \lambda = 2,\ \gamma = 0,\ N/n = 9600 \text{ (model 1)}$$

and $\alpha = 80.9, \kappa_r = -6.39, m = 666, \beta = 0.173, \lambda = 2, \gamma = 3, N/n = 5170$ (model 2).

The results are displayed in the computer plots of Fig.3, in which the

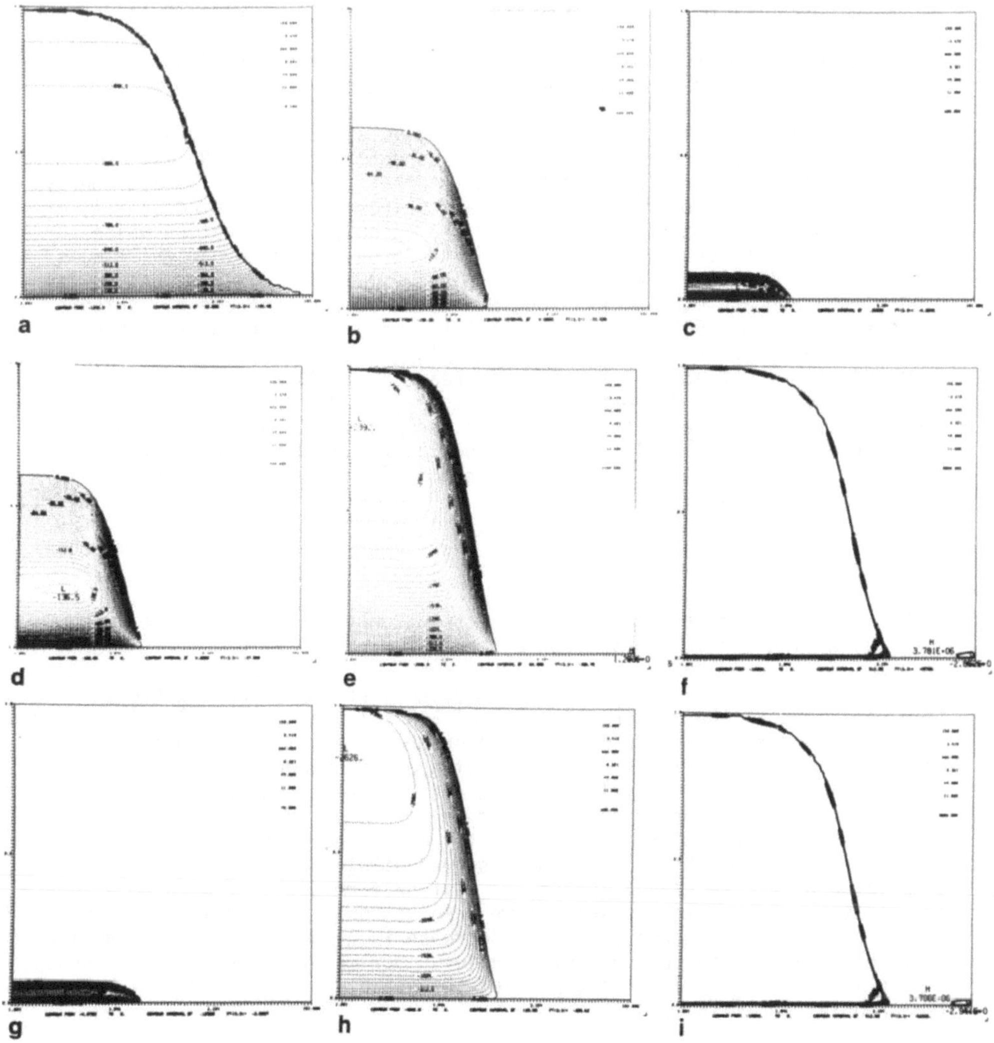

(a) Model 1, $N = 1.1\times10^5$	(b) Model 1, $N = 1.9\times10^6$	(c) Model 1, $N = 5.8\times10^6$
(d) 1, $N = 6.7\times10^6$	(e) 1, $N = 1.1\times10^7$	(f) 1, $N = 7.7\times10^7$
(g) 3. $N = 6.7\times10^5$	(h) 3, $N = 5.8\times10^6$	(i) 3, $N = 7.7\times10^7$

Fig. 3(a) - (i). Contours of energy in the plane of $\log_{10}(\ell - 1)$ and ρ. Contours are drawn only where a crenated form has lower energy than the unperturbed form. N is the number of absorbed molecules.

increase in energy on forming crenations given by (55) is plotted as contours in the plane of $\log_{10}(\ell -1)$ and ρ. Only those contours for which this increase is negative are plotted, and the plots are blank for situations in which the unperturbed form has lower energy than any crenated form.

For the first model (thin spectrin, negative spontaneous curvature), crenated forms appear even for the lowest concentrations n. No minimum appears, and the contours (Fig. 3a,b) suggest that the state of lowest energy is one in which the absorbed molecules are uniformly distributed ($\ell = 1$) or have a reduced concentration ($\ell < 1$) in the spicules, a situation which our plots cannot represent. The fraction of the area covered by spicules ρ is about 0.66 at the lowest concentrations (n = 0.003, N = 29), decreasing to 0.44 at n = 200 and 0.02 at n = 300, N = 2.9×10^6. There are no contours for n = 400 and n = 500, indicating that the unperturbed form is stable. Contours re-appear for n = 600, N = 5.8×10^6, with a minimum at $\rho_\nu = 0.22$, $\rho = 0.04$, $\ell = 1.007$ and $\nu = 2$. As n increases to 1100 (Fig. 3c-e), the sizes of the spicules decrease slightly to $\rho_\nu = 0.20$, the fraction of the surface covered increases to $\rho = 0.82$, ℓ decreases to 1.002 and the number of spicules increases to 43. These trends continue, until (Fig. 3f) at n = 8 000, N = 7.7×10^7, a new deep minimum of energy appears. At n = 10^4 this minimum represents the presence of 8.5×10^5 spicules with $\rho_\nu = 0.00016$, covering only 0.01 of the surface area, with a concentration factor $\ell = 89$. The contour of zero energy links the regions containing the two minima when n $\geq$ 13 000.

For the second model, with thick spectrin, only the first regime, with spicules having $\ell \leq 1$, appears up to n = 10 000, N = 5×10^7.

The first model is satisfactory in showing a region of reversible crenation beginning at the reasonable value of 5.8×10^6 absorbed molecules, with a first-order transition to another region when the number of absorbed molecules reaches 7.7×10^7. The presence of crenation when the number of absorbed molecules is very small, and its absence in the range N = 3.8 - 4.8 $\times 10^6$,are unsatisfactory features. The reason lies in our choice of κ_r, which was based on [25] the value of κ for which the resting form of the normal cell with n = 0 would be the observed dimpled disc. If we choose this value of κ_r, we cannot expect the resting form with n = 0 to be a sphere. The finite number of irregularities we predict for n = 0 are indentations, not spicules. As n increases, these decrease in number, disappear, and are replaced by increasing numbers of spicules. Our model, which is based on the assumption that $\rho_\nu \ll 1$, cannot predict the actual resting form, which has two indentations with $\rho_\nu \simeq 1$.

A satisfactory calculation would combine the analysis [25] of the large-scale deformation of the cell with ours of the small-scale irregularities. The complications do not seem justified at the present time. We content ourselves with an approximation in which κ_r is chosen in such a way that there is no tendency for dimples or pimples to form when n = 0. That is to say, we require that when n = 0, $\partial F_c/\partial \rho = 0$ when $\rho = 0$. This yields

$$\kappa_r = 2 + [(4mG + G^2)^{\frac{1}{2}} - 3G]/2j^2. \tag{64}$$

The contours of energy in the plane of $(\log_{10}(\ell - 1),\rho)$ now represent the observed behaviour of the red cell qualitatively.

The third model is similar to the first, with a thin strong layer of spectrin, but with a spontaneous positive curvature given by (64). Its parameters are

$$\alpha = 150,\ \kappa_r = 3.41,\ m = 666,\ \beta = 0.321,\ \lambda = 2,\ \gamma = 0,\ N/n = 9\,600 \text{ (model 3)}.$$

There are no negative contours, and the unperturbed form is stable, up to $n = 50$, $N = 4.8 \times 10^5$. At $N = 6.7 \times 10^5$, one or two spicules appear with $\rho_\nu = 0.21$, $\ell = 1.008$, covering 0.03 of the surface. As N increases to 5.8×10^6, the radii of the spicules decrease very slightly, their number ν increases to 46, and the fraction of the surface covered increases to $\rho = 0.86$, while ℓ decreases to 1.002 (Figs. 3g,h). These trends continue up to $N = 4.8 \times 10^7$. At $N = 7.7 \times 10^7$ (Fig. 3i), a new deep minimum of energy appears, which at $N = 8.6 \times 10^7$ has 7.6×10^5 spicules with $\rho_\nu = 0.00017$, covering 0.01 of the area with $\ell \simeq 80$. The transition to these states is irreversible, and the model shows a threshold concentration, a range of concentrations within which crenation is reversible, and a range of irreversible crenation.

In fact [15] the spicules in the prolytic state are very fine. They are not, as PONDER believed, and our theory predicts, very numerous. We do not know how the cell crosses the potential barrier between the reversibly crenated state and the prolytic state. It may be that the rounded spicules, some tens in number, of the reversibly crenated state act as nuclei for the fine elongated spicules of the prolytic state, and so determine their number. At this stage the lipid separates from the spectrin in the spicules, our model breaks down, and myelin forms occur as described by DEULING and HELFRICH [37].

The fourth model, with a thick layer of spectrin and positive spontaneous curvature, has the parameters

$$\alpha = 80.9,\ \kappa_r = 2.49,\ m = 666,\ \beta = 0.173,\ \lambda = 2,\ \gamma = 3,\ N/n = 5170 \text{ (model 4)}.$$

It shows no crenation up to $N = 4.7 \times 10^6$, reversible crenation from $N = 5.2 \times 10^6$ to $N = 5.2 \times 10^7$, and no region of irreversible change up to this last concentration. This accords with the observation [38] that the red cells of the camel, which are strongly reinforced to resist hypo-osmotic haemolysis when the camel takes in huge draughts of water, are also very resistant to hyperosmotic crenation.

6. The Behaviour of Diacylglycerol

When blood is stored, the crenation of the red cells is associated with depletion of ATP. Crenation may also be caused by an excess of calcium ions in the cells. ALLAN and MICHELL [29] have shown that the crenation is mediated by an accumulation of 1,2-diacylglycerol inside the cells, with a similar fatty acid composition to that of human erythrocyte phosphatidylcholine. This behaviour fits well with our model, if we assume that diacylglycerol substitutes for phosphatidylcholine rather than being absorbed in a membrane layer as smaller molecules are. Diacylglycerol is a fusogenic lipid which, when applied externally to hen red cells, causes them to fuse to form multinucleated cells. Such lipids show negative deviations from linearity when the mean area per molecule is plotted against concentration for mixtures of the fusogenic lipid with phosphatidylcholine containing cholesterol [39]. The substitution of diacylglycerol for phosphatidylcholine inside the cell leads to a shrinkage of the inner lipid layer, and to the formation of echinocytes, as is observed, while diacylglycerol formed externally leads to invaginations, as is also observed.

More importantly from the point of view of our theory, the molecule of diacylglycerol, having a similar fatty acid composition to the phosphatidylcholine, is almost as long, but has a smaller hydrophilic head. It therefore acts as a negative wedge, and is expected to concentrate in the inner layers of the spicules. ALLAN et al. [30] concentrated the microvesicles budded off from red cells which had been crenated by the introduction of calcium ions. The microvesicle fraction, corresponding to our prolytic branch with $\ell \to \infty$, contained "only a fifth of the cell lipids but almost half of the newly formed diacylglycerol", representing a concentration factor $\ell = 2.5$.

References

1. P.J.Quinn: The Molecular Biology of Cell Membranes (Macmillan, London 1976)

2. A.W.Martin and F.A.Fuhrman: Physiol.Zool. 28, 18-34 (1955)

3. A.C.Burton: Physiology and Biophysics of the Circulation, 2nd ed. (Year Book Medical Publishers, Chicago 1972)

4. R.B.Pennell: In The Red Blood Cell, ed. by D.McN.Surgenor, 2nd ed., Vol. 1 (Academic Press, New York 1974) pp.93-146

5. S.B.Shohet and S.E.Lux: In Hematology of Infancy and Childhood (W.B.Saunders Co., Philadelphia 1974) Chap. 6

6. V.T. Marchesi, E.Steers, T.W.Tillock and S.C.Marchesi: In Red Cell. Membrane, ed. by G.A. Jamieson and T.J.Greenwalt (J.B.Lippincott, Philadelphia 1969) pp.117-130.

7. R.S.Weinstein: ibid. pp.36-76

8. E.A.Evans and P.L.La Celle: Blood 45, 29-43 (1975)

9. E.A.Evans, R.Waugh and C.Melnik: Biophys.J. 16, 585-595 (1976)

10. E.A.Evans and R.M.Hochmuth: In Current Topics in Membranes and Transport, Vol.10 (Academic Press, New York) in the press

11. W.Helfrich: Z.Naturforsch. 28c, 693-703 (1973)

12. F.Brochard and J.F.Lennon: J.Phys.(Paris) 36, 1035-1047 (1975)

13. R.M.Servuss, W.Harbich and W.Helfrich: Biochim.Biophys.Acta 436, 900-903 (1976)

14. E.Ponder: Hemolysis and Related Phenomena (Grune and Stratton, New York 1948)

15. G. Brecher and M.Bessis: Blood 40, 333-344 (1972)

16. B.Chailley, R.I.Weed, R.F Leblond and J.Maigné: Nouv.Rev.Franç. Hematol. 13, 71-87 (1973)

17. J.T.Penniston and D.E.Green: Arch.Biochem.Biophys. 128, 339-350 (1968)

18. B.Deuticke: Biochim.Biophys.Acta 163, 494-500 (1968)

19. M.P.Sheetz and S.J.Singer: Proc.Nat.Acad.Sci.USA 71, 4457-4461 (1974)

20. Y.E.Rahman, E.A.Cerny and C.Peraino: Biochim.Biophys.Acta 321, 526-535 (1973)

21. Y.E.Rahman, B.J.Wright and E.A.Cerny: Mech.Ageing Devel.2, 151-162 (1973)

22. L.Weiss: J.Biophys.Biochem.Cytology 3, 599-610 (1957)

23. E.A.Evans: Biophys.J. 14, 923-931 (1974)

24. F.R.N.Nabarro and J.Kostlan: J.Appl.Phys. 49, 5445-5448 (1978)

25. W.Helfrich and H.J.Deuling: J.Phys.(Paris) 36-C1, 327-329 (1975)

26. P.Seeman: Pharmacolog.Revs. 24, 583-655 (1972)

27. N.Mohandas and C. Féo: Blood Cells 1, 375-384 (1975)

28. E.A.Evans: Biophys.J. 13, 941-954 (1973)

29. D.Allan and R.H.Michell: Nature, Lond. 258, 348-349 (1975)

30. D.Allan, M.M.Billah, J.B.Finean and R.H.Michell: Nature, Lond. 261, 58-60 (1976)

31. H.U.Lutz,Shih-Chan Lui and J.Palek: J.Cell.Biology 73, 548-560 (1977)

32. E.A.Evans and P.F.Leblond: Biorheology 10, 393-404 (1973)

33. A.W.L.Jay: Biophys.J. 15, 205-222 (1975)

34. W.Helfrich: Z.Naturforsch. 29c, 510-515 (1974)

35. H.S.Hendrickson: J.Lipid Res. 17, 393-398 (1976)

36. J.R.Trudell: Biochim.Biophys.Acta 470, 509-510 (1977)

37. H.J.Deuling and W.Helfrich: Blood Cells 3, 713-720 (1977)

38. Reuven Yagil: private communication

39. B.Maggio and J.A.Lucy: Biochem.J. 149, 597-608 (1975)

Lipid-Protein Interaction in Membranes

F. Jähnig

Max-Planck-Institut für Biologie,
D-7400 Tübingen, Federal Republic of Germany

ABSTRACT

Protein molecules distributed homogeneously in a lipid membrane are treated as representing boundary conditions for the lipid long-range orientational order. Within the framework of a Landau theory it is found that the lipid phase transition with increasing protein concentration becomes weaker of first order until a critical point is reached. Response functions such as the specific heat and the compressibility, and also the permeability, are expected to show an increase on approaching the critical point. Available experimental results are discussed.

1. Introduction

If protein molecules are incorporated into a lipid membrane, one is faced with two questions: How do the proteins influence the lipid phase and how does the lipid phase act on protein properties? Before trying to answer mainly the first question, let us briefly recall some basic facts about pure lipid membranes [1].

They show essentially two different phases: An ordered phase at low temperatures with positional and orientational long-range order of the lipid molecules, and a disordered or fluid phase without positional order and less orientational order. The latter does not vanish completely due to the lateral surface pressure exerted by the hydrophobic effect. The transition between the two phases is caused by the spontaneous change of the orientational order, similar to the nematic-isotropic transition. Analogously the transition is of first order. As known from other first order transitions, pretransitional effects may arise in the vicinity of the transition simulating a hypothetical second order transition.

They become manifest in an increase of thermal fluctuations or response functions, such as the specific heat. Since in our case the critical orientational order is coupled to the packing density, the compressibility is also expected to show a critical increase. This has been observed experimentally via the sound velocity [2]. The increase of the compressibility was furthermore utilized to explain the maximum of the permeability found at the phase transition for certain molecules [3].

2. Model for the Lipid-Protein Interaction

If we now study the case of a lipid membrane with incorporated proteins, we restrict ourselves to one type of protein molecules and assume them to be distributed homogeneously in the membrane plane. Since a protein will locally perturb the lipid order, it may be regarded as representing a boundary condition on the lipid order. In the simplest approximation a protein molecule may be described as a cylinder of radius r_0. Then the lipid orientational order parameter S is S_0 at r_0 and either increases or decreases radially to approach the unperturbed bulk value S^u, if no other protein molecules are present (Fig. 1A). Again in the simplest approximation, the radial variation is exponential with a coherence length ξ. Not much is known about the values of S_0 and ξ. Because a protein in the fluid lipid phase represents a

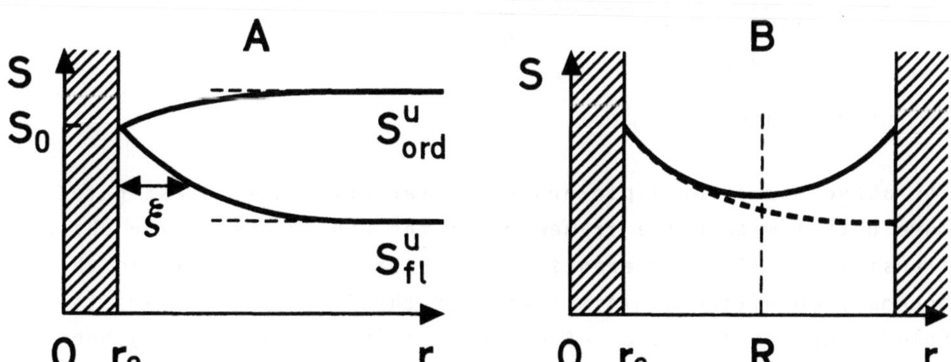

Fig.1 The influence of protein on the lipid orientational order parameter S, for an isolated protein (A) and in the case of overlapping effects from different protein molecules (B).

relatively rigid object, we would expect it to increase the lipid order [4]; in the ordered phase, however, due to its uneven surface it should decrease the lipid order, i.e. $S^u_{fl}<S_o<S^u_{ord}$. The maximal value of ξ, at the phase transition, may be of the order of 10 Å [5].

At finite protein concentration we describe the neighbouring protein molecules of a given protein as forming a cylinder of radius $2R-r_o$ around the given protein, again with the boundary value S_o (Fig. 1B). The spatially averaged lipid order parameter $\bar{S}$ will deviate from the unperturbed value S^u according to the value of S_o. Using for S_o the above relation, $\bar{S}$ is increased in the fluid phase and decreased in the ordered phase. Then the spontaneous change of the order parameter at the phase transition becomes smaller and may even vanish at a critical point (Fig. 2). Such a behaviour was shown to follow from fluorescence anisotropy measurements [6].

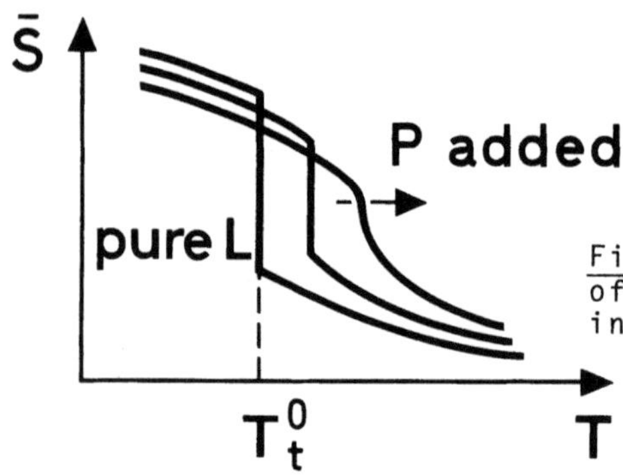

Fig.2 The temperature dependence of the lipid order parameter under increasing protein concentration.

3. Landau Theory

The above model for lipid-protein interaction was investigated in more detail within the framework of the Landau theory for phase transitions. The difference to the usual case consists in the boundary conditions; minimization of the free energy leads to an Euler-Lagrange equation. A computer solution has already been obtained [7]. In order to solve the problem analytically we restrict ourselves to low protein concentrations ($R>>\xi$). For simplicity we furthermore assume the unperturbed order parameter in the fluid phase to vanish ($S^u_{fl}=0$), and also the protein radius to be small (r_o=0). Then the protein induced shift δT_t of the lipid phase transition temperature results as

$$\frac{\delta T_t}{T_t^u - T^*} = \left(\frac{\xi_t}{R}\right)^2 \left(2\frac{S_o}{S_t^u} - 1\right). \quad (1)$$

Here $T_t^u - T^*$ is the difference between the temperatures of the first order transition and the hypothetical second order transition of the unperturbed system; ξ_t and S_t^u are the coherence length and the order parameter at the phase transition in the unperturbed system. The first factor $(\xi_t/R)^2$ makes the protein effect to decrease with decreasing coherence length and protein concentration; the square is a consequence of our two-dimensional geometry and does not appear for one-dimensional geometry, e.g. surface-aligned nematic films [8]. The second factor takes account of the boundary value S_o and determines the sign of the shift δT_t; it vanishes at $S_o = S_t^u/2$, in which case S_o is just the mean value of the unperturbed order parameters of the ordered and fluid phase at T_t^u, and the lipid phase is neither stabilized nor destabilized by the proteins. In the same approximation the relative shift of the latent heat is obtained as

$$\frac{\delta Q}{Q^u} = -\left(\frac{\xi_t}{R}\right)^2 2\frac{S_o}{S_t^u}. \quad (2)$$

δQ being negative means that the latent heat is always decreased by proteins. The critical point is reached if $\delta Q = -Q^u$ thereby fixing a critical value for R

$$\left(\frac{\xi_t}{R_c}\right)^2 = \frac{1}{2S_o/S_t^u}. \quad (3)$$

Releasing the condition r_o=0 and putting $r_o = \xi_t$ we find the improved result

$$\left(\frac{r_o}{R_c}\right)^2 = \frac{2}{9S_o/S_t^u + 1/6}. \quad (4)$$

Using for a typical example $S_o/S_t^u=3/4$ we get $(r_o/R_c)^2=0.24$, which is close to the computer result 0.16 [7] and shows that our approximation $R>>\xi$ may be applied up to the critical point. $(r_o/R)^2$ being the fraction of membrane area covered by protein it may be expressed by the protein molar fraction n and the molecular areas f_P and f_L of protein and lipid. With $f_P/f_L=10$ in the above example we find for the critical protein fraction $n_c=1/40$. This value lies in the range of protein concentrations where the vanishing of the sharp latent heat peak has been observed by calorimetry [9]. In these measurements a much broader peak simultaneously appeared which we interpret as an increase of the specific heat in the vicinity of a critical point. Other pretransitional effects, e.g. in the compressibility and permeability, would be expected but experimental results are lacking.

Calorimetric effects of the same kind have been observed with cholesterol [10], which are interpreted along the same lines. Because the transition temperature is not shifted in this case, we set $S_o/S_t^u=1/2$ and using $f_P/f_L=1$ obtain $n_c=1/3$, in good agreement with the experimental finding. Moreover, a critical slowing down on approaching this concentration was found in dielectric relaxation measurements [11]. And in a nuclear magnetic resonance spin-echo experiment the lateral diffusion constant of the lipid increased due to cholesterol [12]. The proposal of critical phenomena due to cholesterol may open a new way to understand the diverse cholesterol effects.

4. Protein Diffusion

Turning to the influence of the lipid phase on protein properties we confine ourselves to the protein lateral diffusion. In a recent photobleaching experiment the protein diffusion in the ordered lipid phase was found two orders of magnitude faster than the lipid diffusion in the ordered phase [13]. This surprising result finds a simple explanation in our model: In the ordered phase a protein molecule is surrounded by less ordered lipids ($S_o<S_{ord}^u$) and moves essentially in a fluid phase. It translates by melting the ordered lipids, like a wire passing through a block of ice.

1. F. Jähnig, Ann. Phys. (Paris) 3, 291 (1978).
2. S. Mitaku, A. Ikegami, and A. Sakanishi, Biophys. Chem. 8, 295 (1978).
3. S. Doniach, J. Chem. Phys. 68, 4912 (1978).
4. J. Davoust, B.M. Schoot,and P. Devaux, Proc. Natl. Acad. Sci. USA 76, 2755 (1979).
5. S. Marcelja, Biochim. Biophys. Acta 455, 1 (1976).
6. F. Jähnig, Proc. Natl. Acad. Sci. USA 76, 6361 (1979).
7. J.C. Owicki, and H.M. McConnell, Proc. Natl. Acad. Sci. USA 76, 4750 (1979).
8. P. Sheng, Phys. Rev. Lett. 37, 1059 (1976).
9. W. Curatolo, J.D. Sakura, D.M. Small and G.G. Shipley, Biochemistry 16, 2313 (1977).
10. S. Mabrey, P.L. Mateo, and J.M. Sturtevant, Biochemistry 17, 2464 (1978).
11. J.C.W. Shepherd, and G. Büldt, Biochim. Biophys. Acta 558, 41 (1979).
12. A. Kuo, and C.G. Wade, Biochemistry 18, 2300 (1979).
13. W.L.C. Vaz, K. Jacobson, E.S. Wu, and Z. Derzko, Proc. Natl. Acad. Sci. USA 76, 5645 (1979).

Photoreaction of Cholesteryl Cinnamate in Multilayers

Y. Tanaka and M. Suzuki

Research Institute for Polymers and Textiles,
1-1-4 Yatabe-Higashi, Tsukuba, Ibaraki 305, Japan

1. *Introduction*

The designed arrangement of molecules is a valuable tool for research in the molecular sciences. This will enable to construct organized molecular system, which is capable of realizing entirely new functional potentials when the constituents are properly arranged. Thus, the photoreaction of cholesteryl cinnamate in liquid crystalline states has been studied previously [1]. The reaction of the cinnamte irradiated in multilayers is a subject of the present paper. The photoreaction of some derivatives of cinnamic acids has been studied in various solvents and at the solid state, but not in the mesomorphic state nor in multilayers.

2. *Experimental*

Cholesteryl trans-cinnamate was reagent grade and recrystallized repeatedly with benzene as the solvent and methanol as the precipitant. The exciting light was furnished by an irradiator composed of a 500-W xenon lamp as the light source and filters. The light intensity was measured with an Eppley thermopile. To examine the photoreaction of the cinnamate, UV-absorption spectra changes of the sample were measured during reaction on irradiation with filtered light, using a Shimadzu UV-202 spectrophotometer. Using the Langmuir-Blodgett technique [2], multilayers were constructed by depositing a number of monolayers one on top of the other. A similar apparatus to one used by Kuhn [3] was used. This is a special type of trough suitable for the rapid formation and removal of the monolayer on the water-air interface. The reaction mixture was analyzed using a Shimadzu-DuPont LC-2 liquid chromatograph. Cholesteryl cis-cinnamate, truxillic acid and dicholesteryl truxillate were obtained by the similar methods of Vaidya [4] and Green et al. [5].

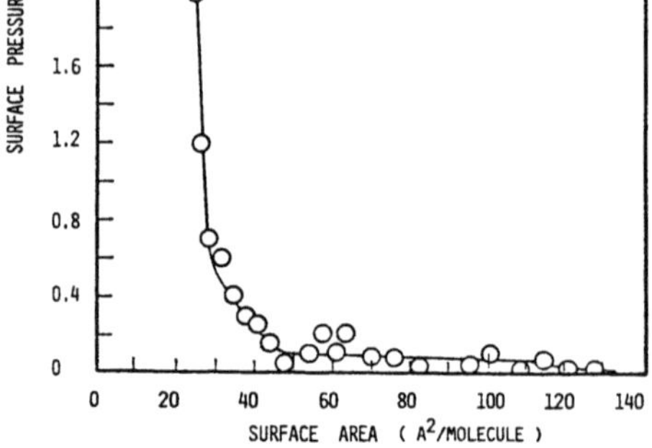

Fig. 1 Surface pressure-area isoherms for cholesteryl trans-cinnamate

3. *Results and discussion*

The monolayer film was deposited on water from a dilute(7.13 x 10^{-4}M)benzene solution. In 20 min after spreading, the π-A curve was obtained as shown in Fig. 1 and a significant decrease in the surface area was observed at all pressures. This is inconsistent with the data on cholesteryl acetate films [6]. The cross-sectional area is obtained by extrapolating the slope to x-axis from the steepest portion of the curve, as about 30 A^2, which is rather smaller than that of cholestrol (41 A^2) [6]. Collapse takes place at ca. 20 dyne/cm.

On the irradiation of the cinnamate in various multilayers at 277 nm, the UV-absorption spectra were changed, as shown in Fig. 2, into those similar to that of photodimers of cinnamic acids and cinnamates [1]. No difference was found among thses multilayers. When the reaction mixtures were dissolved into diethyl ether and then irradiated at 230 nm, the UV-absorption spectrum was changed partly into that of original cholesteryl trans-cinnamate.

Various irradiated samples were analyzed by high-performance liquid chromatography, and the typical result is shown in Fig. 3, which gives the spectra of authentic samples and the reaction mixture. This shows irradiation of the cinnamate in multilayers resulted in formation of a photodimer, dicholesteryl truxillate, as well as cis-isomer, cholesteryl cis-cinnamate. Some amounts of truxillic acid could be also found in the irradiated sample and may be obtained from the hydrolysis of dicholesteryl truxillate or from the photodimerization of cinnamic acid which seems to come from the hydrolysis of the cinnamate.

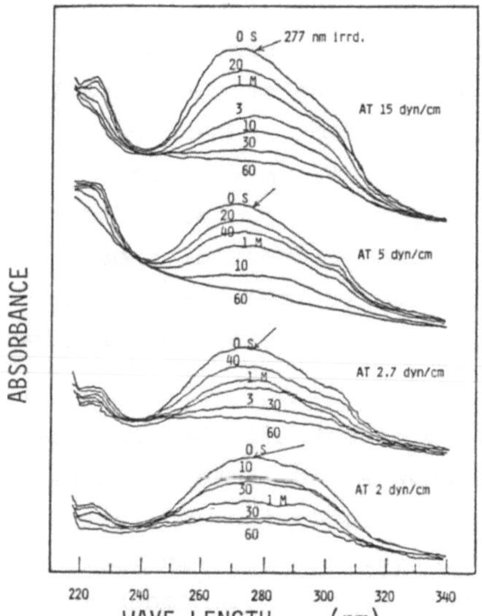

Fig. 2 UV-Spectral changes of cholesteryl trans-cinnamate at r.m. when irradiated in multilayers at 277 nm

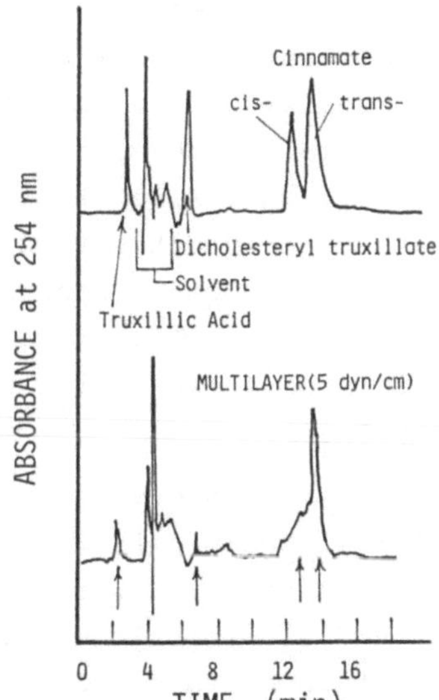

Fig. 3 LC-Spectra of the authentic samples and reaction mixtures column; Shimadzu PCH-05

References: 1. Y. Tanaka & H. Tsuchiya, J. de Phys., 40 (4) C3-41 (1979); 2. K. B. Blodgett, J. Amer. Chem. Sco., 57, 1007 (1935); 3. H. Bücher, O. v. Elsner, D. Möbius, P. Tillmann & J. Wiegand, Z. Phys. Chem. N. F., 65, 152 (1969); 4. B. K. Vaidya, Proc. Roy. Soc. A, 129, 299 (1930); 5. B. S. Green & M. Rejto, J. Org. Chem., 39, 3284 (1974); 6. A. M. Kamel, N. D. Weiner & A. Felmeister, J. Colloid & Inter. Sci., 35 (1) 163 (1971)

Differential Scanning Calorimetric Studies of Mixtures of Cholesterol with Phosphatidylethanolamine and Phosphatidylethanolamine-phosphatidylcholine Mixtures

A. Blume

Institut für Physikalische Chemie,
Albertstr.23a, D-7800 Freiburg, Federal Republic of Germany

The thermotropic behaviour of aqueous dispersions of phosphatidylethanolamine-cholesterol and phosphatidylethanolamine-phosphatidylcholine-cholesterol mixtures was studied by high-sensitivity differential scanning calorimetry. The gel to liquid crystal phase transition of dimyristoyl-phosphatidylethanolamine is lowered and broadened by the addition of cholesterol and disappears at 50 mole% cholesterol. When the cholesterol content is below 25 mole% the calorimetric endotherms seem to consist of two components, a broad one at considerably lower temperature than the original transition and another component at only slightly lower temperature. On the basis of the phase diagram determined by calorimetry we suggest a homogenous distribution of cholesterol in the bilayers of phosphatidylethanolamine. The broader transition at lower temperature may arise from those lipids in contact with only one cholesterol molecule.

Addition of cholesterol to equimolar phosphatidylethanolamine-phosphatidylcholine mixtures, which show either ideal or non-ideal mixing behaviour, also leads to a broadening and lowering of the calorimetric endotherm. However, an unambiguous decomposition of the calorimetric peaks into two components is difficult in these cases. In the phosphatidylethanolamine-phosphatidylcholine mixtures investigated cholesterol showed no preferential affinity for either of the phospholipids.

Ref.: Blume, A. (1980) Biochemistry, submitted for publication

A Two-Dimensional Thermodynamic Field Theory

A. Grauel

Arbeitsgruppe Theoretische Physik, Universität Paderborn,
D-4790 Paderborn, Federal Republic of Germany

1. Introduction

We represent a two-dimensional thermodynamic field theory and we apply this theory on an example, namely a mixture of heat conducting, inviscid fluids in a very thin layer with arbitrary shape. Such a fluid layer can be roughly spoken, considered as a fluid lipid bilayer (the thickness has an order of 10 nm) as well as a monolayer. We consider a semipermeable lipid bilayer with protein films as a model for a biological membrane which consists mainly of an ordered structure (like the smectic phase of liquid crystals). In the following considerations we shall give a description of continuous media and in a first step we shall not consider a polar media with ordered structure.

For the mathematical description, let us consider the fluid surface as a semipermeable surface which shall be in particle and heat exchange with its bulk phases. We assume thermodynamic surface quantities and mechanical properties like surface stress for each fluid on the surface. We discuss field equations and we shall give a representation for the constitutive equations which follows from the principle of material objectivity in space and on surfaces. We show how we can restrict the constitutive equations by an entropy principle. From this entropy principle it follows restrictions for the constitutive equations and surface relations. Finally, we derive a diffusion equation on curved surfaces from the balance of momentum.

2. Theoretical Description

2.1 Thermodynamic Fields

For the thermodynamic properties we must calculate the fields of the density, velocity and temperature for each component of the mixture as function of the particle variable and time. We assume that each constituent in the mixture on the surface has the same temperature and we identify this temperature as the temperature of the mixture.

For the mixture of λ fluids we have to calculate $4\lambda+1$ fields as function of the surface parameter u^1,u^2 and the time t, namely partial densities $\gamma_\delta(u^1,u^2,t)$,

partial velocities $w^k_\delta(u^1,u^2,t)$,

temperature $T_S(u^1,u^2,t)$.

Let us introduce the following definitions which we use in the further considerations. We define the density of the mixture, the velocity of the mixture, the diffusive velocity and the relative velocity by

$$\gamma = \sum_{\delta=1}^{\lambda} \gamma_\delta \quad , \tag{1}$$

$$w^k = \sum_{\delta=1}^{\lambda} \frac{\gamma_\delta}{\gamma} w^k_\delta \quad , \tag{2}$$

$$_S u^k_\delta = w^k_\delta - w^k \quad , \tag{3}$$

$$W^k_\delta = w^k_\delta - w^k_\lambda \quad , \quad \delta = 1,\ldots,\lambda-1. \tag{4}$$

The fluids are characterized by the material properties δ.

2.2 Balance Equations

In a previous paper [1] we have derived balance equations for the densities of mass, momentum and energy. These equations have the following general form

$$\partial_t \gamma_\delta - 2\underset{n}{w_1} K_M(t)\gamma_\delta + (\gamma_\delta \cdot \underset{\tau}{w^A_\delta})_{;A} + [\rho_\sigma(v^j_\sigma - w^j_1)e^j] = \pi_\delta, \tag{5}$$

$$\partial_t(\gamma_\delta \cdot w^k_\delta) - 2\underset{n}{w_1} K_M(t)\gamma_\delta w^k_\delta + (\gamma_\delta \cdot w^k_\delta \cdot \underset{\tau}{w^A_\delta} - {}_S t^{kA}_\delta)_{;A} + [\rho_\sigma v^k_\sigma(v^j_\sigma - w^j_1)e^j - t^{kj}_\sigma e^j] = \tag{6}$$

$$= m^k_\delta + \gamma_\delta \cdot {}_S f^k_\delta$$

$$\partial_t(\gamma \cdot \varepsilon_S) - 2\underset{n}{w_1} K_M(t)\gamma \cdot \varepsilon_S + (\gamma \cdot \varepsilon_S \cdot \underset{\tau}{w^A} + \underset{\tau}{q^A_S})_{;A} - t^{kA}_S w^k_{;A}$$

$$+ [\rho(\varepsilon + \tfrac{1}{2}(v^k - w^k)^2)(v^j - w^j_1)e^j + q^j e^j - t^{kj}(v^k - w^k)e^j] = \gamma \cdot r_S \quad , \tag{7}$$

where we denote the jump of the quantitiy ψ across the surface by $[\psi] = \psi^+ - \psi^-$. Both, ψ^+ and ψ^- are the limit values of $\psi(x^1,t)$ on the sur-

face as x^i from outside or inside of the surface approaches a point $x^i(u^1,u^2,t)$ on the surface. σ takes on different values on each side of the surface.
The semipermeable surface is material for particles of the densitiy γ_1. The quantity $w_1 e^l_n$ is the normal velocity perpendicular to the surface and $K_M = \frac{1}{2}\,\mathrm{tr}(b_{AB})$ is the mean curvature of the deformable surface.

The quantity π_δ is the production of mass due to chemical reactions and m^k_δ , k=1,2,3, are the components of the interaction force. ${}_S f^k_\delta$ are the components of the force density of an external field like gravitation which acts on the fluid particles of the constituent δ in the surface. The radiation density ${}_S r_\delta$ is a supply of internal energy due to radiation which acts on the fluid particle δ in the surface. Let us assume that both, the components of the force density and the radiation density are functions of the surface parameters u^1,u^2 and the time t.
For the time being, let us consider the case that on the fluid surface a mixture of two fluids with the densities γ_0 and γ_{II} exists. Let us assume that

$$\pi_\delta = 0 \ , \ {}_S f^k_\delta = 0 \quad \text{and} \quad r_S = 0 \tag{8}$$

holds in the following considerations. If (8) holds then we shall call the equations (5), (6) and (7) reduced equations of balance. The quantity σ in the jump terms takes on the values σ = I, II inside the closed surface and on the other side we have a single fluid. This fluid is characterized by $\wedge$ or +. For this fluid we assume that the surface is permeable and on the surface the fluid is characterized by an index II.

2.3 Constitutive Assumptions

We obtain a closed system of reduced equations when these equations are supplemented by constitutive equations, namely the internal energy ε_S, the heat flux $\underset{\tau}{q}{}^A_S$, the stress on the surface ${}_S t^{kA}_\delta$ and the interaction force m^k_δ.

These constitutive equations relates the quantities ε_S, $\underset{\tau}{q}{}^A_S$, ${}_S t^{kA}_\delta$ and m^k_δ to the surface fields $\gamma_\delta(u^1,u^2,t)$, $w^k_\delta(u^1,u^2,t)$ and $T_S(u^1,u^2,t)$.
Let us assume that the constitutive quantities on a surface point $P(u^1,u^2,t)$ depend on the values of the quantities $\gamma_\delta(u^1,u^2,t)$, $w^k_\delta(u^1,u^2,t)$, $T_S(u^1,u^2,t)$ and on the density gradients $\gamma_{\delta,A}(u^1,u^2,t)$ and on the temperature gradient $T_{S,A}(u^1,u^2,t)$. We can characterize the curved surface by combinations of geometrical variables, namely the normal vector e^k, surface vectors $x^k_{,A}$ and the derivative of the normal vector $e^k_{,A}$ with respect to the surface parameter. For the time being, for any constitutive quantity we shall write

$$F = \hat{F}(\gamma_\delta, T_S, \gamma_{\delta,A} \ , \ T_{S,A} \ , \ w^k_\delta \ , \ x^k_{,A} \ , \ e^k \ , \ e^k_{,A}) \tag{9}$$

Now we want to restrict the constitutive quantity (9) by additional assumptions. We require that (9) is invariant in form under

1. a change of frame, namely Euclidean transformation,
2. transformation of the surface parameter.

Finally, we obtain further restrictions for the constitutive relations from the entropy principle in the formulation given by MUELLER [2].

2.4 Entropy Principle

Let us assume

1. On the surface there exists a surface entropy η_S and an equation of balance for this quantity

$$\partial_t(\gamma\eta_S)-2\underset{n}{w_o}K_M\gamma\eta_S+(\gamma\eta_S\underset{\tau}{w}^A+\Phi_S^A)_{;A} + [\rho\eta(v^j-w_o^j)e^j+\Phi^j e^j] = \sigma_S. \tag{10}$$

σ_S shall be the production of entropy. The entropy η_S and the flux of entropy Φ_S^A must be given as constitutive quantities. For these quantities the same set of variables as in (9) should be acceptable.

2. On a surface point, the production of entropy σ_S shall not be negative

$$\sigma_S \geqslant 0. \tag{11}$$

for each thermodynamic surface process.

3. The entropy flux will be continuous if the surface temperature is continuous.

The entropy inequality (11) is a restriction for the surface fields $\gamma_\delta(u^1,u^2,t)$, $w_\delta^k(u^1,u^2,t)$ and $T_S(u^1,u^2,t)$ and the requirement 2. is valid only for such fields which are solutions of the field equations. We shall obtain the thermodynamic fields $\gamma_\delta, w_\delta^k$ and T_S out of the inequality (11) if we take the field equations as constraints into consideration. LIU [3] has explored these facts in a lemma and we apply LIU'S lemma for the thermodynamic processes on the surface.

3. Reduced Constitutive Equations

From the entropy principle it follows surface relations and restrictions for the constitutive equations. We obtain the following expressions for the constitutive equations, internal energy: $\overset{I}{\varepsilon}_S = \overset{I}{\varepsilon}_S(\gamma_\delta,T_S)$, where I characterizes the intrinsic part of the internal energy which depends not on the velocities of the diffusive motion,

specific entropy: $\eta_S = \hat{\eta}_S(\gamma_\delta,T_S)$,

heat flux: $\overset{I}{\underset{\tau}{q}}{}_S^A = -(q_1 g^{AB}+q_2 b^{AB})\cdot T_{S,B}+(q_3 g^{AB}+q_4 b^{AB})\cdot \underset{\tau}{w}_B^{II}$,

interaction force:

$$\underset{\tau}{m}{}_S^A = ({}_1m_\delta g^{AB} + {}_2m_\delta b^{AB})\cdot T_{S,B} - ({}_3m_\delta g^{AB} + {}_4m_\delta b^{AB})\cdot \underset{\tau}{w}{}_B^{II} + \sum_{\lambda=II}^{o} g^{AB}\left(\frac{\partial\sigma_\delta}{\partial\gamma_\lambda} - \gamma_\delta \frac{\partial\, {}_I\mu_S^{\delta\,\gamma}}{\partial\gamma_\lambda}\right)\gamma_{\lambda,B} \; ,$$

flux entropy: $\Phi_S^A = \frac{1}{T_S}\left({}^I\underset{\tau}{q}{}_S^A - \sum_\delta \gamma_\delta \cdot {}_S\underset{\tau}{u}{}_\delta^A \cdot {}_I\mu_S^{\gamma_\delta}\right)$,

stress on the surface: ${}_S t_\delta^{BA} = - \sigma_\delta(\gamma_\mu, T_S)\cdot g^{AB}$,

surface tension: $\sigma = \gamma \sum_\delta \gamma_\delta \cdot \frac{\partial}{\partial\gamma_\delta}(\varepsilon_S^I - T_S \eta_S)$,

specific free energy: $\varepsilon_S^I - T_S\,\eta_S = \sum_\delta \frac{\gamma_\delta}{\gamma}\cdot {}_I\mu_S^\delta - \frac{\sigma}{\gamma}$

The scalar coefficients q_1 to q_4 as well as the scalar coefficients ${}_1m_\delta$ to ${}_4m_\delta$ shall be quantities of the density γ_δ and the surface temperature T_S.

4. Diffusion Equation on Curved Surfaces

If we introduce the constitutive expressions for the interaction force ${}_S m_\delta^A$ and the stress on the surface ${}_S t_\delta^{BA}$ into the balance of momentum then we shall obtain the following equation

$$\gamma_{II}\cdot {}_S\underset{\tau}{u}{}_{II}^A = \frac{\gamma_o\gamma_{II}}{\gamma}\, g^{AB} D_W^{-1}\cdot\left(D_T \frac{\partial T_S}{\partial u^B} + \frac{\partial}{\partial u^B}\left({}_I\mu_S^{\gamma_{II}} - {}_I\mu_S^{\gamma_o}\right)\right), \tag{12}$$

where ${}_S\underset{\tau}{u}{}_{II}^A = \underset{\tau}{w}{}_\delta^A - \underset{\tau}{w}{}^A$, $\underset{\tau}{w}{}^A = \sum_{\lambda=0}^{II} \frac{\gamma_\lambda}{\gamma}\cdot \underset{\tau}{w}{}_\lambda^A$ and

$$D_W = - \left(\frac{{}_3m_{II}}{\gamma_{II}} - \frac{{}_3m_o}{\gamma_o}\right) - \left(\frac{{}_4m_{II}}{\gamma_{II}} - \frac{{}_4m_o}{\gamma_o}\right)\cdot b^{AB} g_{AB} \; , \tag{13}$$

$$D_T = \frac{\partial}{\partial T_S}\left(\frac{\sigma_{II}}{\gamma_{II}} - \frac{\sigma_o}{\gamma_o}\right) - \left\{\frac{{}_1m_{II}}{\gamma_{II}} - \frac{{}_1m_o}{\gamma_o} + \left(\frac{{}_2m_{II}}{\gamma_{II}} - \frac{{}_2m_o}{\gamma_o}\right) b^{AB} g_{AB}\right\} - \frac{\partial}{\partial T_S}\left({}_I\mu_S^{\gamma_{II}} - {}_I\mu_S^{\gamma_o}\right) . \tag{14}$$

The metric tensor is defined by $g_{AB} = x^i_{,A} x^i_{,B}$. $b_{AB} = -x^i_{,A} e^i_{,B}$ is the curvature tensor, where A = 1,2.

Equation (12) represents the equation of diffusion on curved surfaces which has a similar structure as the equation of diffusion in space [4]. In (12) we have neglected the acceleration terms. In this case the diffusion flux $\gamma_{II}\, {}_S\underset{\tau}{u}{}_{II}^A$ depends on a temperature gradient as well as on a gradient of the difference of the chemical potential on the surface.

Reference

1 Grauel, A.: On the Thermodynamic of an Interfacial Fluid Membrane, preprint (1979).
2 Müller,I.: Z. Naturforschung 28a (1973) 1801.
3 Liu, I-Shih: Archive for Rational Mechanics and Analysis 46 (1972) 131.
4 Müller, I.: Quaderno del Gruppo Nazionale di Fisica Mathematica del Consiglio Nazionale delle Ricerche, Firenze (1978).

On the Orientation of Liquid Crystals by Monolayers of Amphiphilic Molecules

K. Hiltrop and H. Stegemeyer

Department of Physical Chemistry, University of Paderborn,
D-4790 Paderborn, Federal Republic of Germany

Uniform director field of liquid crystal cells may be induced by surface action which is known as phenomena in many varieties. The hometropic orientation of a liquid crystal may be enforced by a just coherent monolayer of suitable amphiphilic compounds on solid substrates. Polarizing microscopy of nematics and helix unwinding of cholesterics /1/ showed the orientation of the director field in this case to be influenced by the following parameters:
1. temperature; 2. physical properties of the substrate surface (micro geometry, packing density of the amphiphilic monolayer, ...); 3. chemical properties of the substrate (structure of the amphiphilic compounds, ...); 4. structure of the liquid crystal molecules.

Ad 1. Going below a well defined temperature T^* the homeotropic orientation of the director field in a boundary layer near the substrate generally changes to a tilted one and vice versa. The tilt angle increases with decreasing temperature.

Ad 2. a) The instantaneous phase of the monolayer during the transfer from the subphase to a solid substrate severely influences the liquid crystal orientation. Obviously the packing density is frozen at the transfer of the film. b) On changing the packing density from a denser to a looser one the characteristic transition temperature T^* phases through a minimum.

Ad 3. a)Chirality of lecithin molecules does not have any detectable effect on the orientation. b) Increasing chain length of the investigated lecithins results in increased transition temperatures T^*. c) The transition temperatures T^* of monoalkyl lecithins join roughly with those of dialkyl lecithins, if the corresponding mean area values(inverse packing density) are scaled in units per alkyl chain instead of units per lecithin molecule.

Ad 4. The transition between the homeotropic and a tilted orientation for different liquid crystals occurs at different temperatures. This is valid in case of temperature reduction to corresponding order parameters, too. Nevertheless a correlation of T^* to the clearing temperature obviously does exist.

Detailed experimental results and possible models for the orientating action of amphiphilic films on liquid crystals will be described elsewhere /2/.

References

1 M. Brehm, H. Finkelmann, H. Stegemeyer, Ber. Bunsenges. 78, 883 (1974)

2 K. Hiltrop, H. Stegemeyer, to be published in Ber. Bunsenges.

Influence of Phase Transitions of Amphiphilic Monolayers on the Orientation of Liquid Crystals

K. Hiltrop and H. Stegemeyer

Department of Physical Chemistry, University of Paderborn
D-4790 Paderborn, Federal Republic of Germany

Monolayers of different pure lecithins were transferred from aqueous subphases to hydrophilic glass plates at different packing densities of the film. Samples of nematic and induced cholesteric liquid crystals were prepared with these plates. Polarizing microscopy showed that the director field prefers different orientations depending on packing density of the monolayer as well as on temperature of the sample. In particular lowering the temperature of a system substrate/liquid crystal results in the loss of the homeotropic orientation for the benefit of a tilted one. The same alteration of the director field orientation occurs at constant temperature with increasing packing density.

The changes of orientation by variation of temperature and packing density in general may be explained by different models, which are discussed in a forthcoming paper /1/. With one exception all these models are contradicted somewhere by experimental results. The essence of the remaining model is a system consisting of a loose but coherent lecithin monolayer with embedded liquid crystal molecules. Thus the mean orientation of liquid crystal molecules in the bulk of the sample reproduces the mean orientation of micro holes in the lecithin monolayer which are occupied by liquid crystal molecules. The authors suggest the change of liquid crystal orientation to be enforced by phase transitions of this system analogous to those observed by SACKMANN /2/ with phospholipid monolayers on liquid subphases.

References

1 K. Hiltrop, H. Stegemeyer, to be published

2 E. Sackmann, Ber. Bunsenges. 82, 891 (1978)

Infrared Spectroscopic and Ultrastructural Studies of a Hydrophobic Membrane Protein in a Monolayer

F. Kopp* and P.A. Cuendet

Institute of Cell Biology and
Institute of Molecular Biology and Biophysics,
Swiss Federal Institute of Technology,
CH-8093 Zürich, Switzerland

1. Abstract

The arrangement of the molecules of a hydrophobic membrane protein in a monomolecular layer will be presented here as it is obtained by electron microscopical techniques. Infrared spectroscopic data obtained by multiple attenuated total reflection (ATR-IR) allow us, furthermore, to distinguish between monomolecular layers adsorbed to a solid substrate either with their hydrophobic (apolar) or their hydrophilic (polar) surface.

2. The Object

A lipid- and pigment-free polypeptide, isolated from the photosynthetic membrane of the purple bacteria *Rhodospirillum rubrum*, was used to prepare monomolecular layers. *In vivo*, the polypeptide is associated with bacteriochlorophyll and it seems that it represents a subunit of the light-harvesting antenna system. We, therefore, tentatively named it light-harvesting polypeptide (LHP). It has a molecular weight of 14'000 and a comparatively high content of amino acids with apolar side chains (approx. 60%) [1]. An outstanding property of the LHP is its ability to dissolve in several organic solvents such as chloroform/methanol, pyridine and formic acid. From such solvents, it can easily be spread at an air/water interface forming a monomolecular layer [2].

3. Monolayer Preparation

Surface pressure/area isotherms of the protein monolayers were monitored using a LAUDA film balance of the Langmuir type [3]. Monolayers were then transferred, either onto a clean germanium plate for infrared spectroscopy or onto freshly cleaved mica for heavy metal replication and subsequent electron microscopy. The transfer was performed by drawing up the solid (germanium or mica) from the subphase through the compressed monolayer [4]. During the transfer of the monolayer, the surface pressure was kept constant at 20 dynes/cm using a movable barrier. Monolayers transferred in this way adhered to the solid substrate with their hydrophilic surface. For infrared spectroscopy, other monolayers were transferred onto germanium in such a way that their hydrophobic surface adhered to the substrate. Germanium plates used for this purpose had been rendered hydrophobic by treating them with dichlorodimethylsilane before monolayer transfer. The monolayers were then adsorbed to

*Present Address: F. Kopp, Diabetes Institut der Universität Düsseldorf, Auf'm Hennekamp 65, D-4000 Düsseldorf, Fed. Rep. of Germany

X-Ray Diffraction Study of the Mesophase of Octaphenylcyclotetrasiloxane

G. Albertini[1], B. Dubini[1], S. Melone[1], F. Rustichelli[2,3], and G. Torquati[2]
[1] Istituto di Fisica Medica, Facoltà di Medicina e Chirurgia, Ancona, Italy
[2] Dipartimento di Scienze Fisiche, Facoltà di Ingegneria, Ancona, Italy
[3] European Communities J.C.R. Ispra, Varese, Italy

Octaphenylcyclotetrasiloxane (OPCTS) was investigated by KEYES and DANIELS [1], who concluded that this compound presents a plastic phase in the range between 188°C and 205°C during the heating process and before the transition to the isotropic liquid phases. They used differential thermal analysis to observe the appearance of this mesophase and of a solid-solid phase transition at ∿76°C. During the reverse process a supercooling effect was observed: the mesophase exists from 203°C to 155°C. The arguments in favour of the *plastic crystal* nature of the mesophase were the assumed globular shape of the molecule and the observed optical isotropy, which seemed to exclude a *liquid crystal* nature of the mesophase. In fact liquid crystals are in general characterized by optical birefringence and by either rod-like or disc-like molecular shapes.

In spite of this argument neutron diffraction experiments on the OPCTS mesophase suggested to VOLINO and DIANOUX [2] the conclusion that the mesophase could be of "liquid crystal" nature, possibly similar to a smectic A phase. This conclusion was mainly based on the observation of *only one* Bragg peak and pronounced diffuse scattering in the neutron diffraction pattern, as is typical of the molecular arrangement associated to the smectic A phase. However the authors asked for further data by other techniques in order to eventually confirm their assumption. In particular they suggested us to carry out X-ray diffraction experiments on OPCTS.

Then the room temperature structure of OPCTS was obtained from single crystal diffractometry by BRAGA and ZANOTTI [3]. The crystallographic data obtained for the monoclinic cell of OPCTS ($Si_4O_4C_{48}H_{40}$) are: a=21.918(8) Å, b=10.138(6) Å, c=21.743(8) Å, β=116.0(1) deg and volume V_o= 4342.4 $Å^3$, space group $P2_1/c$, Z=4.

At the same time a detailed investigation was performed [4] of the thermodynamic properties of OPCTS: a dependence of the physical behaviour on the thermal history of the sample was observed as is common for either liquid crystal or plastic crystal mesophases.

This note presents the results of an X-ray diffraction study of the OPCTS mesophase, which leads to a conclusion in favour of the *plastic crystal* nature of OPCTS. Furthermore an interpretation is given of the misleading fact connected with the previous interpretation in favour of the *liquid crystal* nature.

The measurements were carried out by using a conventional X-ray powder dif-

the hydrophobized germanium by dipping the plates onto the spread protein film at the air/water interface, the extended plane of the plate being nearly parallel to the film surface.

4. Electron Microscopy

To prevent possible rearrangement of the molecules [5,6], the mica samples were frozen in liquid nitrogen immediately after monolayer transfer. While still under nitrogen, the specimens were placed in a small brass container which was fitted onto a cooled stage in a BALZERS freeze etch unit. The stage permitted rotation of the specimen. After a vacuum better than 10^{-7} Torr was achieved, the container was opened and the exposed specimen replicated by conical shadowing with platinum-carbon under an elevation angle of 9 degrees to an average thickness of 0.28 nm, as measured vertical to the specimen plane with a quartz crystal film thickness monitor. Additional 15 nm of carbon evaporated vertically to the specimen plane reinforced the replicas. The latter were floated off and viewed in a SIEMENS 102 electron microscope. The negatives were directly used for light optical diffractometry and image reconstruction [7].

5. Ultrastructural Results

During shadow-casting, the incoming platinum atoms diffuse laterally on the specimen and condense into small granules (Fig.1,top). Two kinds of decoration have been observed [2]. The mean distance between the platinum granules formed on a protein monolayer has been found to be smaller than between granules that had formed on pure mica. A difference in lateral diffusion due to the "roughness" of the protein monolayer may be responsible for this first decoration effect. Replicas of monolayers transferred under 10 dynes/cm compression showed a random distribution of the platinum granules. At 20 dynes/cm compression, ordered domains in a mosaic-like arrangement were observed (Fig.1,top). This kind of order was again lost when higher surface pressure was exerted during transfer. The order observed is made visible thanks to a second kind of decoration. Some unknown property of the protein molecule itself seems to induce the growth of a certain number of platinum granules with a defined spatial relation between granules and protein molecules. Such granules contributed to the regular spots observed in the optical diffractogram of a regular domain (Fig.1,bottom left). Optical filtration and reconstruction [8] (Fig.1,bottom right) displayed the molecular arrangement in an ordered array. Note the imperfect long range order!

6. Attenuated Total Reflection Infrared Spectroscopy [9]

Infrared spectroscopy of a monomolecular protein layer in the transmission mode would only yield an absorbance in the range of 10^{-4}. We therefore had to take advantage of both the multiple attenuated total reflection technique (ATR) and computer accumulation for getting reasonable spectra. Germanium plates (50x20x1 mm) to which the monolayers had been adsorbed, were used for ATR-IR spectroscopy. The infrared beam, polarized either vertical (v) or parallel (p) to the plane of incidence, enters and leaves the germanium plates through frontal planes inclined at 45 degrees (Fig.2).

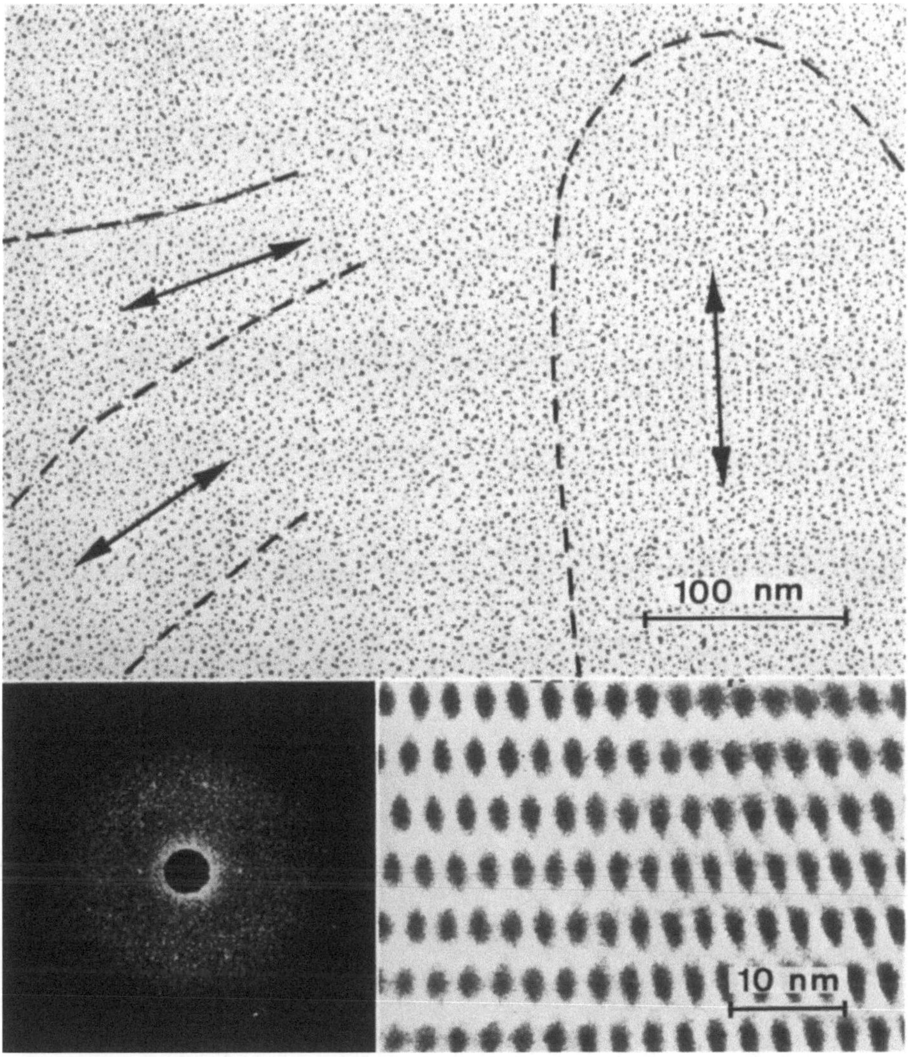

Fig.1 Top: Electron micrograph of a light-harvesting polypeptide monolayer deposited onto mica. The specimen was replicated by rotary platinum shadowing. It exhibits ordered domains in a mosaic-like pattern. Magnification 270 000x. Bottom left: Light optical diffractogram of an ordered domain. Bottom right: Light optical reconstruction of an ordered domain. Magnification 1 500 000x.

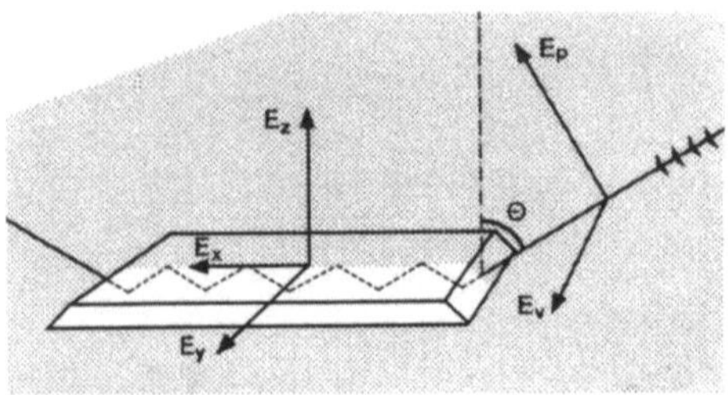

Fig.2 ATR set up. θ: angle of incidence. E_p,E_v: parallel and vertically polarized components to the electric field of incident light. E_x,E_y,E_z are the electric field components with respect to the internal reflection plate. (From ref.5)

The 55 useful reflections obtained in this way resulted in a considerable enhancement of the absorption spectrum. To decrease the noise level, 25 scans were accumulated per spectrum, and the average smoothed and replotted. Examples are given in Fig.3.

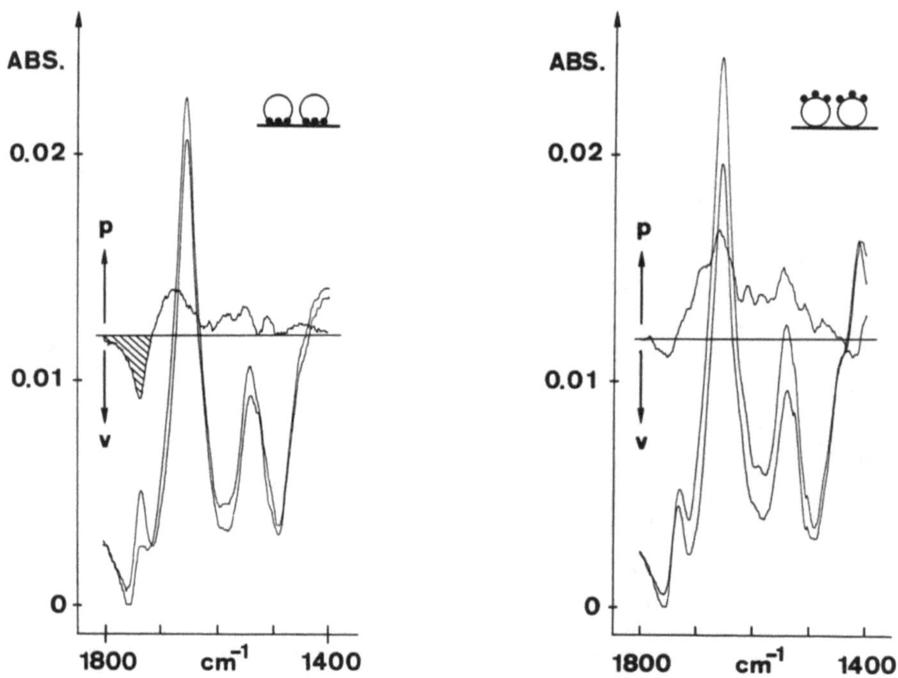

Fig.3 Polarized ATR-IR spectra and their difference spectra (p minus v) of a monomolecular layer of LHP adsorbed with its hydrophilic surface (left) and with its hydrophobic surface (right) to a germanium ATR crystal. Each spectrum is a computer replot of the average of 25 scans. Measuring time: 18 hrs per spectrum. Polarization (p) and (v) as in Fig.2

7. Qualitative Interpretation of IR Spectra (Fig.3)

The position of the amide I (C=O stretching) and the amide II (C-N stretching and N-H bending) bands at 1655 and 1545 cm^{-1} allow for α-helical and random coil conformation [8]. No β-pleated sheet structure has been found (no absorption bands around 1630 and 1520 cm^{-1}). The comparatively narrow half width of the amide I band of less than 40 cm^{-1} found in multilayers of LHP [2], as well as results of deuteration experiments (unpublished) indicate an unusual high helical content of about 80% of this protein. In monomolecular layers however, we observe an increase of the half width of the amide I band dependent on polarization. It amounts to approximately 45 cm^{-1}(v) and 50 cm^{-1} (p). This enlarged half width is found in monolayers adsorbed with either their hydrophilic or hydrophobic surface. We attribute this effect to a deformation of the helix due to the adsorption to the substrate. The influence of the two different kinds of monolayer adsorption on the infrared spectra can most easily be recognized by comparing the difference spectra given in Fig.3.

Interaction between the protein and the germanium substrate is strongly indicated by the behaviour of the C=O stretching band of the carboxyl groups of the polypeptide side chains at 1730 cm^{-1}. A strong vertical polarization (relative dichroic ratio R'= 0.4) is observed (Fig.3,left) if the carboxyl groups exposed at the hydrophilic surface of the monolayer are adsorbed to the germanium substrate. If the monolayer is adsorbed with its apolar surface, the free carboxyl side groups exhibit a more random orientation, i.e., no appreciable polarization of the C=O stretching band at 1730 cm^{-1} is observed.

8. Final Remarks

In this report, we intended to show that the combination of film balance, ATR-IR spectroscopy and electron microscopy allows the concomitant investigation of chemistry, molecular conformation and ultrastructural organization within a twodimensional specimen only one molecule thick.

The original aim of our work was to construct membrane models using defined membrane proteins and lipids. With such models we planned to investigate the changes that occur in biogenic material when such specimens are prepared for electron microscopy. Conformational and chemical alterations have to be expected to occur in such specimens due to treatments such as adsorption to a solid support, chemical fixation, chemical staining with heavy metal compounds or decoration with evaporated heavy metal or other substances.

The combination of the experimental techniques we used could be of some interest to those who would like to investigate the first layer (or layers) at solid/liquid crystal interfaces.

9. References

1. P.A.Cuendet, H.Zuber: FEBS Lett. 79, 96-100 (1977)
2. F.Kopp, P.A.Cuendet, K.Mühlethaler, H.Zuber: Biochim. Biophys. Acta 553, 438-449 (1979)

3. G.L. Gaines: Insoluble Monolayers at Liquid-Gas Interfaces. (New York: Interscience Publishers 1966)
4. K.B.Blodgett: J. Amer. Chem. Soc. 57, 1007-1022 (1935)
5. F.Kopp, U.P.Fringeli, K.Mühlethaler, Hs.H.Günthard: Biophys. Struct. Mechanism 1, 75-96 (1975)
6. F.Kopp, U.P.Fringeli, K.Mühlethaler, Hs.H. Günthard: Z. Naturforsch. 30c, 711-717 (1975)
7. B.E.P.Beeston, R.W.Horn, R.Markham: Electron Diffraction and Optical Diffraction Techniques (North Holland/American Elsevier 1973)
8. T.Miyazawa, E.R.Blout: J. Am. Chem. Soc. 83, 712-719 (1961)
9. N.J.Harrick: Internal Reflection Spectroscopy (New York: Interscience Publishers 1967)

Dependence of the Optical Contrast of Vesicle Walls on Lamellarity and Curvature[1]

R.M. Servuss and E. Boroske

Freie Universität Berlin, Institut für Atom- und Festkörperphysik
Königin-Luise-Str.28/30, D-1000 Berlin 33, Federal Republic of Germany

Giant phospholipid vesicles with diameters up to 100 µm could be swollen from a lump of lecithin in water and be observed with a phase-contrast microscope. The contrast of the contour of the vesicle in the image plane depends on the difference of refractive indices of the lipid and the surrounding medium, the thickness of the wall (i.e. lamellarity), and the curvature of the wall in the object plane and perpendicular to it. The intensity along a scan perpendicular to the vesicle's contour in the image plane was measured with a step-motor driven photometer. Due to the influence of the annular phase-plate in the focal plane of the microscope the image of an extended object reveals structures that do not correspond to real properties of the object but must be understood by the complex analysis of the modification of the diffraction patterns.

The contrast of egg-lecithin spheres and tubes with diameters between 2 and 20 µm and having their axis in the object plane was recorded for several diameters and thicknesses of walls. For a given curvature perpendicular to the obiect plane it was easy to discern the number of lamellas in the wall at least for small lamellarity and thus to single out the most interesting vesicles with unilamellar walls. Further it was shown that for a given diameter the contrast of a tube differs from that of a comparable sphere mainly due to a stronger halo in the image of the wall of the tube. This could be understood from the analysis of the influence of the curvature of the wall in the object plane on the formation of the image. Photometric phase-contrast microscopy is thus an undestructive method to single out unilamellar vesicles with diameters down to the resolving power of the microscope and to get information about the curvature perpendicular to the optical axis by analysis of the two-dimensional image.

[1]) To be published in Chem. Phys. Lipids (1980)

Osmotic Shrinkage of Giant Egg-Lecithin Vesicles [1]

M. Elwenspoek and E. Boroske[2]

Freie Universität Berlin, Institut für Atom- und Festkörperphysik
Königin-Luise-Str.28/30, D-1000 Berlin 33, Germany

W. Helfrich

Freie Universität Berlin, Institut für Theorie der Kondensierten Materie,
Arnimallee 3, D-1000 Berlin 33, Germany

Osmotic shrinkage of giant egg-lecithin vesicles was observed by phase-contrast microscopy. The vesicles remained or became spherical when shrinking. Small and thick-walled vesicles formed visible fingers attached to the sphere. The water permeability of the single bilayer was found to be 41 µm/s. A variety of observations indicate that osmosis induces a parallel lipid flow between the monolayers of the bilayer, leading to a strong positive spontaneous curvature. They also suggest the formation of mostly submicroscopic daughter vesicles. The estimated coupling constant, $2 \cdot 10^{-6}$ mole/mole, is large enough to be biologically significant.

1) Submitted to Biophysical Journal
2) To whom correspondence should be sent

Direct X-Ray Study of the Molecular Tilt in Dipalmitoy Lecithin Bilayers

M. Hentschel[1,2], R. Hosemann[1], and W. Helfrich[2]

[1] Fritz-Haber-Institut der MPG, Teilinstitut für Strukturforschung Faradayweg 4-6, D-1000 Berlin 33, Germany

[2] WE 1B, FB Physik, Freie Universität Berlin, Königin-Luise-Str.28/30, D-1000 Berlin 33, Germany

Well-ordered multilayer systems of dipalmitoyl lecithin and water containing enough material for X-ray diffraction were prepared. They consisted of 100 subsystems alternating with mylar films, each film and subsystem being 5 µm thick. The tilt of the hydrocarbon chains away from the layer normal was measured in the water-saturated gel phase (L_{β}') as a function of temperature. The angle varied between 15° and 8° with the maximum at 10°C. The tilt was directed along one of the basic vectors of the quasihexagonal lattice of the hydrocarbon chains.

submitted to Z. Naturforsch., Part a

Part IX

Mesophases of Disk-Like Molecules

Carbonaceous Mesophase and Disk-Like Molecules

H. Gasparoux

Centre de Recherche Paul Pascal, Université de Bordeaux I,
Domaine Universitaire, F-33405 Talence, France

In a first part, the formation of the carbonaceous mesophase will be described and some physical properties of this material reviewed. We would like to show in a second part in what way the synthesis of disc-like molecules provides systems which exhibit striking analogies with the carbonaceous mesophase.

I. Carbonaceous Mesophase

The chemistry of the liquid-phase carbonization is very complex but the understanding of the formation of microstructure in cokes and graphite was substantially improved since the role of the mesophase transformation was recognized first by BROOKS and TAYLOR (1). During carbonization the transformation takes place in the graphitizable organic materials at temperatures comprised between 350°C and 550°C. During this transformation, the large planar molecules formed by the reactions of thermal cracking and aromatic polymerization, orientate more or less parallel to form an optically anisotropic phase called the carbonaceous mesophase. The life time of this mesophase is limited by its hardening to a semi-coke but the "mesomorphic" organization of the flat molecules is essential for thermal graphitizability of the pyrolysis product.

1 Formation

In the earty stages of nucleation and growth the carbonaceous mesophase appears as small spherules suspended in the optically isotropic matrix (fig.1).

Fig. 1 : Spherules of mesophases

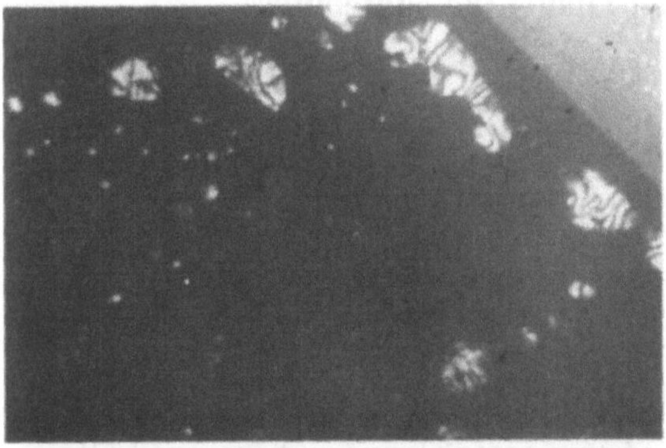

Fig.2 : Coalescence of spherules which leads to an important amount of mesophase in the medium (fig. 3)

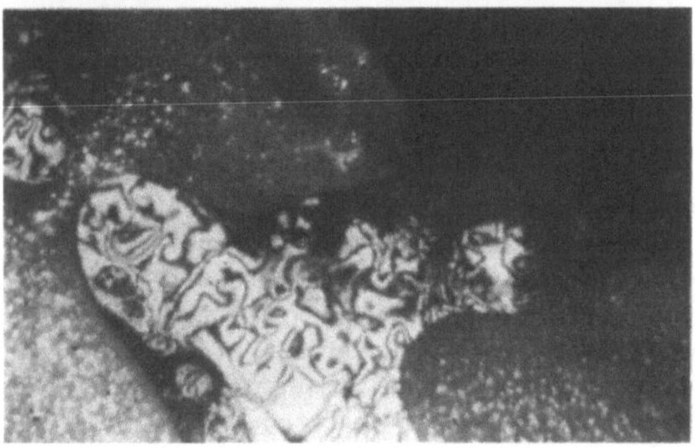

Fig.3 : Mesophase at the end of the coalescence process

As pyrolysis proceeds the spherules grow and begin to sink through the less dense matrix.When spherules meet,coalescence may occur to produce larger droplets (fig.2).

2 *General Characters*

The structural features of these mesophase spherule were described by BROOKS and TAYLOR (2) by applying selected area electron diffraction to thin sections of spherules

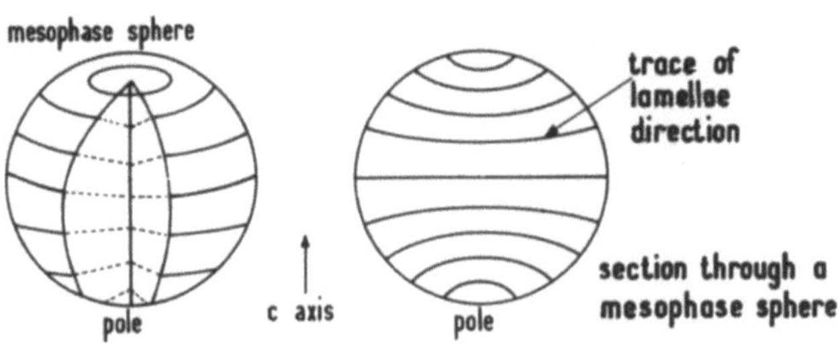

Fig.4 : Diagram of structure within a mesophase spherule

Recently, MOCHIDA and MARSH (3) described large anisotropic spherules of mesophase and proposed two types of structure for these spherules : the onion-like and the concentric structures (fig. 5).

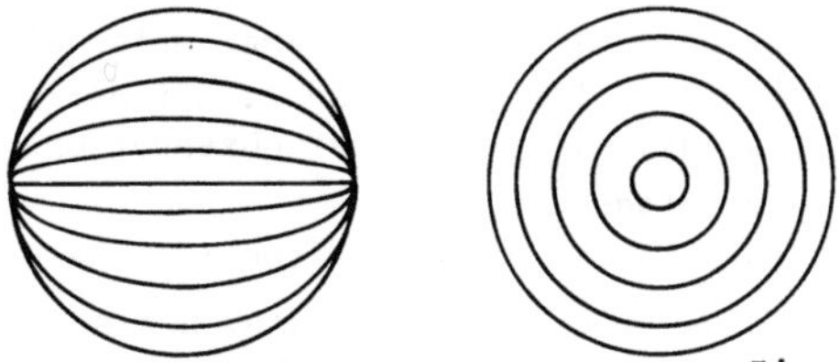

Fig. 5 : Onion-like and concentric structures

The various optical textures of cokes indicate that many parameters drive this microscopic aspect. We have to notice : the composition of the pyrolyzed substances, the range of temperature over which the mesophase is fluid, the viscosity values and the physical disturbance within the carbonization system caused by volatile compounds and convection flows. Thus, the approach of WHITE et al. (4)(5) is to characterize the "defects-structures" found in the industrial prepared cokes with respect to the factors which influence the amount and the dynamics of the mesophase.

This mesophase exhibits many analogies with the classical nematic liquid-crystal, in particular the dynamic behaviour of some disclinations. In addition, it is possible by using polarized-light micrography to inter-relate the defects of the mesophase with those of a classical nematic. WHITE develops, as done in the nematics, the use of co-rotating and counter-rotating nodes and crosses which are labelled as U,Y,O and X respectively (2,2,4,6)(fig.6).

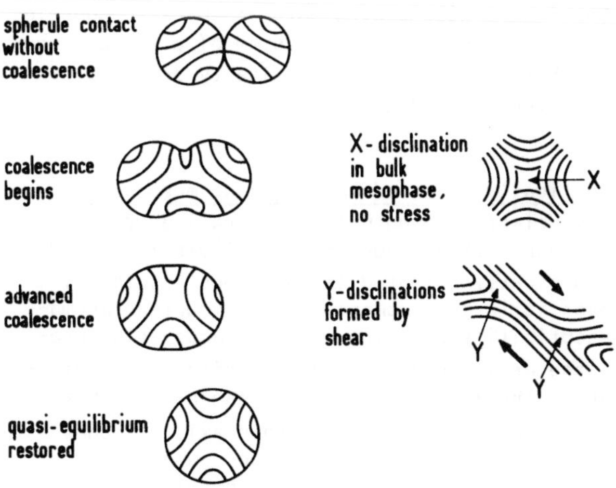

Fig.6: Schematic models for formation and deformation of disclination structures

The density of the carbonaceous mesophase ranges from 1.4 to 1.6 (respectively 1.25 and 1.32 for the isotropic medium).

It has been believed for a long time that the carbonaceous mesophase was insoluble in most solvents and particulary in quinoleine and pyridine which were commonly used in its separation from the isotropic medium. Recent results (7) show that the mesophase solubility depends on the original material from which it is formed and that the percentage of the mesophase is not necessarily the same as the one of insolubles.Thus, direct optical observations seem to be the way of evaluation of this percentage.

It is difficult to precise the exact chemical composition of the mesophase, it depends on the pyrolized compound and is continually changing during the heat treatment. Mass spectrometry studies have shown the coexistence of numerous chemical substances. The average molecular weight of the mesophase is in the 2000 range,much larger than in the isotropic medium, this value increases with temperature and heat treatment time. The C/H ratio is large, which indicates an important aromaticity percentage. Moreover,some recent results suggest the existence of small aromatic molecules or large molecules composed of small aromatic units linked together with alkyl bonds.

3 Industrial Importance of the Carbonaceous Mesophase

Graphite is usually manufactured from a coke prepared from a petroleum pitch and a binder such as a coal tar pitch. The size, growth and coalescence characteristics of the mesophase spherules influence very closely the physical properties of the final products ; so graphite industries are interested in those stages of pyrolysis where mesophase growth and interactions are important.For example, a recent technology is applied in which mesophase still retaining some fluidity is moulded or formed into specific shapes and then graphitized.

Currently the major source of carbon fibers is from synthetic textiles among which polyacrilonitrile is the most important. Economically it is much more attractive to use as raw materials petroleum or coal tar pitch or coal extracts. In order to get carbon fibers with convenient mechanical properties the best results are obtained by spinning a bi-phase system of pitch and mesophase. The process being such that the mesophase is suitably oriented by the shear forces generated during the spinning.

4 Some Recent Advances

Physical properties of a magnetically oriented carbonaceous mesophase

Generally the presence of many defects prevents from generating large single domains, so in order to approach such a structure many investigations concern the properties of a magnetically oriented carbonaceous mesophase. As previously seen, the mesophase is largely composed of polynuclear aromatic molecules the magnetic susceptibility of which is much larger (in absolute value) perpendicularly to the aromatic planes. By rotating the mesophase sample around an axis normal to the magnetic field, we can assume that the aromatic parts lie, on average, perpendicularly to this axis. I summarize here some results recently published (8,9) which concern a quenched mesophase. These studies have shown that it is possible to obtain a highly oriented quenched sample by rotating the mesophase in a magnetic field of about 10 KG.The diamagnetic susceptibility and anisotropy results suggest the presence of either small aromatic molecules or larger molecules with small aromatic regions probably connected by aliphatic bridges ($\Delta\chi \simeq -0{,}80 . 10^{-6}$ uem CGS g^{-1}).

We can remark that the specific heat results indicate the existence of a pregraphitic organization already in the mesophase.

X-ray measurements

Concerning the anthracene carbonization from 400 to 800°C, X-ray diffraction experimental diagrams have been recently compared with theoretical curves which have been computed using several types of models. The results show (10) that the aromatic system at 400°C corresponds approximately to the tetramer of the anthracene. From 400°C to 600°C the growth of these systems is not visible and we can suppose that the aromatic species, bridged by atom groups (not necessarily aromatic) are organized in type of "cobweb". The evolution of the material between 400°C to 600°C can be described as the development of this cobweb structure without changing the size of the elementary aromatic species.

Rheological Investigations

The rheological behaviour is important for controlled growth and coalescence of mesophase spherules. A few measurements deal with rheological characteristics of coal tar pitches during their transformation to mesophase. Mesophase microstructure can be correlated with the flow behaviour followed by rotational viscometry (11,12). A typical curve shows a maximum of the viscosity around 400°C. Flow curves give evidence that the pitches are newtonian liquids at low temperature but a non newtonian character appears above $\sim$ 400°C. COLLET and RAND suggest that the system can be regarded as emulsions and the maximum in the apparent viscosity corresponds to the phase inversion point. It is also possible to find similarities with the evolution of the viscosity of thermotropic liquid crystals, for example, with the nematic-isotropic phase transition (13).

If we sum up the information about the carbonaceous mesophase we can conclude that this material presents numerous analogies with a nematic phase : optical texture, orientation effect by a magnetic field ($\Delta\chi<0$), low viscosity, rheological behaviour ..

II. Disc Like Molecules

All the measurements and observations on the carbonaceous mesophase are difficult because of the high formation temperature of the mesophase joined to gaseous releases which restrict the in situ experiments and at last because of the chemical complexity of the medium obtained by carbonization either of petroleum pitch or of pure aromatic compounds. It is now evident that there was clearly a need for a well defined chemically stable system which could serve as a model for the carbonaceous mesophase. We can define this model as an anisotropic mesophase formed of a pure disc like aromatic compound (starting point of a "cobweb structure" obtained at rather low temperature, without chemical transformations.

1 Columnar Phases

Now several disc-like mesogens are known. The Bangalore group first (4) showed the existence of one type of mesophase with the hexa-n-alcanoyloxy benzene series, then in Paris and in Bordeaux new materials were synthetized and published a few months later : the series of hexa-ester (fig. 8) and hexa-ether (fig. 7) of triphenylene (15,16,17).

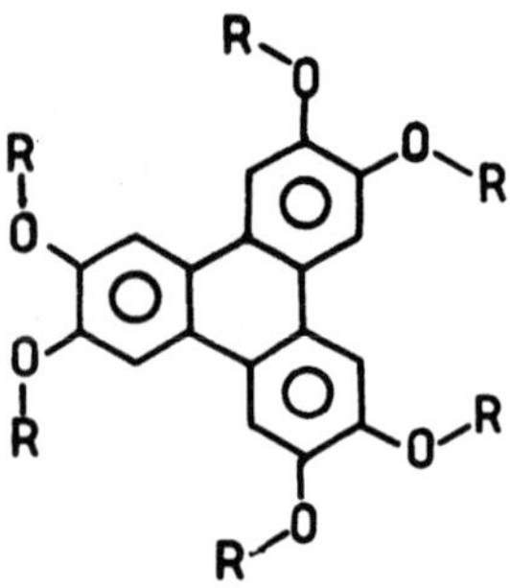

Fig.7 : Hexa-ether of triphenylene

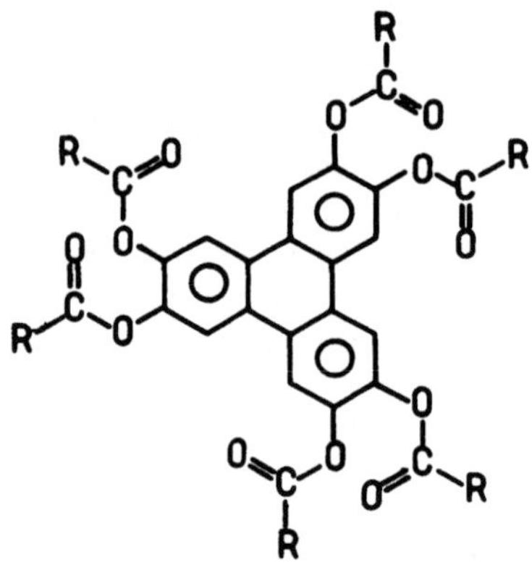

Fig.8:Hexa-ester of triphenylene

The evidence of a polymorphism has been shown by the Bordeaux group (18) and we can propose a classification for these disc-like mesogens starting from X-ray measurements and from systematic microscopic observations. These phases are labelled D_{ho}, D_{hd} or D_{rd} with respect to an "ordered" (o) or "disordered" (d) stacking of disc-molecules constituting columns and with respect to the symmetry of the two dimensional lattice of the columns : hexagonal (h) or rectangular (r) (19).

Nevertheless the columnar lattice proposed by X-ray experiments for all these systems seems not to agree with the structure of the carbonaceous mesophase. Moreover, if we observe the textures of the different columnar phases it is obvious that there is no textural analogy with the micrographies of the carbonaceous mesophase.

2 Nematic Phase

In order to obtain a fluid mesophase it was necessary to find a molecular modification which induces the destruction of the columnar packing.In this way, two projects have been developped. The substitution of a disc-like core with asymmetric substituents has unfortunately leed to the columnar phases (24) On the other hand (20), the substitution of the triphenylene core with highly sterical hindrance substituents enables us to reveal the new type of fluid mesophase we were looking for in the series of hexa-alkyl or alcoxy benzoates of triphenylene (fig.9).

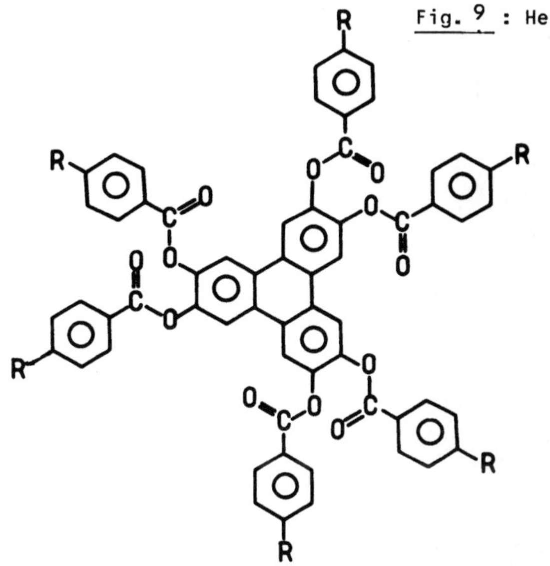

Fig. 9 : Hexa benzoate of triphenylene

An example of the polymorphism of these compounds is presented in the following table.

R	K		D		N_D		I
$C_5H_{11}O$	•	224	–		•	298	•
$C_6H_{13}O$	•	186	•	193	•	274	•
$C_9H_{19}O$	•	154	•	183	•	227	•
$C_{10}H_{21}O$	•	142	•	191	•	212	•
$C_{11}H_{23}O$	•	145	•	179	•	185	•
C_7H_{15}	•	130	•	210	–		•
C_8H_{17}	•	179	–		•	192	•

Polymorphism of the hexabenzoate series

K = crystal D = columnar phase N_D = nematic disc-like phase
I - isotropic

The new fluid phase is labelled N_D because it is quite a nematic phase according to the following arguments (21) :

- the microscopic observations show a great fluidity of this anisotropic phase and give an evidence for schlieren textures similar to classical nematics (Fig. 10 and 11).

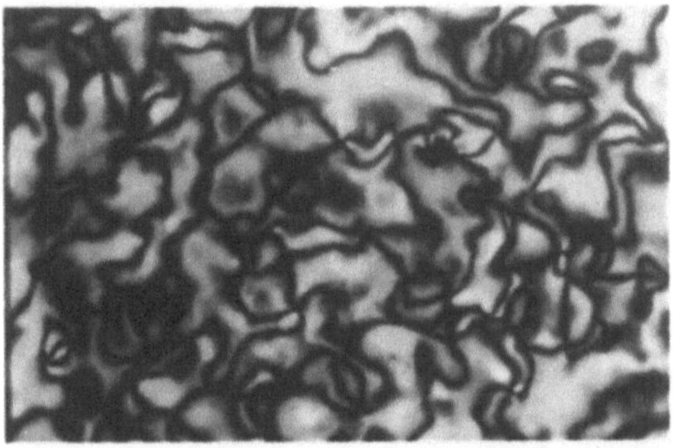

Fig.10 : Nematic phase N_D

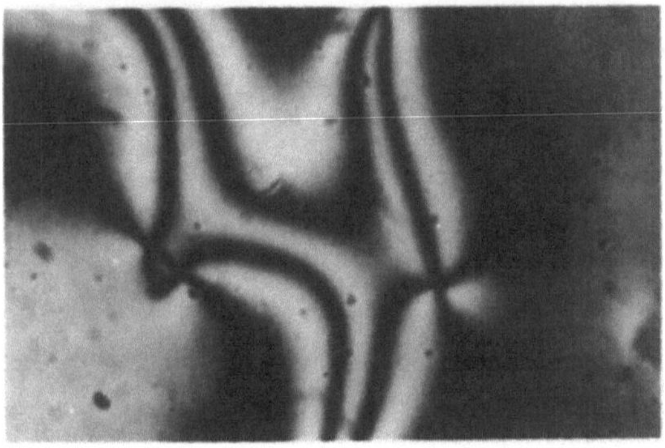

Fig.11 : A detail of the optical texture of the N_D phase

- the X-ray powder diagrams present two diffuse rings which ensure the liquid nature of this phase. In addition, the distortions of these rings in presence of a magnetic field, observed for one compound, involve an orientating effect of this external field. On this aligned sample the diffraction patterns have an infinite rotational symmetry axis perpendicular to the direction of the aromatic planes. Moreover, the structure of the diffuse scatterings at small angles is similar to the diffraction patterns of a nematic phase of elongated molecules with "skewed cybotactic groups"

- the thermal evolution of the magnetic sysceptibility χ corroborates the existence of an orientating effect of the field at the isotropic-fluid mesophase transition (i.e. a sharp decrease of the absolute value of the diamagnetic susceptibility measured in the direction of the magnetic field).It is important to note that contrarily to the columnar phase of the triphenylene ether this effect is saturated for $H \geqslant 5KG$. In addition, the study of the behaviour of a sample in an homogeneous rotating magnetic field of 11 KG shows that the fluid mesophase is isotropic in the rotation plane of the magnetic field, thus this medium is magnetically uniaxial (the director is perpendicular to the magnetic field rotation plane).

The magnetic anisotropy is defined as $\Delta\chi = \chi_{\parallel} - \chi_{\perp}$

$\chi_{\parallel}$: magnetic susceptibility in a direction parallel to the director

$\chi_{\perp}$: magnetic susceptibility in a direction perpendicular to the director

$\chi_{\parallel}$ and $\chi_{\perp}$ are negative (diamagnetic molecules) and for such "disc-like" aromatic molecules $|\chi_{\parallel}| > |\chi_{\perp}|$ in the fluid mesophase, thus $\Delta\chi < 0$.

From these arguments we can claim that this fluid mesophase uniaxial and without translational order is a disc-like nematic phase (N_D for short). As for series of rod-like molecules the low temperature ordered phase which succeeds to the nematic phase can exhibit different structures according to the length of the substituents.

Conclusion

The nematic disc-like mesophases we have just described present many characteristics of the carbonaceous mesophase (22) and we can expect that the understanding of this complex system will be facilitated by the investigations on pure compounds exhibiting disc-like nematic phases.

It is now important to obtain new series with such phases based on cores of species participating in the carbonaceous mixture. For this purpose, we have undertaken the synthesis of substituted truxene and decacyclene. (fig. 12). At the same time, the study of the carbonization of the corresponding hexasubstituted core would provide interesting comparison.

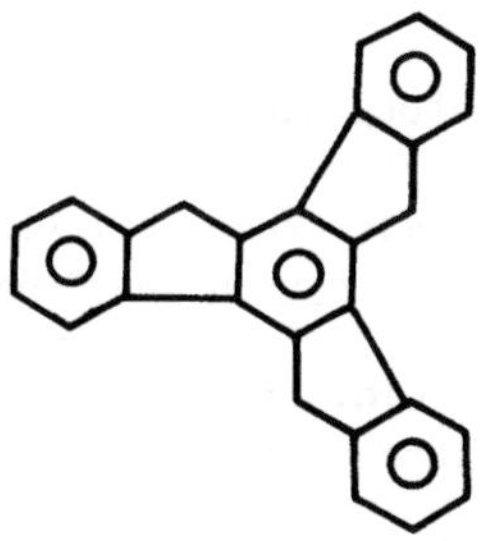

Truxene

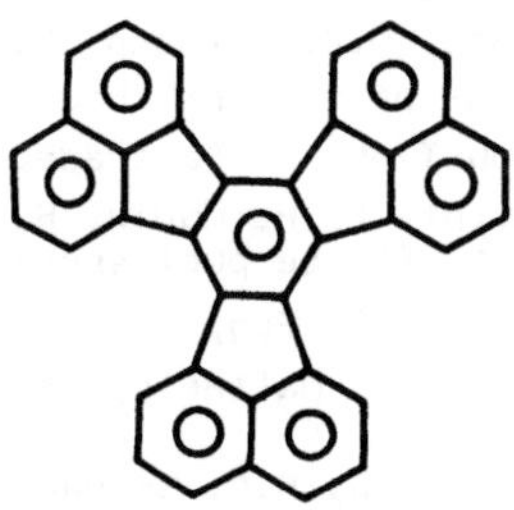

Decacyclene

Fig. 12

Besides this expected clarification of the carbonaceous mesophase problems, the analysis of this disc-like nematic in itself is very stimulating especially because we are in the presence of a new fluid phase which is able to orientate in an external field ... and we know what happened during the last years with orientated rod like mesophases. I wish to thank M.F. ACHARD, C. DESTRADE, F. HARDOUIN, A. JAUNAIT, J. PROST , G. SIGAUD and NGUYEN HUU TINH for valuable discussions.

Bibliography

1 - J.D. Brooks and G.H. Taylor, Nature 226, 697 (1965)
Carbon 3, 185 (1965)

2 - J.D. Brooks and G.H. Taylor, Chem. Phys. Carbon 4, 243 (1968)

3 - I. Mochida and H. Marsh, 14th Biennal Conf. on Carbon-Penn State Univ. (1979)

4 - J.L. White, G.L. Guthrie, J.O. Garner, Carbon 5, 517 (1967)

5 - J.L. White, Prog. Sol. State Chem. 9, 59 (1975)
6 - H. Honda, H. Kimura, Y. Sanada, Carbon 9, 695 (1971)
7 - S. Chevastiak and I.C. Lewis, Carbon 16, 156 (1978)
8 - L.S. Singer and R.T. Lewis, 11th Biennal Conf. on Carbon, (1973)
9 - P. Delhaes, J.C. Rouillon, G. Fug, L.S. Singer, Carbon 17, 435 (1979)
10 - J.M. Guet, F. Rousseaux, D. Tchoubar, 14th Biennal Conf. on Carbon, Penn. State Univ. (1979)
11 - G.W. Collet and B. Rand, Fuel 57, 162 (1978)
12 - R. Baldaha and E. Fitzer, 14th Biennal Conf. on Carbon, Penn. State Univ. (1979)
13 - R.S. Porter, J.F. Johnson, Rheology Vol. 4, Acad. Press. p. 317 (1967)
14 - S. Chandrasekhar, B.K. Sadashiva, K.A. Suresh, Pramana 9, 471 (1977)
15 - Nguyen Huu Tinh, J.C. Dubois, J. Malthete, C. Destrade, C.R. Acad. Sci. C, 286 (1978)
16 - J. Billard, J.C. Dubois, Nguyen Huu Tinh, A. Zann, Nouv. J. de Chim.2, 535 (1978)
17 - C. Destrade, M.C. Mondon, J. Malthete, J. Phys. 40 C 3, 17 (1979)
18 - C. Destrade, M.C. Mondon, Nguyen Huu Tinh, Mol. Cryst. Liq. Cryst. Liq. Lett. 49, 169 (1979)
19 - C. Destrade, M.C. Bernaud, H. Gasparoux, A.M. Levelut, Nguyen Huu Tinh Proceedings of the Int. Liq. Cryst. Conf. Bangalore (1979)
20 - Nguyen Huu Tinh, C. Destrade, H. Gasparoux, Phys. Lett. 72 A, 251 (1971)
21 - A.M. Levelut, F. Hardouin, H. Gasparoux, C. Destrade, Nguyen Huu Tinh, J. de Phys. (to be published)
22 - H. Gasparoux, G. Fug, C. Destrade, Mol. Cryst. and Liq. Cryst.
23 - C. Destrade, J. Malthete, Nguyen Huu Tinh, H. Gasparoux (to be published)
24 - Nguyen Huu Tinh, M.C. Bernaud and C. Destrade, 3d Congrès of Liq. Cryst. of Socialist Countries, Budapest 1979

Discotic Mesophases: A Review

J. Billard

Laboratoire de Physique de la Matière Condensée*, Collège de France,
F-75231 Paris Cédex 05, France and
Laboratoire de Dynamique des Cristaux Moléculaires*, Université de Lille,
F-59655 Villeneuve d'Ascq Cédex, France

1. Introduction

The different nematic and smectic mesophases correspond to some fashions to arrange, with a partial order, elongated and flexible molecules [1] that is to say calamitic molecules (from χαλαμοσ : reed). Other mesophases, the plastic crystals, are formed, with globular molecules [2]. Some people have wondered if we would obtain another sort of mesophases. A possibility is to consider molecules with a different simple general form. The more simple form after the elongated ellipsoid and the sphere is the flat ellipsoid. If we can obtain also mesophases with these discoid molecules we can name these ones *discotic* (from δισχοξ : quoit) [3]. Many modes to arrange disks with partial order are conceivable [3]. These considerations have been experimentally confirmed, firstly on some hexahydroxybenzene-n-alkanoates [4] and, secondly, on hexaalkoxytriphenylenes [3]. Observed with a polarizing microscope equipped with a heating stage these derivatives exhibit birefringent and non rigid phases.

Some precursors can be found :

- the hexahydroxybenzene heptanoate was first synthesized in 1937 [5], but the mesophase hadn't been detected,
- the mesophase of the di *iso*butylsilane diol elaborated in 1952 [6] and studied in 1955 [7] is discotic [8],
- different models are proposed [9 to 11] for mesophases of anhydrous soaps where the polar groups are located in disks arranged in parallel and equidistant planes and the paraffinic chains disordered,
- the model imagined [12] for the carbonaceous mesophase is in fact a discotic mesophase (for a new evidence by high resolution electron microscopy see [14]).

During the last years, many papers and communications are devoted to this subject. With this rapid increasing of the scientific activity over this type of mesophases we can conclude we have now a new field in the mesomorphy.

2. Molecular Architecture

The first molecules elaborated exhibiting a discotic mesophase are a transposition of the more common calamitic mesogens. These molecules possess a rigid central part and flexible side chains. Contrary to the calamitic mesogens where we can have only two-fold axis parallel or perpendicular to the director, it is possible for the discogens, to have a more diversified symmetry : Fig.1.

*Equipes associées au C. N. R. S.

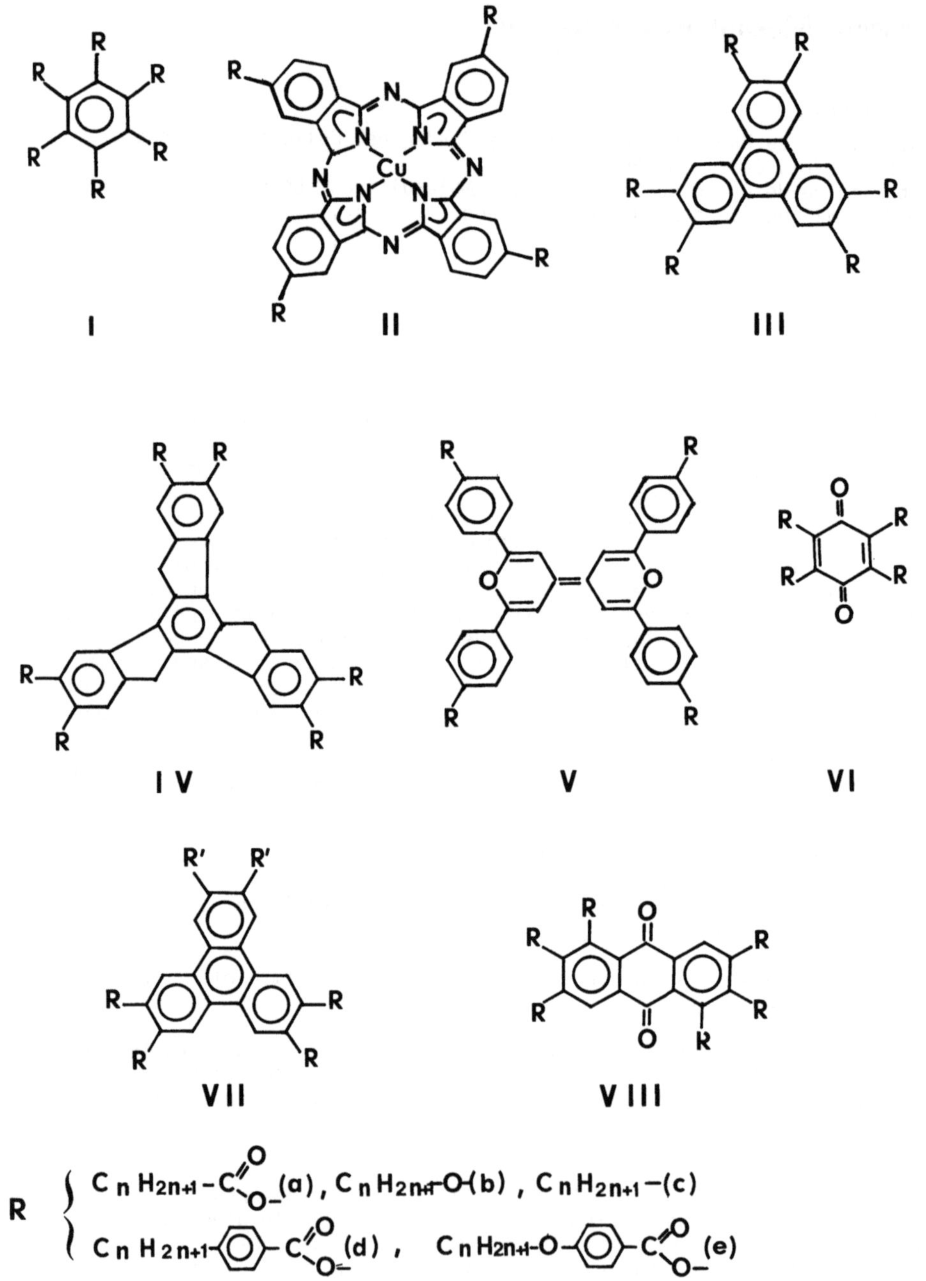

Fig. 1 Some known discogenic compounds (I to VIII) with different side chains R (a to e)

The symmetries are the hexagonal 6/m m m for I_[4], tetragonal 4/m for the only lyotropic discogen II [14], the trigonal $\bar{3}$ m for III [3,15,16] and 3 for IV [17], the binary m m m for V [18] and VI [19] (but these last compounds exhibit no stable discotic mesophase), 2 m m for VII [20] and 2/m for VIII [21]. We can conclude that many chemical series of disk-like molecules with different molecular symmetries can exhibit the discotic mesomorphism.

Many sorts of central cores and side chains can be imagined. But if we combine only two by two the today used parts we obtain twenty possibilities. Only seven combinations are tested : I a 6 and 7 [4,22,23] , III a 6 to 13 [24] , III b 4 to 10 [3,16] , III d 7 and 8 [24] , III e 5, 6 and 9 to 11 [24] , V c 9 and 12 [18] and VIII c 9 and 12 [21] . In this list and later the compounds are described by a roman numeral for the core type, followed by a letter for the side parts sort and, endly, by the number n to specify the chain length. Are not mentioned here the Ringsdorf's polymers with negative unaxial mesophases [25] , the di *iso*butylsilane diol [8], the lyotropic salt of the 4,4',4'',4''' tetracarboxylated copperphtalocyanine [14] , the truxene derivatives [17] and the molecules with non equal side chains [20].

The list of the studied discogenic compounds is short and it is difficult to make interesting comparisons. For the I a, III a, III b series, versus the chain length n, the clearing point is a decreasing function and the melting temperature exhibits a minimum [23] . The effect of the central core can be studied on I a 7, III a 7 and VIII a 7 compounds : the more extended mesomorphic range (60°C) and the lower melting point (66°C) are obtained with III a 7. For this triphenylene III core we can observe the influence of the side groups. The larger mesomorphic ranges are obtained with III a 8, III b 4, III e 6 and III d 7. These two last series are the most interesting. This is to rapproach to the fact that the acids corresponding to the side parts are calamitic mesogens [26,27] ; that is to say with interactions which give a tendency to the molecular parallelism.

There is no general remark over the synthesis of these derivatives. Only, often, the purification is difficult.

3. Microscopical Observations

Observed with a polarizing microscope equipped with a heating stage the mesophases of the III b compounds obtained by the melting of the crystals have a confuse birefringence. By cooling of the liquid phase, the mesophase appears in domains with digitized contours. After a total transformation it exists many areas with irregularly curved limits. By a slow cooling are obtained areas totally black between crossed polars and rare birefringent defects. The homogeneous parts observed in convergent light appear uniaxial with the optical axis perpendicular to the slide. We name them : *normally oriented areas*. The optical sign is negative.

In each defect observed without analyzer a *rectilinear axis* appears. This axis is not visible if it is parallel to the electrical field of the light. The two optical neutral lines are respectively parallel and perpendicular to the rectilinear axis. The observation with a compensator furnishes the orientations of the optical axis near a defect. The optical axis are contained in planes perpendicular to the axis of the defect. The surfaces perpendicular to the optical axis (the *normal surfaces* [28]) are cylinders parallel to the

Fig. 2

Discophase obtained by cooling of the liquid phase of 2,3,6,7,10,11 hexa heptyloxytriphenylene: a) usual texture ; b) defects in normally oriented areas (courtesy J.C. Dubois and A. Zann) axis of the defect. They are surfaces with a single curvature. Consequently, the curvature have no effect on the distance between two points of a normal surface. The axis of the defect is an intersection line for the normal surfaces [3] . We can define a unit polar vector $\vec{n}$ parallel in each point to the optical axis and with an arbitrary choice of the sense but definited by continuity outside the defect axis : *director vector*. This vectors field has a conservative flow : $\vec{\nabla}.\vec{n} = 0$; there is no splay. The circulation of $\vec{n}$ around a closed path surrounding the defect axis is non null : $\vec{\nabla} \times \vec{n} \neq 0$; we have $\vec{n}.(\vec{\nabla} \times \vec{n}) = 0$: there is no twist. We have only bend [1] . This is the experimental basis of the theory from M. KLEMAN [29] . We can't have $\vec{n}$ equal to a gradient and the structure is not layered.

If we press over the cover slip of a microscopical preparation with normally oriented areas the number of defects is increased, and the axis of defects are locally parallel, practically any relaxation of the defects is observed : the mesophase is too viscous [3] . For preliminary theoretical studies of the hydrodynamical properties see [30] to [32] .

The microscopical observations can furnish also information about the symmetry of the mesophase. In such a supercooled mesophase the crystals appear in fine rectilinear needles. Arrived to a limit between two different normally oriented areas these needles curve and take again their rectilinear growth. Progressively a network is obtained. In a single normally oriented area the needles made between them angles multiple of thirty degrees. We have only six directions perpendicular to the optical axis for the easy crystal growth. Consequently around the optical axis all the directions are not equivalent. This optical axis is only a three or sixfold symmetry axis [3] . This symmetry is confirmed by the observation of the slow transformation of the liquid in mesophase : the contours of a normally oriented areas have clearly a sixfold symmetry [33] . But with these observations it is not possible to distinguish between a trigonal and an hexagonal symmetry.

The mesophases of I a compounds are also birefringent and viscous [4] . By cooling the liquids give mesomorphic domains without digitation. The growth speed is highly anisotropic and occurs practically in only one direction. Many defects appear with rectilinear axis. The mesophase obtained by melting of a single crystalin needle have defects with rectilinear axis parallel to the original needle length [23] . With the disk-like compounds the supercooling is important and monotropic discotic mesophases can be so observed [4] .

By cooling of the liquid phases two V c compounds exhibit highly viscous and birefringent mesophases. By pressing over the cover slip, defects with rectilinear axis are obtained. No relaxation is observed. By a slow cooling, domains with digit-like contours admitting a fourfold symmetry appear. By cooling of a single domain the crystalline needles formed are oriented in two perpendicular directions. These microscopical observations conduct to attribute a two or fourfold symmetry at these mesophases formed by molecules admitting a binary symmetry [18] .

The III a compounds exhibit also mesophases [16] with rectilinear axis. In this series, for the first time, a discotic polymorphism has been observed [16,34] . At the transition texture changes are observed. In normally oriented areas appear, by cooling, *rectangular areas* (Fig.7 in [21]) exhibiting a more intense birefringence, but with the same extinctions directions as the field around. The thermal measurements confirm the existence of this polymorphism (see *infra*).

Similar observations can be made on VIII a 7 : by cooling the liquid, appears a viscous birefringent mesophase with digitized contours. After the total change of state the texture exhibits defects with rectilinear axis. By suitable cooling a new texture with supplementary bars is produced. This is a new example of discotic polymorphism. Information about the symmetry of the mesophase stable at more elevated temperature can be obtained by microscopical observations. If the cooling is slow, the mesophase appears from the liquid in domains with, apparently, a sixfold symmetry. By pressing over the mesophase the unmber of defects with rectilinear axis increases. The greatest part of these axis is parallel in an homogeneous area [21] . This molecules with a binary symmetry are assembled in a mesophase which exhibits a six, three or twofold symmetry.

With the III d and III e compounds another discotic polymorphism is observed [24] . By cooling, the liquids give spherical droplets of a birefringent and *fluid* phase. After the total transformation, the texture is homeotropic or schlieren-like, very similar to these of the nematic calamitic mesophases with curved $s = \pm 1/2$ and $s = \pm 1$ disclination lines [17] . By pressing over the mesophase the number of defects increases. These one relax spontaneously in a short time [24] . This phase is optically uniaxial [35] . An intense thermal agitation in the orientation of the optical axis is easy to observe [17] . The coherence length [1] is similar to this one of the nematic mesophase. The main optical difference between these two sorts of mesophases is the optical sign : positive in nematics [36] and negative for the corresponding discotic mesophase [35] . By suitable cooling the fluid mesophases of III e 6 and III d 8 give mesophases with *mosaic textures*. For III d 8 some domains with the optical axis perpendicular to the slide can be observed. This is an uniaxial or slightly biaxial mesophase. By pressing a fragmentation and deformations of the domains can be obtained, but without formation of birefringent defects. By increasing the temperature the birefringence decreases at the transition from the mosaic-like to the fluid mesophase : it is a lowering of the order parameter. The normally oriented domains give homeotropic fluid areas with only some disclination lines. From an oblique mosaic grain appear fluid regions with practically uniform extinctions. Inversely by decreasing the temperature the germs of the mosaic-like mesophase appearing in an uniformly oriented fluid area conduct to mosaic grains with the same orientation. We can conclude to a reciprocal orientation of these two mesophases [35] similar to this one observed with nematic and smectic B calamitic mesophases [37] .

4. Thermodynamical Observations

Sometimes it is very difficult to observe the phase transitions with a polarizing microscope, a more sure method to detect these phenomenons is the differential scanning calorimetry. This technic provides also a good purity criterion.

Measurements has been made for I a [4,22,23] , III b [3,16] , III a [16] , III e [24] , VIII a 7 [21] and V c [18] . Generally, the melting enthalpy changes are larger compared to the discotic-discotic and to the discotic-liquid values. Only for V c derivatives the molar enthalpy changes are larger for mesophase-isotropic transitions than for the meltings. This situation has exceptionally been observed with only one calamitic mesogen [38] . These calorimetric measurements conjointly with the microscopical texture observations establish the discotic polymorphism. For the I a 5 and I a 6 compounds the heat capacities under constant pressure are measured with a good accuracy between

13 and 393°K [39,40] . A rich solid polymorphism and post-transitional thermal effects are to note. Curiously in the mesophase the heat capacity is lower than in the adjacent solid and liquid phases.

To build the isobaric binary phase diagrams the microscopical contact method [41] can be used. But there are here some particular difficulties : the big supercoolings and the small texture changes at the melting or at the transitions between two highly viscous discophases. To distinguish the solid and viscous discotic mesophases we can press over the cover slip and observe the deformations. Without any exception until now, the disk-like compounds are totally miscible in the liquid state and do not form continuous solid solutions. Generally, two compounds of a same series are totally miscible in their discotic state and the observed eutectic point is in accordance with the calculations including only the temperatures and the enthalpy changes of the pure components [42,43] . Often the spindle between the mesophase and the liquid exhibits a minimum due to the non ideality of these mixtures, similarly to several instances for smectogens [44] . But if the chain lengths are too different there is no total miscibility in the discotic state [16] . Today, only one example of total miscibility in a discotic state for members from two different series is known : III a 7 and VIII a 7 [21] . In this case, the two components have dissimilar molecular symmetries. This result proves we can use for discotic mesophases the miscibility criteria similarly to the calamitic mesophases [21] .

If two mesogens are totally miscible in a mesomorphic state these mesophases have the same type. Until now no exception has been found to the four following miscibility rules :

1) The reciprocal of the definition does not exist : two compounds exhibiting mesophases of the same type are not always totally miscible in this state.

2) If two mesophases can be in equilibrium in a two phases system these two mesophases have different types.

3) For pure compounds the inversion in the succession order for the mesophases versus the temperature (reentrant polymorphism) is exceptional.

4) Two emantionners are totally miscible in all their non-solid states.

These rules applied to the discotic mesophases conduct to attribute the same name to the stable mesophases of III a 7 and VIII a 7. This III a 7 is totally miscible in the discotic state with III a 9 itself totally miscible with the III a 10 in its discophase stable at lower temperatures [16] . We can use a notation similar to the one used for the smectic mesophases : $\mathcal{D}$ for discotic and a latin letter in inferior index. The two mesophases of III a 10 separated by a first order transition are called $\mathcal{D}_A$ and $\mathcal{D}_B$. So the III a 9, III a 7 and VIII a 7 have each a $\mathcal{D}_B$ phase. The monotropic mesophase of VIII a 7 is different from the preceding, we name this $\mathcal{D}_C$ [21] . For the fluid discophase of III d and III e I suggest the denomination $\mathcal{D}_F$ with F for fluid. Other phase diagrams are necessary to classify all the known discophases.

More than the identification by isomorphy the isobaric binary phase diagrams can give other information. For example adiscogen and a nematogen or a smectogen are totally miscible in the liquid state, but in the mesophases the mutual miscibilities are very small [3,24,45,46] and the addition of a chiral calamitic mesogen cannot induce an observable twist in a $\mathcal{D}_F$ phase [24] .

Similarly to the case of the calamitic mesogens [47] no significative particularity is observed in the phase diagrams for monotropic or virtual phase transitions involving a discophase [23]. It is possible, by the isobaric binary phase diagram method, to determine the virtual discotic-liquid temperatures of the hydrocarbons corresponding to some possible cores : 55°C for the triphenylene, - 80°C for the pyrene [48] and - 24°C for the biphenylene. But + 50°C for the anthraquinone, that is to say a value near this one of the triphenylene and this data was an argument to chose the rufigallol derivatives VIII to synthesize [21] .

To study by this way the discotic potentiality of the benzene as central core, benzene has mixed with I a 5. In the phase diagram appears a narrow stability field for the discophase. This is the first stable lyotropic discotic mesophase found [23] . The crystallization process is difficult for the corresponding mixtures. This fact explains the observations previously made on benzenic solutions evaporated at room temperature [22] . The solutions in water of the sodium salt of II with R = $-coo^-$ exhibit also a lyotropic discophase [14] . It is to note that this mesophase is oriented between two slides rubbed in one direction [14] .

The formation of discotic mesophase by mixing with a volatile solvent is interesting to observe the textures : after the departure of the solvent the texture is more net that this are obtained by cooling of the pure liquid [22, 21] . It is so possible to observe a texture of a mesophase without memory from the texture of a mesophase stable at a more elevated temperature : by evaporation of a suitable volatile solvent at one temperature where the mesophase to observe exists [21] .

The thermobarograms of the pure disk-like compounds are rare : only for I a 7 and I a 5 [22] . For I a 7, the discophase disappears at a triple point corresponding to 1.4 k bar. By use of the Clapeyron's relation the molar volume changes are obtained : 12.3 cm^3/mole for the melting and 3.5 cm^3/mole for the clearing under zero pressure. Contrary for I a 5 the discophase is stable only under a pressure over 160 bars [22] . The extrapolation of the discophase-liquid coexisting curve conducts to a virtual transition under zero pressure in an excellent accordance with this one obtained by the isobaric binary phase diagram method [23] .

5. Structures

We have seen that some information on the structures can be obtained by the microscopical observation. The binary phase diagrams can establish the isomorphism. To obtain more rich structural information the X-ray diffraction is used. This method permits also to detect the practically second order phase transitions without clear texture change.

From the first X-ray observations are deduced for the mesophases of the I a compounds the irregular stacking of the molecular disks in columns, the perpendicularity of the molecular planes to the column axis and the hexagonal packing for the columns [4,22] . Similar results are obtained for III b with details about the orientational disorder of the molecules around their director in the columns, the absence of correlation for the positions about the axis of the columns for the molecules from two neighbouring columns, the bending of the columns and the disorder in the side chains [49] . For the III a 11 compound a similar structure is obtained for the discophase $\mathcal{D}_2$ stable in the more elevated temperature range [50] . If III a 11 and III a 10 have the same discotic polymorphism, this phase $\mathcal{D}_2$ is a $\mathcal{D}_A$ phase. For the $\mathcal{D}_B$ phases of III a 7 and III a 11 the structure is also in columns perpendicular to the molecular planes, but the lattice of these columns is rectangular [51] . The columns don't admit any infinitefold symmetry axis, only twofold axis and a mirror plane perpendicular to their axis. The transition between $\mathcal{D}_B$ and $\mathcal{D}_A$ is similar to a smectic E smectic B phase change. For the studied $\mathcal{D}_B$ and $\mathcal{D}_A$ phases the distribution of the molecules along the column axis and the side chains are more disorder than in the discophases of I a [50] .

For the $\mathcal{D}_F$ phases of III d and III e compounds the order is only in the average parallelism for the directors [17,52] . The III e 11 compound exhibits, at lower temperatures, another discophase with texture similar to these of the $\mathcal{D}_B$ phases of III a [52]. With III e 6 under $\mathcal{D}_F$ is obtained a discophase with a mosaic texture. The structure of this phase is columnar. The lattice of the columns is rectangular and the average planes of the molecules are not perpendicular to the column axis, this is a tilted mesophase : $\mathcal{D}_L$ [52] .

These X-ray structural determinations can't distinct between centered and non-centered structures. By these methods is obtained only the maximal symmetry for the phases. The actual symmetry can be a meriedry of the collected symmetry.

With the datas from microscopical examinations, binary phase diagrams and X-ray observations we obtain the following maximal symmetries :
2/m for $\mathcal{D}_L$ obtained with molecules having the $\bar{3}$ m symmetry,
2/m m m for $\mathcal{D}_B$ formed by molecules having the $\bar{3}$ m or 2/m symmetries,
6/m m m for $\mathcal{D}_A$ formed by molecules having the $\bar{3}$ m symmetry,
∞/m m m for $\mathcal{D}_F$ obtained with molecules exhibiting the $\bar{3}$ m symmetry,
for $\mathcal{D}_C$ it is no sufficient data available today.

The known succession order for the discophases versus the temperature is reported Fig.3. With the data now collected we know only the order $\mathcal{D}_C$ $\mathcal{D}_B$ $\mathcal{D}_A$

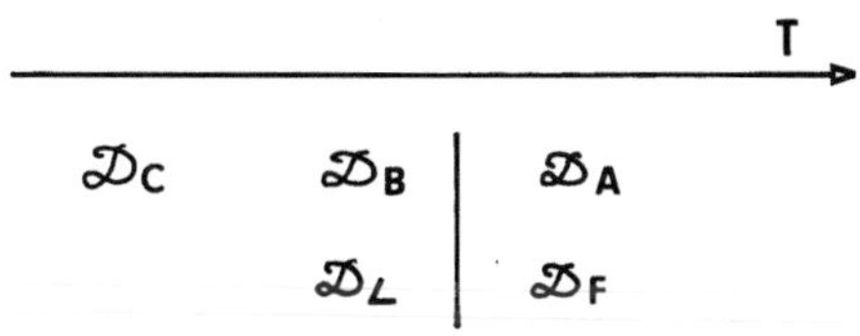

Fig.3 Succession order of the discophases versus the temperature

by increasing temperature, the position of $\mathcal{D}_F$ over $\mathcal{D}_B$ and of $\mathcal{D}_L$ under $\mathcal{D}_F$

With the structural information we can understand some aspects of the textures. For the mesophases with columns perpendicular to the molecular planes the disorder along the column axis is in accordance with the non layered organization obtained from the defects observations. The absence of strong molecular correlations between the molecules of two adjacent columns is in accordance with an easy bending. The compact packing of the columns explains the difficulty to change the distance between two points from a surface normal to the optical axis and the single curvature of this surface. For the molecules around a defect with rectilinear axis the average molecular planes contain the defect axis, they are similar to the pages of an opened book : Fig.4.

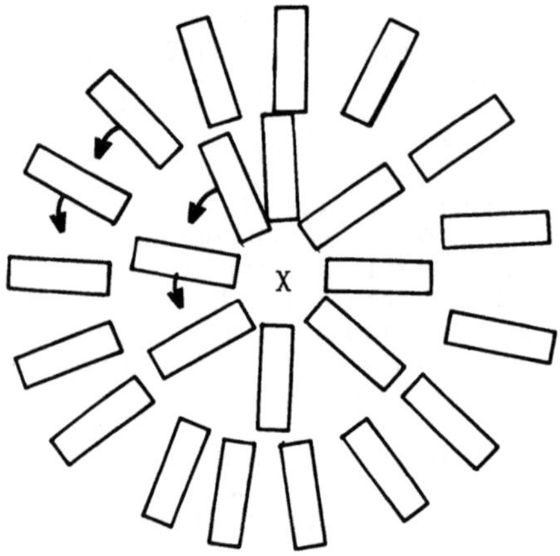

Fig.4 Disposition of the molecules near a defect with rectilinear axis (X). The arrows represent the directors

The mosaic texture of the tilted $\mathcal{D}_t$ mesophase proves the difficulty to bend the columns, in fact the tilt conducts to stronger correlations for the positions along the column axis of the molecules from two neighbouring columns : Fig.5.

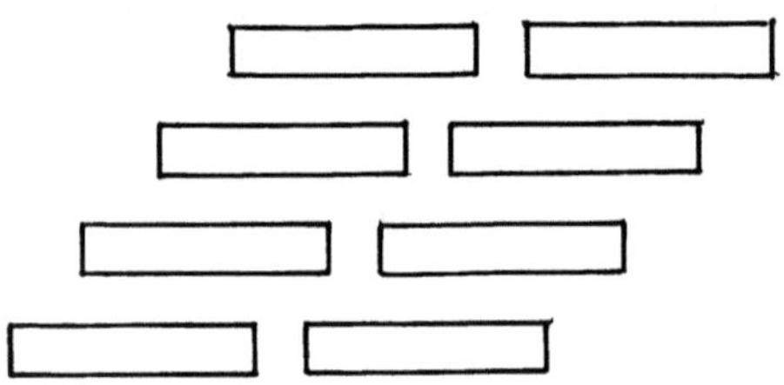

Fig.5 Correlations for the positions along the column axis of the molecules from two neighbouring columns in a tilted discophase.

In the $\mathcal{D}_F$ phase there is no columnar organization and bend, twist and splay deformations are possible.

The structural information about the solid phases of the discogenic compounds is rare : preliminary datas for I a 7 [4] and the crystal structures for three III a 3, III a 4 and III a 5 compounds [53] . But none of these compounds exhibit stable discophase.

6. Other Physical Properties

We have seen that the discophases exhibit continuous tensorial properties (the refractive index for example) and discontinuous tensorial properties [54] (the orientation of crystals-growing in an orthogonal columnar discophase for example).

Some other physical properties experimentally studied are to mention briefly :

- the infra-red absorption spectra of I a 5 and I a 6 compounds [39,40] ,
- the magnetic susceptibilities [55] and the magnetic orientation of the discophases from III derivatives [55,52] ,
- the formation of cybotactic groups in the $\mathcal{D}_F$ phase [52] ,
- the distances between the disk-like molecules in a discophase incorporating non mesogenic molecules [46] .

The first specifical application of the discophases is the study of the carbonaceous mesophase [56] . Another possible interest is as model to study the chloroplasts of the plants with chlorophyl.

7. Conclusion

There are many works to made about the today known discogenic compounds. It is necessary to elaborate more diversified compounds to elucidate the relations between the molecular structure and the thermodynamical stability of the discophases. With the some available discogens a rich polymorphism is obtained. Probably it remains other discophases to discover.

For the calamitic mesogens the first cholesteric has been studied in 1888 [57] and a clear view was obtained only in 1922 [36] . The existence of the discophases has been established in 1977 and we can say that, with the knowledge previously accumulated, has been possible to made similar progress during only two years.

The discophases constitut the first state of the matter intentionally discovered [3] . This success proves that we have made a step forward. After the synthesis of compounds having such chemical or physical properties, molecules which assemble in a fashion previously wanted have been built. The proof would be more clear if we could create voluntarily other partially ordered states of the matter.

Acknowledgements

I am much indebted to many colleagues for valuable discussions and information particularly Professors P.G. de Gennes and S. Chandrasekhar and Doctors J.C. Dubois, M. Dvolaïtzky, P. Le Barny, A.M. Levelut, J. Malthète, Nguyen Huu Tinh, H. Strzelecka and A. Zann.

I thank the University of Lille for the break of the discophase study prohibition.

References

1. P.G. de Gennes : "The Physics of Liquid Crystals", (Clarendon, Oxford 1974)
2. J. Timmermans : J. Chim. Phys. *35*, 331 (1938)
3. J. Billard, J.C. Dubois, Nguyen Huu Tinh, A. Zann : Nouv. J. de Chi., *2*, 535 (1978)
4. S. Chandrasekhar, B.K. Sadashiva, K.A. Suresh : Pramàna, *9*, 471 (1977)
5. H.J. Backer, S.J. van der Baan : Rec. Trav. Chim. Pays-Bas, *56*, 1161 (1937)
6. C. Eaborn : J. Chem. Soc, 2840 (1952)
7. C. Eaborn, H. Hartshorne : J. Chem. Soc, 549 (1955)
8. J.D. Bunning, J.W. Goodby, G.W. Gray, J.E. Lydon these proceedings, p. 397
9. A.E. Skoulios, V. Luzzati : Acta Cryst., *14*, 278 (1961)
10. P.A. Spegt, A.E. Skoulios : Acta Cryst., *17*, 198 (1964)

11. B. Gallot, A.E. Skoulios : Kolloïd Z., *210*, 143 (1966)
12. J.D. Brooks, G.H. Taylor : Carbon, *3*, 185 (1965)
13. D. Augié, M. Oberlin, A. Oberlin, P. Hyvernat : Carbon, to appear
14. S. Gaspard, A. Hochapfel, R. Viovy : C. R. Acad. Sci. Paris, *289C*, 387 (1979)
15. Tinh Nguyen Huu, J.C. Dubois, J. Malthète, C. Destrade : C. R. Acad. Sci. Paris, *286C*, 463 (1978)
16. C. Destrade, M.C. Mondon, J. Malthète : J. de Phys., *40C3*, 17 (1979)
17. C. Destrade, M.C. Mondon-Bernaud, H. Gasparoux, A.M. Levelut, Nguyen Huu Tinh : Proc. Int. Liquid Crystals Conf., Bangalore, 1979 (Heyden and Son, London, to appear)
18. R. Fuguitto, H. Strzelecka, A. Zann, J.C. Dubois, J. Billard : Chem. Comm., to appear
19. S. Chandrasekhar, B.K. Sadashiva, K.A. Suresh : Proc. Int. Liquid Crystals Conf., Bangalore, 1979 (Heyden and Son, London, to appear)
20. Nguyen Huu Tinh, M.C. Bernaud, C. Destrade : Proc. Conf. Liquid Crystals, Budapest, 1979, to appear
21. A. Queguiner, A. Zann, J.C. Dubois, J. Billard : Proc. Int. Liquid Crystals Conf., Bangalore, 1979 (Heyden and Son, London, to appear)
22. S. Chandrasekhar, B.K. Sadashiva, K.A. Suresh, N.V. Madhusudana, S. Kumar, R. Shashidhar, G. Venkatesh : J. de Phys., *40C3*, 120 (1979)
23. J. Billard, B.K. Sadashiva : Pramāṇa, *13*, 309 (1979)
24. Nguyen Huu Tinh, C. Destrade, H. Gasparoux : Phys. Let., *72A*, 251 (1979)
25. H. Kelker, U.G. Wirzing : Molec. Cryst. Let., *49*, 175 (1979)
26. C. Weygand, R. Gabler : Z. phys. Chem., *46B*, 270 (1940)
27. D. Demus, H. Demus, H. Zaschke : "Flüssige Kristalle in Tabellen", (Deutscher Verlag f. Grundstoffindustrie, Leipzig 1974)
28. F. Grandjean : Bull. Soc. fr. Min., *42*, 42 (1919)
29. M. Kléman : these proceedings, p. 97 .
30. L. Ricard, J. Prost : Proc. Conf. Liquid Crystals, Budapest, 1979, to appear
31. J. Prost, N.A. Clark : Proc. Int. Liquid Crystals Conf., Bangalore, 1979 (Heyden and Son, London, to appear)
32. J. Prost, N.A. Clark : these proceedings, part X, p. 409 ff
33. Y. Bouligand, O. Rechou : "Discotic Liquid Crystals", (Institut de Cinématographie scientifique, Paris, 1980)
34. C. Destrade, M.C. Mondon-Bernaud, Nguyen Huu Tinh : Molec. Cryst., *49*, 169 (1979)
35. A. Zann, P. Le Barny, J.C. Dubois, J. Billard : to be publish
36. G. Friedel : Ann. de Phys., *18*, 237 (1922)
37. M. Warenghem : these Proceedings, **part X, p. 409 ff**
38. A. de Vries, D.L. Fishel : Molec. Cryst., *16*, 311 (1972)
39. M. Sorai, K. Tsuji, S. Seki : Molec. Cryst., to appear
40. M. Sorai, K. Tsuji, H. Suga, S. Seki : Proc. Int. Liquid Crystals Conf., Bangalore, 1979 (Heyden and Son, London, to appear)
41. L. Kofler, A. Kofler : "Thermomikromethoden", (Verlag Chemie, Weinheim 1954)
42. J. Malthète, M. Leclercq, J. Gabard, J. Billard, J. Jacques : C. R. Acad. Sci. Paris, *273C*, 265 (1971)
43. J. Malthète, M. Leclercq, M. Dvolaïtzky, J. Gabard, J. Billard, V. Pontikis, J. Jacques : Molec. Cryst., *23*, 233 (1973)
44. M. Domon, J. Billard : J. de Phys., *40C3*, 413 (1979)
45. F. Hardouin, G. Sigaud, M.F. Achard, H. Gasparoux : Molec. Cryst., to appear
46. R.E. Goozner, M.M. Labes : Molec. Cryst. Let., *56*, 75 (1979)
47. M. Domon, J. Billard : Proc. Int. Liquid Crystals Conf., Bangalore, (1973) Pramàna Suppl. p. 131

48. A. Béguin, J. Billard, J.C. Dubois, Nguyen Huu Tinh, A. Zann : J. de Phys., *40C3*, 15 (1979)
49. A.M. Levelut : J. de Phys. Let., *40*, 81 (1979)
50. A.M. Levelut : J. de Phys., to appear
51. C. Destrade, J. Malthête, A.M. Levelut, Nguyen Huu Tinh : Proc. Int. Liquid Crystals Conf., Budapest, (1979) to appear
52. A.M. Levelut, F. Hardouin, H. Gasparoux, C. Destrade, Nguyen Huu Tinh : these proceedings, p. 396
53. M. Cotrait, P. Marsau, C. Destrade, J. Malthète : J. de Phys. Let. *40*, 519 (1979)
54. G. Friedel : "Leçons de Cristallographie", (Berger-Levrault, Nancy 1926)
55. G. Sigaud, M.F. Achard, C. Destrade, Nguyen Huu Tinh: these proceedings, p. 403
56. H. Gasparoux : these proceedings, p. 373
57. H. Kelker : Molec. Cryst., *21*, 1 (1973)

X-Ray Investigations and Magnetic Field Effects on a Fluid Mesophase of Disk-Like Molecules[1]

A.M. Levelut and F. Hardouin[2]

Laboratoire de Physique des Solides associé au CNRS, Université de Paris-Sud, Bât.510, F-91405 Orsay, France

H. Gasparoux, C. Destrade, Nguyen Huu Tinh

Centre de Recherche Paul Pascal, Université de Bordeaux I, F-33405 Talence, France

Abstract

The X-ray diffraction patterns of the fluid phase of two alkoxy benzoates of triphenylene have been investigated : for non oriented sample, this pattern is very similar to the isotropic liquid one, since we observed two broad diffraction rings ; the outer ring is near $1/d \simeq 1/4.5$ Å^{-} while the inner rings corresponds respectively to d = 22 Å and 27 Å for the hexyloxy and the undecyloxy derivative.

In order to check the anisotropy of this phase we have performed magnetic measurements. By Faraday method we observe that a static magnetic field has an orienting effect on this phase. The orientational order is improved by the rotation of sample with respect to the magnetic field. Measurements in a rotating field show that the samples are uniaxial, the axis being perpendicular to the magnetic field and the Faraday method allows us to measure the diamagnetic anisotropy $\Delta\chi = \chi_{//} - \chi_{\perp}$ where $\chi_{//}$ and $\chi_{\perp}$ are respectively parallel and perpendicular to the director :

$\Delta\chi$ = - 0,49 10 uem csgg^{-1} for the C_6 derivative

$\Delta\chi$ = - 0,21 10 uem cgsg^{-1} for the C_{11} derivative

The nature of the orientational order can be checked by performing X-ray diffraction experiment on oriented samples. The compounds held in a Lindemann glass capillary tube are put into a magnetic field of .3T perpendicular to the incident beam. We have obtained oriented samples of the C_6 derivative but not of the C_{11}. For the best oriented sample the inner ring splits into for broad spots while the 4.5 Å outer rings split into two discs as in the type II lyotropic nematic phase. The structure of the inner ring is similar to the pattern of skewed cybotatic nematic and the four spots are indicative of a pretransitional order. This order is reminiscent of the low temperature columnar phase order, since the columnar phase of the C_6 derivative is a phase formed of columns in which the cores of the molecules are tilted with respect to the plane normal to the column axis with a 55° angle. The lattice of the column is centred rectangular. On the contrary the columnar phase of the C_{11} derivative exhibits a non-tilted herringbone rectangular lattice similar to the columnar phase of the short chains paraffinic esters of the triphenylene.

---

[1]Will be published in Journal de Physique

[2]Permanent address: C.R.P.P., 33405 Talence (France)

The Classification of Mesophase of Di–i–butylsilanediol

J.D. Bunning[2], J.W. Goodby[1], G.W. Gray[1] and J.E. Lydon[2]

[1] Department of Chemistry, The University, Hull, HU6 7RX, England

[2] Astbury Department of Biophysics, The University, Leeds, LS2 9JT, England

1. Introduction

The classification of the thermotropic mesophase of di*iso*butylsilanediol (DIIBSD) has remained an unsolved problem for the past 25 years. This compound was prepared in 1952 by EABORN [1] and he noticed that it had a "double melting point". A widespread programme of synthesis and examination of other alkyl-silanediols was carried out subsequently but it failed to reveal any other meso-genic compounds and it would appear therefore that the di*iso*butyl compound is unique. Examination of DIIBSD by hot stage optical microscopy confirmed that a mesophase was formed over the temperature range 89.5° - 101.5° (our values differ slightly from these and are indicated in Fig.1). The optical textures seen were unlike any previously encountered and in their paper of 1955 EABORN and HARTSHORNE [2] described these in detail but were unable to offer a satis-factory model for the molecular arrangement in the mesophase.

Prior to this investigation, an X-ray crystal structure determination had been carried out on the diethylsilanediol by KAKUDO and WATASE [3] and it was reported that in the crystalline solid of this compound the molecules are held together in chains by the association of opposed OH dipoles thus -

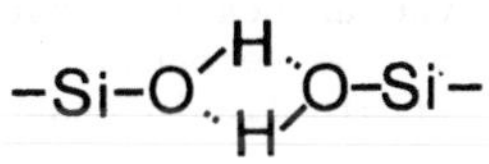

rather than by conventional hydrogen bonds.

A preliminary X-ray investigation of the crystalline solid of DIIBSD was carried out by BERNAL *et al.* (see EABORN and HARTSHORNE [2]). The crystals were found to be triclinic with a = 17.79Å, b = 5.06Å, c = 28.82Å, $\alpha \simeq 90^{\circ}$, $\beta \simeq 121^{\circ}$, $\gamma \simeq 96^{\circ}$. A Patterson projection showed a pattern of peaks compatible with a structure similar to that reported for the diethyl compound. EABORN and HARTSHORNE therefore accepted this model for the structure of solid DIIBSD and sought an explanation for the optical properties of the mesophase in terms of structures consisting of either chains or sheets of chains bonded as shown above.

The main points raised by the observations of EABORN and HARTSHORNE may be listed:

1) The molecular structure offers no clue as to the type of mesophase formed and there is no apparent reason why mesogenic properties should be restricted to the di*iso*butyl compound.

2) The crystal structure of DIIBSD does not resemble that of typical nema-togenic or smectogenic compounds.

3) The mesophase shows unique optical textures, one of which contains large, optically negative, homeotropic areas and it is difficult to reconcile this with a structure for the mesophase consisting of molecular chains of the type thought to be present in the solid.

Since 1955 no relevant publications have appeared, except for a recent report of the crystal structure determination of diphenylsilanediol [4]. In this case only conventional hydrogen bonds were found and there were no interactions of the opposed dipole type.

We have attempted a comprehensive structural study of the DIIBSD mesophase using a range of physical techniques. In addition to repeating the optical microscopy we have extended the X-ray investigation to a study of the mesophase itself and we have undertaken differential thermal analysis and miscibility tests.

2. Experimental

2.1 Differential thermal analysis

DTA was used to confirm the transition temperatures obtained from hot stage optical microscopy and to determine the enthalpies of the crystalline solid → mesophase and mesophase → isotropic liquid transitions. The DTA trace is shown in Fig. 1. Note that the ΔH for the mesophase → isotropic transition is unusually large - the inference being that a considerable fraction of the intermolecular bonding in the crystalline solid is retained in the mesophase.

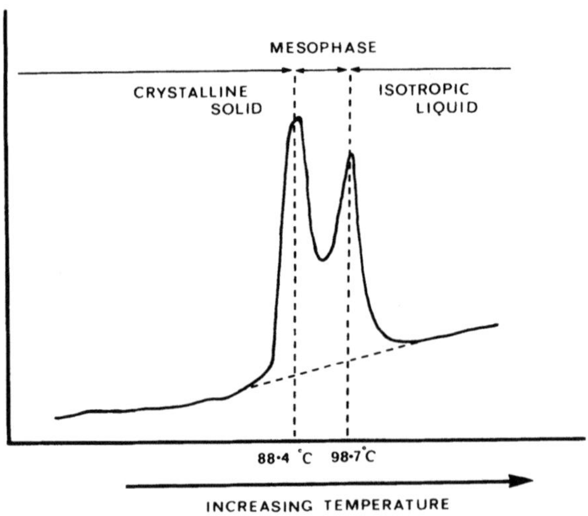

Fig.1 The DTA trace for DIIBSD on heating. Note the unusually large size of the mesophase → isotropic liquid peak

It was also found that, in the trace obtained for cooling, the peak corresponding to the isotropic → mesophase transition is considerably broadened - indicating that appreciable decomposition occurs for even short periods at these temperatures.

2.2 X-ray Diffraction

The X-ray diffraction pattern of the mesophase of DIIBSD is particularly simple. Only two factors are apparent: an outer diffuse ring corresponding to a repeat distance of 4.7Å and an inner ring corresponding to a repeat distance of about 11Å. This inner ring is of an intermediate type - being more diffuse than the low angle reflections found for smectic phases but less diffuse than those of nematic phases.

2.3 Optical Microscopy

Isotropic Liquid → Mesophase Transition

Rapid cooling of the isotropic liquid causes the mesophase to appear as rounded dendritic growths as shown in Fig. 2. These regions are homeotropic and conoscopic investigation showed them to be optically negative. They can be seen most easily in polarised light - with the polariser in and the analyser out (rather than with crossed polars) by virtue of the Becké line effect at their edges. Occasional regions develop which are birefringent - presumably where the dendrite does not lie in the plane of the slide.

As the dendrites increase in size and coalesce to form the bulk mesophase, long, perfectly straight, lines of disclination appear giving the "rod" texture shown in Fig. 3.

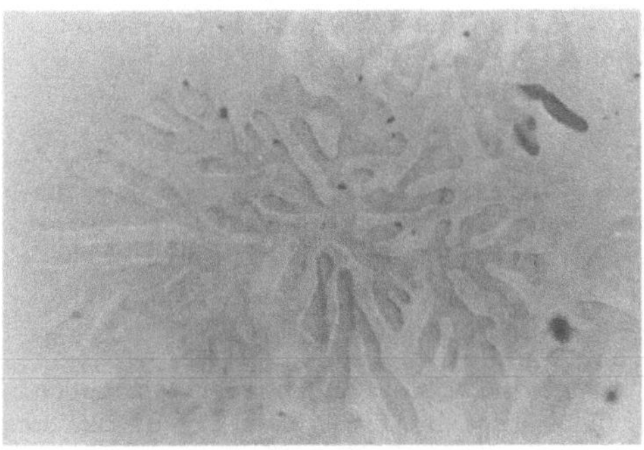

Fig. 2 A homeotropic dendritic growth of the DIIBSD mesophase (surrounded by the isotropic liquid). Polarised light, x100

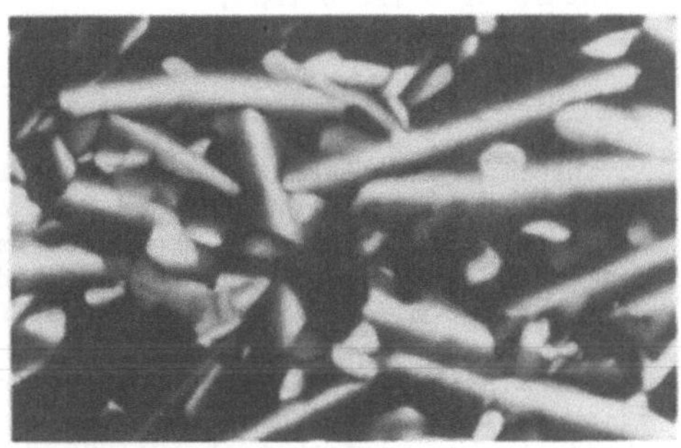

Fig. 3 The "rod" texture of the DIIBSD mesophase. Crossed polars, x100

Under crossed polars, these "rods" appear as bright lines shading gradually into the dark background and they extinguish every 90°. Under polarised light, with the polariser in and the analyser out, they are also discernable as discontinuities in the structure and the contrast with the surrounding regions is lost every 180°.

We have also observed a more confused petal-like fan texture which can be produced by slower cooling.

Crystal → Mesophase Transition

When the crystalline solid is produced by rapid cooling (via the mesophase) or by evaporation of a solution, spherulites are formed. When these are heated, the radial array of needles is replaced by a tangential array of mesophase domains as shown in Figs. 4 and 5. Note that these domains are often tilted out of a strictly parallel alignment to give oblique chevron-like patterns.

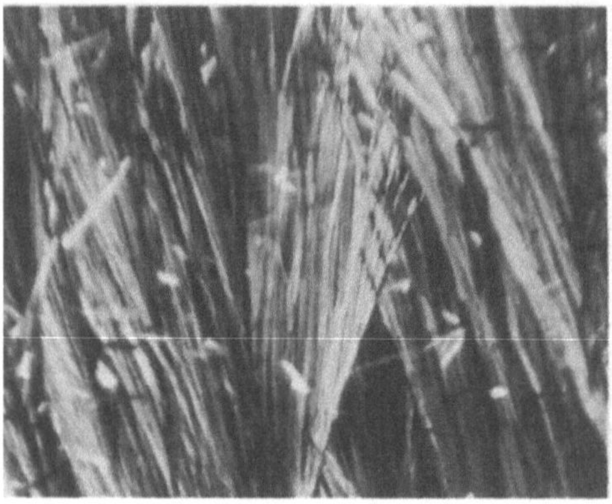

Fig. 4 A spherulitic array of crystals of solid DIIBSD. Crossed polars, ×75

Fig. 5 The same area of sample as Fig. 4 when heated to form the mesophase. Crossed polars, ×75

2.4 Miscibility Studies

At the time when EABORN and HARTSHORNE were investigating this mesophase, no comparable optical textures had been observed. There are, however, a number of points of similarity with the textures seen recently for the discotic phase of benzene-hexa-n-heptanoate as described by CHANDRASEKHAR *et al.* [5] . This prompted our miscibility study of DIIBSD and BHH.

The miscibility diagram of state for binary mixtures of these two compounds is shown in Fig. 6 and we take the continuous region of miscibility of the mesophases to be proof that the DIIBSD mesophase is indeed discotic.

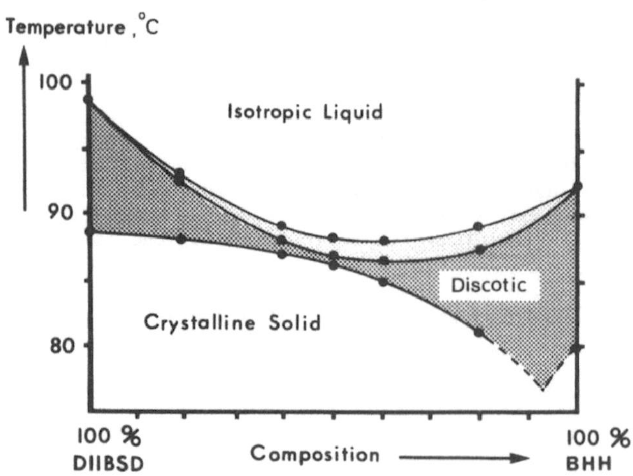

Fig. 6 The miscibility diagram of state for DIIBSD and BHH

The mixing would appear, however, to be far from thermodynamically ideal because of the extent of the depression of the isotropic liquid → mesophase transition temperatures.

3. Discussion

In retrospect, the bonding scheme proposed by KAKUDO and WATASE looks very suspect. We are not aware of any other crystal structure where this type of molecular association has been found and it certainly does not exist in crystals of diphenylsilane diol (the only other silane diol to have been subjected to a full crystal structure determination).

If space-filling models of DIIBSD are examined, it appears that their geometry is favourable for dimer formation by conventional hydrogen bonding as shown in Fig. 7. The *iso*butyl groups give the dimer the profile of an approximately square slab. This, coupled with the miscibility test, leads us to postulate that the DIIBSD mesophase is discotic and that the basic unit is the dimer. Model building also suggests that these dimers can stack in a tilted column enabling further hydrogen bonding to occur along the stack axis. The transition from crystal → mesophase is pictured as involving a weakening of the bonding along the stacks and a realignment of the dimers from a tilted to a normal pattern. This would explain the dramatic texture changes at the crystal → mesophase transition. The untilting of the dimers causes the structure to contract along the stack axis and expand about a perpendicular direction. This expansion cannot be accommodated without some buckling of the structure giving the chevron-like patterns mentioned above. A reappraisal of the single crystal X-ray diffraction data also makes a discotic model appear attractive. If additional lattice points are added half-way along the c axis, as shown in Fig. 8, this produces a close approximation to a hexagonal array of the type suggested for the BHH mesophase.

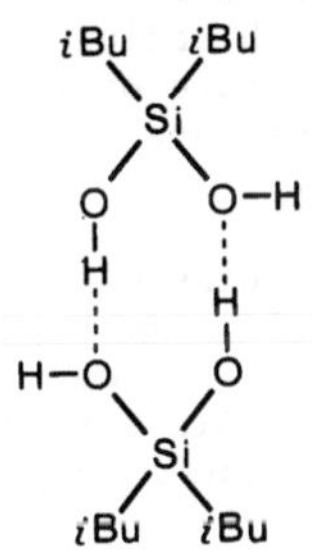

Fig.7 The DIIBSD dimer which we postulate is the basic unit of the mesophase

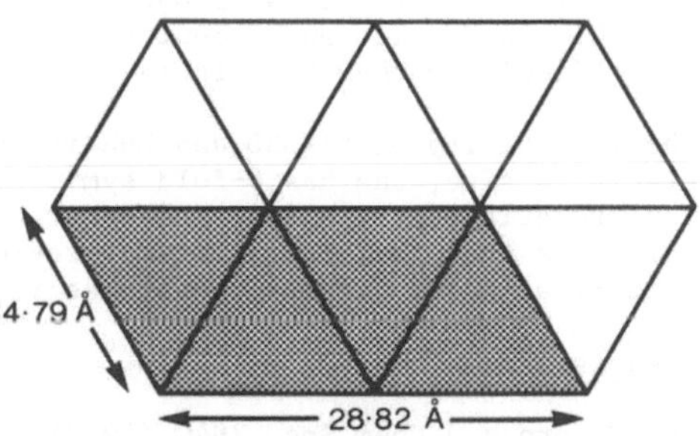

Fig.8 The unit cell of the crystalline solid of DIIBSD with extra lattice points at c = ½ forming an approximately hexagonal lattice

The crystal → mesophase transition involves breaking the inter-dimer hydrogen bonds and the mesophase → isotropic liquid transition involves breaking the intra-dimer hydrogen bonding. Since these are present in equal numbers this would explain why the two transition enthalpies have approximately equal magnitudes.

The discotic model for the mesophase also offers satisfactory explanations for the optical textures observed. The dendritic mesophase growths as shown in Fig. 2 are single domain regions of liquid crystal and because they are being viewed down the 6-fold axis they are pseudo-homeotropic. The pattern of branching is a manifestation of the 6-fold symmetry analogous to that of the snowflake. Dendritic mesophase growths were not reported for BHH but they have been found recently for other discotic systems by BILLARD [6].

Straight line disclinations have also been found recently by BILLARD [6] in discotic mesophase textures. We propose the structure shown in Fig. 9.

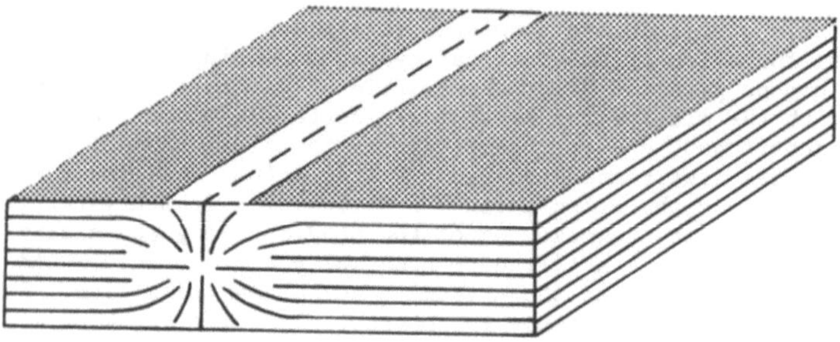

Fig. 9 The molecular arrangements in the rod texture. The planes drawn here indicate the alignments of the discotic molecules. It is not intended to imply that the structure is layered. The molecular stacks lie perpendicular to these construction planes

The straightness of the lines must owe its origin to the relative values of the elastic constants of the discotic phase, bending being much easier than splay. This gives the phase properties which are similar to those of a sheet of flexible card. It can be bent easily in one direction, but once it is bent, it is difficult to then bend it in a perpendicular direction.

The miscibility of DIIBSD and BHH is surprising. The molecules differ considerably in size, one has 6-fold symmetry and the other 4-fold symmetry and one has an aromatic centre whilst the other has sites for hydrogen bonding. The corollary is that the miscibility criterion appears to be as valid for the characterisation of discotic phases as it has proved to be for smectic phases.

References

1. C. Eaborn : J.Chem.Soc. 2840 (1952)
2. C. Eaborn, N.H.Hartshorne : J.Chem.Soc. 549 (1955)
3. M. Kakudo, T. Watase : Technol.Reports Osaka Univ. 50, 2, 247 (1952)
4. L. Párkányi : Cryst.Struct.Comm. 7, 337 (1978)
5. S. Chandrasekhar, B.K.Sadashiva, K.A.Suresh : Pramana 9, No 5, 471 (1977)
6. J. Billard : these proceedings, p. 383

Magnetic Susceptibility of Mesophases of Disk-Like Molecules

G. Sigaud, M.F. Achard, C. Destrade, and Nguyen Huu Tinh

Centre de Recherche Paul Pascal, Université de Bordeaux I,
F-33405 Talence, France

Since the discovery of the first disc-like molecule [1] exhibiting a thermotropic mesomorphism, several series have been synthesized [2,3,4,5,6] and some structural properties of these new compounds were reported essentially by the way of X-ray experiments [7,8,9]. For further knowledge of these materials, it seems fruitful in particular to precise their behavior under an external magnetic field. Thus we have performed the thermal variations of the diamagnetic susceptibility χ (measured in the direction of the magnetic field, Faraday method [10]) of several compounds of two series synthesized in our laboratory [3,5] and with the general formula :

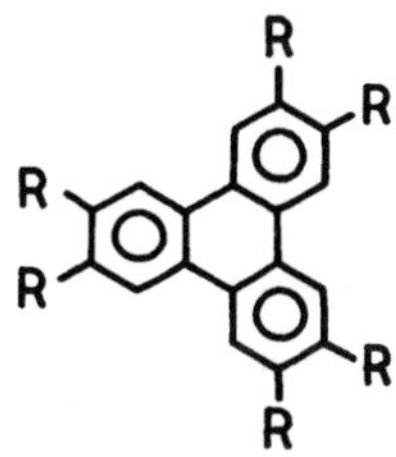

with R $= OC_nH_{2n+1}$: hexaalkoxytriphenylenes

with R $= OCOC_nH_{2n+1}$: hexaalcanoyloxytriphenylenes

The experiments are performed from the isotropic phase by slow cooling of the sample under magnetic field. The magnetic susceptibility measured in the isotropic phase is noted $\bar{\chi}$.

1. The Hexaalkoxytriphenylene series

The homologues of these series exhibit only one mesomorphic state [5]. The structure of this mesophase has been studied by X-ray measurements [7] :the observations reveal a rather regular stacking of molecules (i.e. ordered) in parallel columns (interspacing of the molecules in each column ∿ 3.6 A) and a hexagonal arrangement of the columns which nevertheless remain uncorrelated each one from another (figure 1). This phase has been labelled D_{ho} : "h" for hexagonal bidimensional lattice, "O" for ordered molecular spacing in each column [11,12].

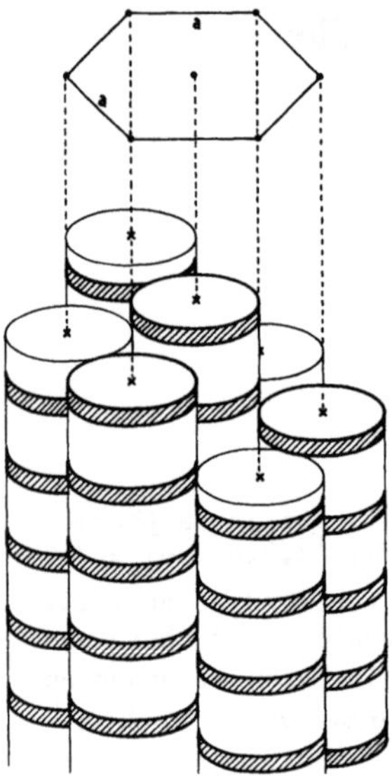

Fig. 1 : Structure of the Dh_o phase of hexa-alkoxytriphenylenes.

Magnetic behavior :

As shown on figure 2 (n=5,8), at the isotropic → columnar phase transition the diamagnetism decreases : this indicates an orientating effect of the magnetic field.

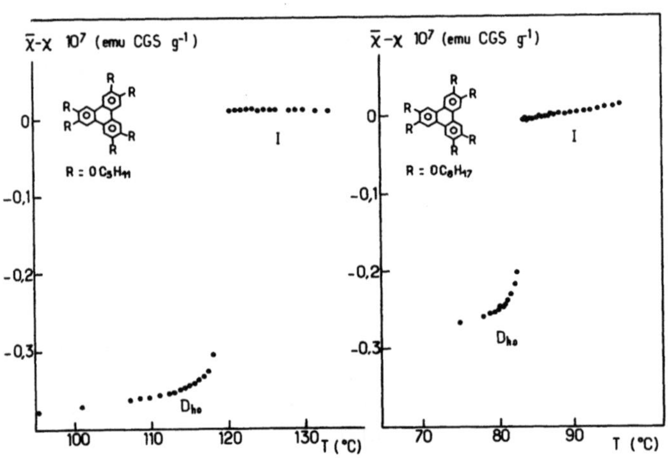

Fig. 2

This slight decrease of the diamagnetic susceptibility measured in the direction of the magnetic field means that the molecules tend to orientate with their aromatic core parallel to the magnetic field because the diamagnetism is essentially due to the aromatic part of the molecule. Nevertheless the orientation is not complete. Indeed we have performed a magnetic field variation on the n=5 compound and measured the magnetic susceptibility discontinuity at the isotropic - columnar phase transition : stronger the field is, better the orientating effect (figure 3), but nevertheless it is not yet saturated with the maximum value of the magnetic field (17KG).

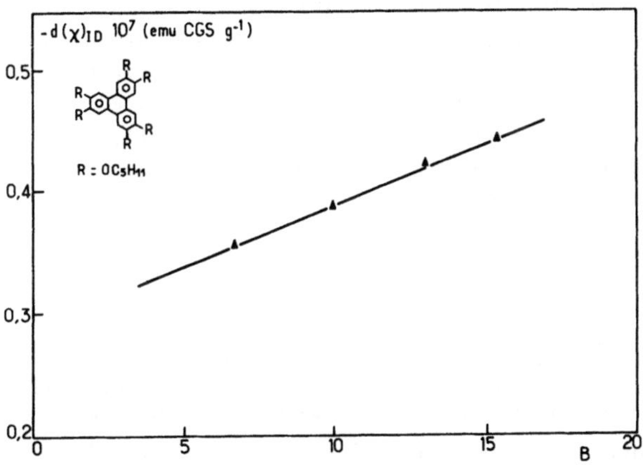

Fig. 3

Now, if we compare the thermal variations of the diamagnetic susceptibility for the n=5 and n=8 homologues (figure 2), the I-D discontinuity is weaker for the long chain compound.

This effect can be attributed in part to the difference in molecular weight but this seems not to be sufficient. From n=5 to n=8, the variation of the ratio of the aromatic part with respect to the aliphatic part (which modifies the values of the molecular anisotropy $\Delta\chi_0$) has also an influence in the same way. At last the difference in the degrees of order at the transition has certainly a non negligible effect which can be appreciated by the comparison of the I-D transition enthalpies of the two compounds:

n=5 $\Delta H = 2.04$ Kcal.mole^{-1}

n=8 $\Delta H = 1.04$ Kcal.mole^{-1}

the greater enthalpy corresponds to the larger magnetic susceptibility discontinuity.

2. The Hexaalcanoyloxytriphenylene Series

In this series, we have to distinguish the lower homologues (with n⩽9) which present only one mesomorphic state, from the long chain compounds for which a polymorphism has been revealed [5]. In this case, optical observations lead to differentiate three columnar phases, called D_2, D_1 and D_o [13], when D.S.C. measurements and X-ray diffraction study show only two mesophases D_2 and D_1 [9]. These disc-like phase analysed by X-ray experiments present the following characteristics [9] :

- D_2 corresponds to an hexagonal lattice of columns : the molecules are more or less irregularly located in each column (i.e. "disordered", and the columns are uncorrelated each one from another. This phase could be called D_{hd} : "h" for hexagonal, "d" for disordered [11,12].
- D_1 presents a rectangular lattice of columns : the molecules are also irregularly stacked in a column and there is no correlation between neighbouring columns. The abbreviation D_{rd} ("r" : rectangular, "d" = disordered) summarizes the essential structural characteristics of this phase [11,12].

Note also that the short-chains homologue mesophase is of D_1 type and that the D_2-D_1 transition for the superior homologues can be interpreted as a buckling of the molecules.

Magnetic behavior :

- n ⩽ 9 (n = 6,7 figure 4).

At the opposite of the first series, at the isotropic-D_1 transition, no decrease of diamagnetism observed: thus, the magnetic field has no orientating effect on this phase D_1. At the opposite a weak increase of diamagnetism is measured : it can be probably due to the boundary effect of the cell walls.

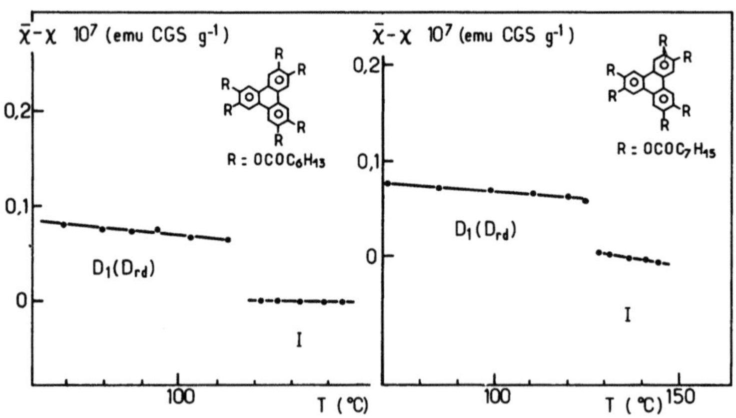

Fig. 4

- 10<n≤13 (n=11,12 figure 5)

At the isotropic-D_2 transition the diamagnetism decreases : the field orientates to a certain extent the sample.

Then when the D_2 to D_1 change occurs, the sharp decrease of the diamagnetism observed indicates that the D_1 phase remains more or less organized.

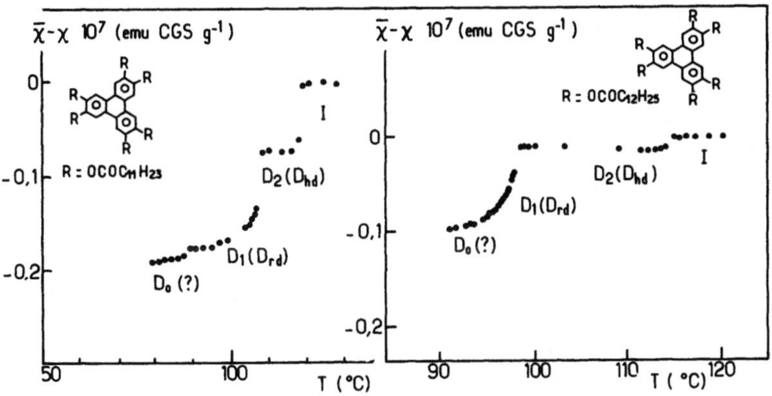

Fig. 5

At last, the D_1-D_o transition remains almost undetected in the limit of accuracy of our experiment although it could be discerned for the n=11 homologue.

Thus in this series, the magnetic field has an orientating effect only when it appears a polymorphism and in fact only when an uniaxial D_2 phase exists before the D_1 biaxial one. In this case, the D_1 phase remains "orientated" whereas it is unorientated when it transforms directly from the isotropic phase.

This can be compared to what is magnetically observed in the case of transitions between smectic mesophases of rod-like molecules : indeed for transition between an orthogonal uniaxial smectic B and an orthogonal biaxial smectic E to which the D_2-D_1 transition could be analogous [9] the orientational order is slightly improved [14].

3. Conclusion

At last, if we compare the behavior of the two series we can conclude that only the uniaxial columnar phases organize more or less in a magnetic field (as observed for uniaxial smectic mesophases of rod-like molecules) but this order is not yet saturated by a 17 KG field. In addition, it is clear that the diamagnetic susceptibility measurement is a good detector of the isotropic - columnar or columnar - columnar transitions.

References

1. S. Chandrasekhar, B.K. Sadashiva, K.A. Suresh, Pramana 9, 471 (1977)
2. J.C. Dubois, Ann. Phys. 3, 135 (1978)
3. Nguyen Huu Tinh, J.C. Dubois, J. Malthete, C. Destrade, C.R. Acad. Sci. C 286, 463 (1978)
4. J.C. Billard, J.C. Dubois, Nguyen Huu Tinh, A. Zann, Nouv. Journ. Chim. 2, 535 (1978)
5. C. Destrade, M.C. Mondon, J. Malthete, Journ. Phys. 40 C3-17 (1979)
6. A. Queguiner, A. Zann, J.C. Dubois, J. Billard, to be published
7. A.M. Levelut, J. Phys. Lett. 40, L-81 (1979)
8. S. Chandrasekhar, B.K. Sadashiva, K.A. Suresh, N.V. Madhusudana, S. Kumar, R. Shashidhar, G. Venkatesh, Journal Phys. 40, C3-120 (1979)
9. A.M. Levelut, Int. Liq. Cryst. Conf., Raman Res. Inst. Bangalore 1979
10. H. Gasparoux, B. Regaya, J. Prost, C.R. Acad. Sci. Paris, 72B, 1168 (1971)
11. C. Destrade, J. Malthete, A.M. Levelut, Nguyen Huu Tinh, IIIrd Liquid Crystal Conf. of Socialist Countries Budapest (1979)
12. C. Destrade, M.C. Bernaud, H. Gasparoux, A.M. Levelut, Nguyen Huu Tinh Proceedings of the Int. Liq. Cryst. Conf. Bangalore 1979
13. C. Destrade, M.C. Mondon-Bernaud, Nguyen Huu Tinh, Mol. Cryst. Liq. Cryst. Lett. 49, 169 (1979)
14. F. Hardouin, M.F. Achard, G. Sigaud, H. Gasparoux, Mol. Cryst. Liq. Cryst. 39, 241 (1977)

Part X

Further Contributions

AC-MECHANICAL IMPEDANCE OF SMECTIC LIQUID CRYSTALS UNDER SHEAR OR COMPRESSION

G. Durand*
Université Paris-Sud, Bât. 510, F-11405 Orsay, France

CONFIRMATIONS OF A RECENT STRUCTURAL CLASSIFICATION OF LAYERED SMECTIC PHASES

A. de Vries
Liquid Crystal Institute, Kent State University, Kent, Ohio 44242, U.S.A.

UNIDENTIFIED SMECTIC PHASES IN THE BIS-(4'-n-ALKOXYBENZYLIDENE)-1,4-PHENYLENEDIAMINES

J.W. Goodby[1], G.W. Gray[1], and E.M. Barrall II[2]
1) Department of Chemistry, The University, Hull, HU6 7RX, U.K.
2) Intern. Business Machines Corporation, San José, California 95193, USA

MOLECULAR ORDERING AND DYNAMICS IN THE SMECTIC PHASES OF IBPBAC AS STUDIED BY ^{14}N NQR AND PROTON NMR

M. Vilfan, J. Seliger, V. Zagar, and R. Blinc
J. Stefan Institute, E. Kardelj University of Ljubljana, 61000 Ljubljana, Yugoslavia

GAS-LIQUID CHROMATOGRAPHY AS A TECHNIC FOR THERMODYNAMIC STUDIES OF LIQUID CRYSTAL SOLUTIONS

J.F. Bocquet, C. Pommier
Physical Chemistry Laboratory, Université Paris XIII, F-93430 Villetaneuse, France

STUDY IN THE SMECTIC C AND THE CHIRAL SMECTIC C PHASES OF COFOCAL DOMAINS OBTAINED IN A SMECTIC A PHASE

L. Bourdon
Laboratoire de Physique des Solides, Bât. 510, F-91405 Orsay, France

THE ORDER-DISORDER PHASE TRANSITION IN LIQUID CRYSTALS AS A FUNCTION OF MOLECULAR STRUCTURE: THE DIRECT ISOTROPIC-SMECTIC A PHASE TRANSITION

H.J. Coles, C. Strazielle[1]
Dept. of Physics, Schuster Laboratory, University of Manchester, Manchester, M13 9PL, U.K.
1) C.R.M., C.N.R.S., 6, rue Boussingault, Strasbourg-Cedex, France

---

* invited lecturer

NEUTRON-SPIN-ECHO MEASUREMENT OF THE CRITICAL DYNAMICS OF THE SMECTIC-A TO NEMATIC PHASE TRANSITION IN CBOOA

H.M. Conrad, F. Mezei[1]
Institut für Festkörperforschung der Kernforschungsanlage Jülich GmbH, D-5170 Jülich, Federal Republic of Germany
1) Institut Laue-Langevin, F-3800 Grenoble, France

TRICRITICAL BEHAVIOUR OF CHOLESTEROGENES AT HIGH PRESSURES

P. Pollmann, G. Scherer, H. Stegemeyer
Physikalische Chemie, Gesamthochschule Paderborn, D-4790 Paderborn, Federal Republic of Germany

^{13}C NMR STUDY OF SOME FERROELECTRIC AND NON-FERROELECTRIC SMECTIC PHASES

M. Luzar, J. Seliger, V. Rutar, R. Blinc
J. Stefan Institute, E. Kardelj University of Ljubljana, 61000 Ljubljana, Yugoslavia

PERMEATION BOUNDARY LAYER IN A SMECTIC A LIQUID CRYSTAL

R. Bartolino and G. Durand
Université Paris-Sud, Bât. 510, F-91405 Orsay, France

DYNAMIC SHEAR STUDY OF A SMECTIC A - SMECTIC B PHASE TRANSITION

S. Bhattacharya, S.V. Letcher
University of Rhode Island, Kingston, R.I. 02881, USA
K. Cawthon, W. Pickens, B.A. Lowry
Erskine College, Due West, SC 29639, U.S.A.
F. Bitter
National Magnet Laboratory, MIT, Cambridge, MA 02139, USA

INSTABILITIES IN SMECTICS-A SUBMITTED TO AN ALTERNATIVE SHEAR FLOW

J. Marignan, O. Parodi
Groupe Dynamique des Phases Condensées, Laboratoire de Minéralogie Crystallographie. U.S.T.L., F-34060 Montpellier Cedex, France
E. Dubois-Violette
Laboratoire de Physique des Solides, Université Paris-Sud, F-91405 Orsay, France

COFOCAL CONICS IN CHIRAL SMECTICS C

A. Perez, M. Brunet, O. Parodi
Groupe de Dynamique des Pahses Condensées, Laboratoire de Mineralogie, U.S.T.L., F-34060 Montpellier Cedex, France

RATIONAL VISCOSITY OF A REENTRANT NEMATIC NEAR THE N-A and N*-A Phase Transitions

J.A. Stroscio, S. Bhattacharya, S.V. Letcher
Department of Physics, University of Rhode Island, Kingston, R.I. 02881, USA

DYNAMICS OF FIELD INDUCED NEMATIC-CHOLESTERIC TRANSITIONS

A. Saupe
Liquid Crystal Institute, Kent State University, Kent, 44242, USA

EXPERIMENTAL STUDIES OF THE TEMPERATURE DEPENDENCE OF THE HELICAL TWISTING POWER

A. Göbl-Wunsch, G. Heppke, F. Oestreicher
Inst. für Anorganische u. Analytische Chemie der Technischen Universität Berlin, Straße des 17. Juni 135, D-1000 Berlin 12, Germany

FLEXIBLE POLYMERS IN NEMATIC SOLVENTS

F. Brochard
Collège de France, Physique de la Matière Condensée,
11 place Marcelin-Berthelot, F-75231 Paris Cedex 05, France

POLYMERIZATION OF SUBSTITUTED BUTADIENES AT THE GAS-WATER INTERFACE

H. Ringsdorf, H. Schupp
Institut für Organische Chemie, Universität Mainz, Federal Republic of Germany

DETERMINATION OF THE SKIN-THICKNESS OF LECITHIN-(DLPC)-LIPSOMES BY NEUTRON SMALL-ANGLE SCATTERING

H.M. Conrad, K. Salm[1], J. Schelten, W. Stoffel[1]
Inst.f.Festkörperforschung der Kernforschungsanlage Jülich GmbH,
D-5170 Jülich, F.R.G.
1) Inst. f. Phyisolog. Chemie der Universität Köln,
Josef-Stelzmann-Str. 52, D-5000 Köln, Federal Republic of Germany

CONCERTED INTERMOLECULAR REACTION IN LIPID-BILAYER ON SEMICONDUCTOR ELECTRODE

P. Fromherz, W. Arden
Max-Planck-Institut für biophysikalische Chemie, D-3400 Göttingen-Nikolausberg
Federal Republic of Germany

HYDRODYNAMICS OF TWO DIMENSIONAL SYSTEMS

D. Langevin
CNRS Lab. Spectroscopie Hertzienne, Paris Cedex, France

FLEXOELECTRIC AND STERIC INTERACTIONS BETWEEN TWO BILAYER LIPID MEMBRANES RESULTING FROM THEIR CURVATURE FLUCTUATIONS

I. Bivas, A.G. Petrov
Bulgarian Academy of Sciences, Sofia 1113, Bulgaria

PORE FORMATION IN LIPID BILAYERS AS INVERTED MICELLATION

M.D. Mitov, A.G. Petrov
Liquid Crystal Group, Institute of Solid State Physics,
Bulgarian Academy of Sciences, Sofia 1113, Bulgaria

SMECTIC B FACIES AND MUTUAL ORIENTATION WITH NEMATICS

M. Warenghem
Laboratoire de Physique des Stases Anisotropes, Université des Sciences et Techniques de Lille, F-59655 Villeneuve d'Ascq Cedes, France

NEUTRON SMALL ANGLE SCATTERING ON MODEL MEMBRANES

W. Knoll, E. Sackmann, H.B. Stuhrmann[1]
Abteilung für Biophysik, Universität Ulm, F.R.G.
1) E.M.B.L., c/o DESY, Hamburg, Federal Republic of Germany

HYDRONAMIC PROPERTIES OF TWO DIMENSIONALLY ORDERED LIQUID CRYSTALS

J. Prost, N.A. Clark[1]
Centre de Recherche Paul Pascal, Domaine Universitaire, F-33405 Talence, France
1) University of Colorado, Department of Physics and Astrophysics, Boulder, CO 80309, USA

MISCIBILITY STUDIES CONCERNING THE TWO SMECTIC A PHASES OF REENTRANT NEMATICS

A. Göbl-Wunsch, G. Heppke, R. Hopf
Inst. f. Anorg. u. Analyt. Chemie der Technischen Universität Berlin, Straße des 17. Juni 135, D-1000 Berlin 12, Germany

Index of Contributors

Display Devices

Editor: J. I. Pankove

1980. 140 figures, 15 tables.
XIII, 252 pages
(Topics in Applied Physics, Volume 40)
ISBN 3-540-09868-2

Contents:
J. I. Pankove: Introduction. – *C. J. Nuese,* J. I. Pankove: Light-Emitting Diodes – LEDs. – *T. N. Criscimagna, P. Pleshko:* AC Plasma Display. – *D. J. Channin, A. Sussman:* Liquid-Crystal Displays, LCD. – *B. W. Faughnan, R. S. Crandall:* Electrochromic Displays Based on WO_3. – *A. L. Dalisa:* Electrophoretic Displays. – *E. O. Johnson:* Electric Displays.

Semiconductor Devices for Optical Communication

Editor: H. Kressel

1980. 186 figures, 6 tables. XIV, 289 pages
(Topics in Applied Physics, Volume 39)
ISBN 3-540-09636-1

Contents:
H. Kressel: Introduction. – *H. Kressel, M. Ettenberg, J. P. Wittke, I. Ladany:* Laser Diodes and LEDs for Fiber Optical Communication. – *D. P. Schinke, R. G. Smith, A. R. Hartman:* Photodetectors. – *R. G. Smith, S. D. Personick:* Receiver Design for Optical Fiber Communication Systems. – *P. W. Shumate, Jr., M. DiDomenico:* Lightwave Transmitters. – *M. K. Barnoski:* Fiber Couplers. – *G. Arnold, P. Russer, K. Petermann:* Modulation of Laser Diodes. – *J. K. Butler:* The Effect of Junction Heating on Laser Linearity and Harmonic Distortion. – *J. H. Mullins:* An Illustrative Optical Communication System.

Electroluminescence

Editor: J. I. Pankove

1977. 127 figures, 16 tables. XI, 212 pages
(Topics in Applied Physics, Volume 17)
ISBN 3-540-08127-5

Contents:
J. I. Pankove: Introduction. – *Y. M. Tairov, Y. A. Vodakov:* Group IV Materials (Mainly SiC). – *P. J. Dean:* III-V Compound Semiconductors. – *Y. S. Park, B. K. Shin:* Recent Advances in Injection Luminescence in II-VI Compounds. – *S. Wagner:* Chalcopyrites. – *T. Inoguchi, S. Mito:* Phosphor Films.

I. Kohonen

Associative Memory

A System-Theoretical Approach

Corrected printing 1978. 54 figures,
7 tables. IX, 176 pages
(Communication and Cybernetics, Volume 17)
ISBN 3-540-08017-1

Contents:
Introduction. – Associative Search Methods. – Adaptive Formation in Optimal Associative Mappings. – On Biological Associative Memory.

Springer-Verlag
Berlin
Heidelberg
New York

A monthly journal

Applied Physics

Board of Editors:
S. Amelinckx, Mol; **V. P. Chebotayev,** Novosibirsk; **R. Gomer,** Chicago, IL; **P. Hautojärvi,** Espoo; **H. Ibach,** Jülich; **K. L. Kompa,** Garching; **V. S. Letokhov,** Moskau; **H. K. V. Lotsch,** Heidelberg; **H. J. Queisser,** Stuttgart; **F. P. Schäfer,** Göttingen; **R. Ulrich,** Stuttgart; **W. T. Welford,** London; **H. P. J. Wijn,** Endhoven; **T. Yajima,** Tokyo

Coverage:
Application-oriented experimental and theoretical physics:
Solid-State Physics
Surface Science
Solar Energy Physics
Microwave Acoustics
Electrophysics
Quantum Electronics
Laser Spectroscopy
Photophysical Chemistry
Optical Physics
Optical Communications

Special Features:
Rapid publication (3-4 months)
No page charges for concise reports
Microform edition available

Languages:
Mostly English

Articles:
Original reports, and short communications review and/or tutorial papers

Manuscripts:
To Springer-Verlag (Attn. H. Lotsch), P.O. Box 105280, D-6900 Heidelberg 1, FRG

Please ask for your sample copy.
Send your order or request to your bookseller or directly to:
Springer-Verlag, Promotion Department, P.O. Box 105280, D-6900 Heidelberg 1, FRG
North America: Springer-Verlag New York Inc., Journal Sales Dept., 44 Hartz Way, Secaucus, NJ 07094, USA

Springer-Verlag
Berlin
Heidelberg
New York

MIX
Papier aus verantwortungsvollen Quellen
Paper from responsible sources
FSC® C105338

If you have any concerns about our products,
you can contact us on
ProductSafety@springernature.com

In case Publisher is established outside the EU,
the EU authorized representative is:
Springer Nature Customer Service Center GmbH
Europaplatz 3, 69115 Heidelberg, Germany

Printed by Libri Plureos GmbH
in Hamburg, Germany